Operation, Maintenance, and Management of Irrigation Systems

Operation, Maintenance, and Management of Irrigation Systems serves as a comprehensive guide to irrigation system governance across global contexts, from state-controlled systems in developing countries to privately managed districts in developed nations. This book systematically covers water delivery methods, automation systems, hydraulic structures, drainage principles, equipment maintenance, financial management, and policy frameworks, while addressing modern concerns such as water quality, safety protocols, and wildlife enhancement. It serves as an essential resource for irrigation professionals, agricultural engineers, water resource managers, and policymakers seeking to optimize system performance across diverse geographical and governance environments.

Key features:

- Combines engineering principles with socio-economic, managerial, and stakeholder considerations for holistic irrigation system management.
- Features real-world case studies, advanced automation systems, water delivery methods, hydraulic structures, and maintenance strategies.

Operation, Maintenance, and Management of Irrigation Systems

Vijay P. Singh and Qiong Su

CRC Press
Taylor & Francis Group
Boca Raton London New York

CRC Press is an imprint of the
Taylor & Francis Group, an **informa** business

Designed cover image: Shutterstock

First edition published 2027
by CRC Press
2385 NW Executive Center Drive, Suite 320, Boca Raton FL 33431

and by CRC Press
4 Park Square, Milton Park, Abingdon, Oxon, OX14 4RN

CRC Press is an imprint of Taylor & Francis Group, LLC

ISBN: 978-1-041-25608-3 (hbk)
ISBN: 978-1-041-25607-6 (pbk)
ISBN: 978-1-003-74744-4 (ebk)

DOI: 10.1201/9781003747444

Typeset in Times
by codeMantra

VPS: (late) wife Anita; son: Vinay; daughter: Arti; daughter-in-law: Sonali; son-in-law: Vamsi Reddy; grandsons: Ronin, Kayden, Davin, and Vayu; granddaughter: Alivia

QS: In loving memory of my father, Zhongke; my mother, Guizhen; and my husband, Yun.

Contents

SECTION 2 Operation of Irrigation Systems

SECTION 3 Maintenance of Irrigation Systems

Preface

Irrigation systems are of different types, depending on the method of irrigation, source of water, quality of water, type of landscape, type of climate, type of soil, types of crops, size of agricultural land, state and federal laws and regulations, manpower, and finances. The organization of an irrigation system depends on the type of system and the system of governance prevailing at the place. It requires the integration of technical, socio-economic, managerial, and stakeholder considerations. There are different types of organizations in different countries and in different parts of the same country. In many countries, especially developing countries, such as India, irrigation systems are under the control of state governments, and stakeholders have no role to play in their operation and management. On the other hand, in developed countries, irrigation systems are operated and managed by private organizations, often referred to as irrigation districts. Regardless of who has the authority to operate and manage irrigation systems, the underlying issues are more or less the same. This book discusses the operation, maintenance, and management of irrigation systems.

The subject matter of the book comprises 23 chapters, which are grouped into three sections. Section I contains five chapters discussing preliminaries. Chapter 1 provides an integrated view of irrigation systems. Beginning with a discussion of the physical system, it goes on to discussing source system, conveyance system, delivery system, receiver system, open canals, pipe systems, canal structures, classification of irrigation systems, disciplinary considerations, scale considerations, and irrigation enterprise. Chapter 2 discusses a brief history of irrigation systems, including irrigation practices in ancient civilizations, irrigation from groundwater, irrigation development, growth of irrigation systems, and operation and management.

On-farm irrigation practices are dealt with in Chapter 3, which include surface irrigation, sprinkler irrigation, and micro-irrigation. Chapter 4 discusses the organization for operation and management, encompassing a range of activities from water acquisition, conveyance, and delivery to the development and maintenance of crop-growing areas, as well as potentially managing drainage and other water supply works. It includes planning, implementing, and maintaining the infrastructure needed to supply water to agricultural lands efficiently and sustainably. Chapter 5 examines three primary administration structures: community-led, government-led, and hybrid systems. Each administrative structure is evaluated for its advantages and disadvantages, with examples from various regions, such as Irrigation Associations in the United States, Public Irrigation Schemes in Spain and Kenya, and mixed-control systems in Asia.

Section 2 on operation of irrigation systems comprises four chapters. Chapter 6 evaluates four water delivery methods, i.e., continuous, rotation, demand, and arranged supply, and their adaptation to diverse crops, climate, soil, landscape, and infrastructure conditions. Advanced automation systems and hydraulic structures are evaluated for their precision in flow regulation and canal management. Chapter 7 explores principles, methodologies, and real-world applications of managing irrigation systems, with case studies from irrigation districts.

Chapter 8 explores essential methods for measuring water to support efficient agricultural water management. It covers techniques for both groundwater and surface water, including propeller, ultrasonic, Doppler, and magnetic flow meters for wells, as well as weirs, flumes, and orifices for open channels. Chapter 9 provides a comprehensive overview of drainage principles and practices essential for sustainable irrigation management and soil productivity. It introduces basic drainage terminology, including ditches, drains, channels, and outlets, and explains the function of drainage systems in controlling groundwater levels and managing soil salinity.

Section 3 on maintenance of irrigation systems contains 14 chapters. Chapter 10 outlines key principles and practical strategies for maintaining irrigation systems, including inspection, feedback integration, and implementing normal, catch-up, essential, and preventive maintenance programs.

Chapter 11 discusses the equipment used for the effective operation and maintenance of irrigation systems. It outlines the factors influencing equipment needs, including system type, scale, and issues related to sediment or weeds. Key equipment types, such as excavators, tractors, graders, and communication systems, are discussed. Chapter 12 addresses the comprehensive maintenance requirements for dams and conveyance structures in irrigation and water management systems. This chapter covers service and repair practices for key components, including reservoirs, headworks, canals, structures, bridges, and communication systems.

Chapter 13 provides a comprehensive guide for the maintenance of canal systems, covers key maintenance activities, and addresses the removal of aquatic and semi-aquatic vegetation, the control of burrowing animals, and the protection of blowing soils. Water quality is a critical factor in sustainable irrigation, influencing crop productivity, system longevity, and environmental compliance. Chapter 14 examines key water quality challenges and their management under frameworks, such as the Clean Water Act. It provides guidelines for assessing irrigation water, mitigating infiltration issues, and controlling salinity through leaching and drainage.

Weed control is a critical component of the operation and maintenance of irrigation systems. Chapter 15 provides a comprehensive overview of the classification, impacts, and management strategies for weeds in irrigation canals, drains, reservoirs, and pipelines. Chapter 16 provides a structured framework for the rehabilitation and betterment of irrigation systems. This chapter explores a wide range of water conservation strategies, including alternative irrigation methods, return flow reuse, and pollution control.

Chapter 17 provides a comprehensive overview of financial management in irrigation enterprises, emphasizing the critical role of sound fiscal planning, budgeting, and cost recovery strategies in sustaining system operations. It explores various cost allocation methods and presents financing mechanisms, such as bonds, sinking funds, and federal aid. Chapter 18 covers key policy areas, including water conservation, drought contingency planning, water allocation, and delivery regulations. It also addresses operational protocols for canal maintenance, user rights and responsibilities, complaint resolution, and enforcement mechanisms.

Chapter 19 introduces legal frameworks that shape the operation of irrigation enterprises in the United States. Federal authorities cover navigation, pollution control, and federal lands, whereas states govern allocation through doctrines.

Groundwater law varies from absolute ownership to reasonable use and correlative rights. Irrigation systems pose a variety of hazards to the public, operating personnel, livestock, and systems themselves. Chapter 20 provides a comprehensive framework for safety management in irrigation systems, covering four key domains: public safety, operation and maintenance personnel safety, livestock and wildlife protection, and system safety. It also emphasizes emergency preparedness, response planning, and recordkeeping as foundational elements for continuous improvement.

Chapter 21 highlights the growing role of recreation and fish and wildlife enhancement in irrigation projects, reflecting rising public demand for water-based leisure and ecological stewardship. It explores how reservoirs, canals, and multipurpose facilities provide opportunities for recreation. Legal and financial considerations in providing recreational facilities are addressed. Case studies highlight real-world approaches, such as habitat enhancement at California's Imperial Irrigation District and integrated recreation–wildlife initiatives at Texas's Delta Lake.

Chapter 22 discusses public relations that shape how enterprises interact with water users, employees, agencies, and the broader public. It emphasizes that effective communication fosters transparency, builds trust, and ensures that management actions are accepted and supported. Irrigation enterprises operate amid constant change: shifting crops and markets, climate variability, urban encroachment, regulatory reform, technological disruption, and evolving social expectations. Chapter 23 offers a practical framework for anticipating, evaluating, and managing change at the project and district scales.

Vijay P. Singh
College Station, Texas
Qiong Su
Clemson, South Carolina

Acknowledgments

The authors would like to express their gratitude to irrigation organizations, including government departments (national and foreign), and irrigation districts that are engaged in managing irrigation systems for allowing access to their reports and publications.

About the Authors

Vijay P. Singh is a Distinguished Professor, a Regents Professor, and the inaugural holder of the Caroline and William N. Lehrer Distinguished Chair in Water Engineering at Texas A&M University. He received his B.Sc. & Tech., M.S., Ph.D., and D.Sc. degrees, all in engineering. He is a registered professional engineer, a registered professional hydrologist, a Board-Certified Water Resources Engineer (ASCE-AAWRE), and an Honorary Diplomate of the European Water Resources Association. He is a member of the National Academy of Engineering (NAE), a member of the Canadian Academy of Engineering, an academician of the Georgia Fazisi Academy, a member of the European Academy of Sciences and Arts, and a fellow/member of 11 other international science/engineering academies. He is a Distinguished Member of ASCE, an Honorary Distinguished Member of IWRA, a Distinguished Fellow of AGGS, an Honorary Member of AWRA, and a Fellow of AGU, IAHR, EWRI, ASCE, IAH, ISAE, IWRS, and IASWC. He has published extensively in the areas of hydrology, irrigation engineering, hydraulics, groundwater, water quality, food-water-energy-environment nexus, water resources, entropy theory, copula theory, and climate change impacts, with more than 1680 refereed journal articles, 41 books, 102 edited reference books, 134 book chapters, and 330 conference papers. His Google Scholar citations are more than 133,095 with an h-index of 150 and an I10-index of 1,547. He has received four honorary doctorates and more than 122 national and international awards. He has served as President of the American Institute of Hydrology (AIH); President of American Academy of Water Resources Engineers; President of International Association for Water, Environment, Energy, and Society; President of International Society for Smart Construction and Production; and Chair of Watershed Council of American Society of Civil Engineers. He has served as the editor-in-chief of five journals and two book series, and he has served/serves on editorial boards of more than 40 journals and three book series. He has given more than 210 keynote lectures and 300 invited seminars all over the world, and he has organized more than 27 international conferences. Research.Com has ranked Professor Singh number 2 in hydrology in the world and number 13 among all earth scientists in the world, whereas Scholar GPS has ranked him number 1 in hydrology, number 1 in drought, and number 4 in water resources in the world.

Qiong Su is an Assistant Professor in the Department of Agricultural Sciences at Clemson University. She earned a B.S. in Environmental Science and Computer Science (Ocean University of China), an M.S. in Environmental Science (Peking University), and a Ph.D. in Water Management and Hydrological Science (Texas A&M University). Her research bridges fundamental and applied irrigation management, advancing water-use efficiency and climate-resilient crop production by integrating field measurements, process-based models, geospatial big data, remote sensing, and artificial intelligence. She has published 40 peer-reviewed journal articles, a textbook on irrigation engineering, and multiple book chapters. Her contributions include the calibration-free PT_SS method for reference evapotranspiration, a seasonal forecast downscaling tool for regional crop modeling and irrigation management, and idCROP, a sensor- and crop-model-based irrigation decision app. She teaches irrigation system design and land surveying, serves as a reviewer for more than 20 journals, and is an active member of the American Society of Agricultural and Biological Engineers (ASABE) and the American Geophysical Union (AGU).

Section 1

Preliminaries

1 An Integrated View of Irrigation Systems

NOTATION

z side slope of the canal
H height of the canal (m, ft)
W horizontal distance of the canal banks to bottom width (m, ft)
s bottom slope of the canal
Q discharge in the channel (m^3/s, cfs)

1.1 INTRODUCTION

It is well known that nature does not distribute precipitation evenly over our globe (spatial distribution). Neither is the precipitation evenly distributed throughout the year (temporal distribution). Therefore, there are humid areas, there are deserts, and there are many "in between". Even in humid areas, there may not be adequate rainfall to support the growth of crops for certain periods during growing seasons.

Much of our land resource lies in areas that are arid, semi-arid, or in areas where the time distribution of precipitation is not satisfactory for optimum crop production. What has to be done in these areas? Water is provided by irrigation to supplement the normal precipitation. The amount of irrigated water in these areas depends on the type of crops, soil, climate, etc. Then, it may be said that the purpose of an irrigation system is to supplement rainfall by supplying water: (1) in the proper amount, (2) at the proper time, (3) of the desired quality, and (4) at the proper place efficiently and economically so that the land resource may be put to use for agriculture. Therefore, irrigation may be defined as the application of water to soil in order to supplement the natural water and thus provide the moisture essential for plant growth. There are broader definitions that include providing crop insurance against short-duration droughts and leaching of or dilution of salts in the soil. There are different types of systems that are employed to do irrigation. These systems vary with the method of water delivery in conjunction with method of irrigation and type of landscape and climate. For example, there is the canal system, which is employed for irrigating large areas. This system is operated, managed, and maintained by an organization—whether public or private. On the other hand, there is the drip system that is employed by individual farmers who themselves operate, manage, and maintain the system. If the landscape is undulating and hilly, sprinkler or drip irrigation systems are the only viable systems. Similarly, if the climate is wet and it rains a lot, then there is no reason to have canals or tubewells.

Operation, management, and maintenance of irrigation systems are important functions and are technical jobs when sophisticated irrigation systems of today are considered. Strangely enough, even today, these jobs do not receive the attention

DOI: 10.1201/9781003747444-2

they deserve and are frequently passed on to those people who may or may not have received the training for discharge of the duties in the most efficient manner.

The success or failure of an irrigation enterprise is generally believed to be related to management. Management, in the establishment of irrigation and drainage projects, includes planning, organizing, equipping, and staffing for the direction and control of a complex business venture. Many factors are involved in the management of natural resources, water, and the effective use and conservation of that resource utilizing efficient operation and maintenance as an economic outlay.

The management of an irrigation and drainage system depends not only on people, as do many business enterprises, but also on equipment, procedures, practices, and crops to be grown, which are important considerations. In arid and semi-arid areas, where irrigation and general drainage are essential to the people, a strong, dependable, and capable administrative organization is indispensable. Civilizations have followed the development of irrigation and have risen on irrigated lands; they also have decayed and disintegrated in irrigated regions. Most men who are well informed are certain of its perpetuity, as long it is intelligently practiced.

Although irrigation is defined in terms of the delivery of water to the land or farms, the system that makes the delivery possible is quite complicated. This may also include all of the physical improvements necessary to transfer the water in time and space from the source to the farm. Therefore, the management and operation of an irrigation system require an understanding of the system in full detail. The objective of this chapter is to present an integrated view of the irrigation system.

1.2 PHYSICAL SYSTEM

An irrigation system may be composed of four components, each constituting an independent system: (1) source system, (2) conveyance system, (3) delivery system, and (4) receiver system. The major components of a typical irrigation system are shown in Figure 1.1. In this figure, the source of supply is surface water derived from

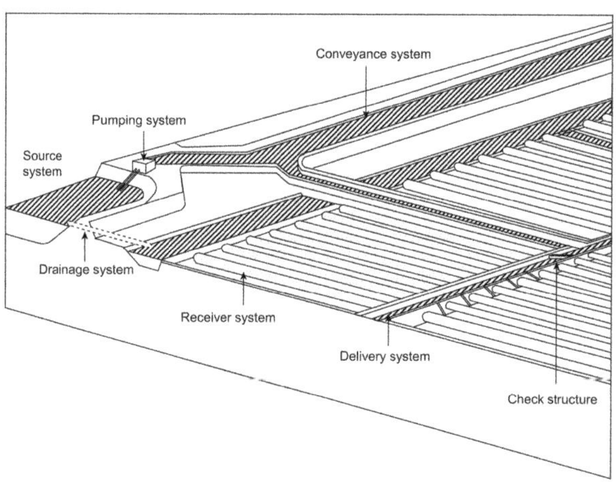

FIGURE 1.1 Major components of an irrigation system.

a stream. The source might also include groundwater used exclusively or in conjunction with the surface supply. Storage is provided by an on-stream reservoir located higher on the stream than the project land. The stream is used as a conveyance, for reasons of economy, to a point near the beginning of the project. The water is then diverted from the stream at a diversion dam and conveyed to the land area by a system of canals and laterals. No attempt has been made to show the numerous minor structures in the system. In a particular irrigation system, conveyance subsystems and/or delivery subsystems may not exist. Examples of different irrigation systems will be presented later in the chapter.

1.2.1 Source System

The source system is defined as the system that houses or stores water or from which water is drawn. The source of water may exist above or below the ground. The source may be surface water, groundwater, or a combination of the two. Thus, depending upon its existence, the source system may be defined as surficial or subsurficial. The systems existing on the ground are called surficial. Examples of such systems are dams, reservoirs, lakes, ponds, rivers, and springs. The source systems that exist below the land surface are referred to as subsurficial and are exemplified by aquifers, wells, etc.

The source can be natural or man-made. For example, lakes, reservoirs, rivers, springs, aquifers, and natural lakes are natural sources of water. Man-made source systems are exemplified by tanks, dams, reservoirs, artificial lakes, ponds, etc.

A source system may be either static or dynamic. Examples of static systems are ponds, tanks, reservoirs, and lakes, and examples of dynamic systems are rivers, canals, springs, and aquifers. Different source systems have many different functions. For example, dams and reservoirs store water that can be used whenever needed, especially during dry periods. Rivers have many functions, including water supply, water transfer, irrigation, transportation, navigation, purification, habitat, power generation, recreation, and fishing. Canals are used for the conveyance of water for irrigation, drainage, water supply, etc. Lakes are also used for a variety of functions, such as storage or as regulating reservoirs, irrigation, recreation, and fishing.

Irrigation systems can also be classified, based on scale: microirrigation systems, mesoirrigation systems, macroirrigation systems, and megairrigation systems. Examples of microirrigation systems are drip irrigation networks, gated pipe systems, and sprinkler systems. Mesoirrigation systems can be tubewells, and small canal irrigation systems receiving water from small dams. Macroirrigation systems can be medium canal irrigation systems receiving water from medium dams and a network of tubewells. Megairrigation systems are large irrigation canal systems, such as those in India, China, Brazil, and Pakistan, receiving water from large dams.

1.2.2 Conveyance System

A conveyance system is defined as a system that receives water from a source and conveys water to a delivery system or a receiver system. As the name suggests, a conveyance system conveys water and may comprise flumes, tunnels, canals aqueducts,

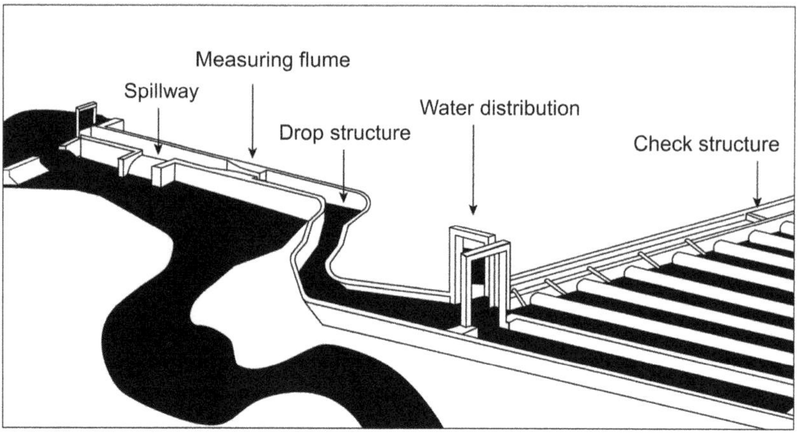

FIGURE 1.2 Multi-branch conveyance irrigation system.

ditches, channels, and rivers. The system transports large discharges over long distances. Conveyance systems can be natural or man-made. Rivers, streams, brooks, bayous, and creeks are examples of natural conveyance systems, whereas canals, man-made channels, and ditches exemplify (artificial) conveyance systems. A conveyance system can be a branch or a multi-branch system, as shown in Figure 1.2. A conveyance system may be lined or unlined. It may extend from a few kilometers to tens of kilometers. A conveyance system may be of limited extent and simple, or it may encompass many miles of canals and laterals and may be quite complicated with various types of structures.

1.2.3 Delivery System

A delivery system is a system that connects a receiver system to a conveyance system. The conveyance or primary canal originates at the source and provides water to the receiver system like a secondary canal. The secondary canal receives water from the primary canal and branches to cover more agricultural fields. The tertiary canals contain turnouts and deliver water to individual fields. On some occasions, turnouts can be installed in primary and secondary canals. However, turnout density is higher in tertiary canals. The system is generally organized at the secondary or tertiary canal level.

Examples of delivery systems are aqueducts, grassed channels, grassed waterways, pipes, and tubes, or a source system. Usually, a delivery canal system may be lined or unlined, grassed or earthen. It may run from a few meters to a few kilometers. A man-made system is exemplified by a pipe or tube.

The manner in which water will be delivered to the farms during the operation of the system must be contemplated in the planning and design stages. Changes in procedures from those considered in the design stage may cause difficulties in operation, depending on the flexibility designed into the system. Project irrigation efficiency is affected to a pronounced degree by the method of water delivery. It is important that the system be designed and constructed to permit flexible operation, because alternate operational procedures must frequently be used to attain reasonable delivery system efficiency.

1.2.4 Receiver System

A receiver system is defined as a system that receives water for irrigation. That is where irrigation takes place. Agricultural land, orchards, forest lands, etc. constitute receiver systems. Irrigation is done in different ways: (1) basin irrigation, (2) border irrigation, (3) furrow irrigation, (4) sprinkler irrigation, and (5) drip irrigation.

The application of water to land has a significant effect on operation. Farm practices are usually not the responsibility of the irrigation system management. However, the system must be operated and managed in conformity with farming practices. Although the activities on the farm are normally the responsibility of the irrigator, many of these practices must be considered in the operation of the system. It must be kept in mind that the design of the system governs the operation to a major degree. Changes in farm practices, including cropping patterns and methods of application of water to the soil, from those anticipated in design, always cause difficulties in operational procedures.

1.3 OPEN CANALS

Water can be transported by open canals or pipes. Open canals and channels are used to transport water from one place to another by gravity flow. Field ditches have smaller dimensions and are used to convey water inside the farm. Drainage ditches also have small dimensions and are used to receive drainage water or return flows from the farm fields.

The shape of canals can be rectangular, trapezoidal, circular, parabolic, triangular, regular, or irregular. In irrigation systems, the main shape is the trapezoidal cross-section. A sketch of a canal is shown in Figure 1.3. The freeboard of the canal is the vertical distance between the maximum anticipated water surface elevation in the design and the top of retaining banks, pipeline vents, or other structures, provided to prevent overtopping due to unforeseen conditions.

The side slope of the canal (z) is the ratio between the height of the canal (H) and its horizontal distance (W). For example, the side slope can be expressed as 1:2, which means the horizontal distance is two times the vertical height of the canal (as shown in Figure 1.3b).

The bottom slope (s) of the canal can be shown from the side view (as shown in Figure 1.3c). The bottom slope of the canal in the unit of percent is expressed as the height difference of the canal and its horizontal distance.

$$\text{Bottom slope }(\%) = \frac{\text{Height difference (m)}}{\text{Horizontal distance (m)}} \times 100 \tag{1.1}$$

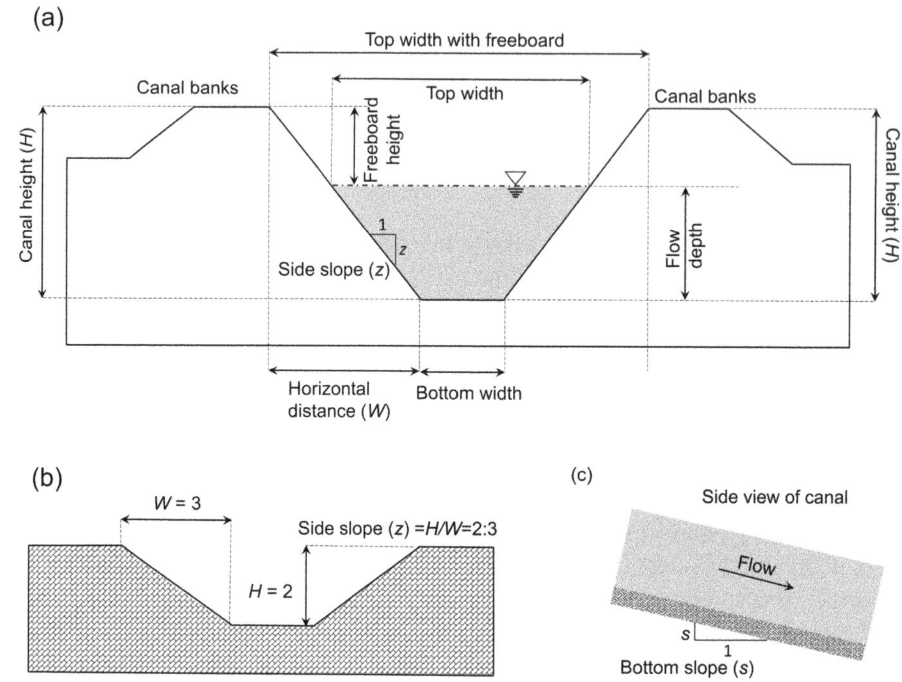

FIGURE 1.3 Examples of (a) a trapezoidal cross-section, (b) the side slope, and (c) the bottom slope of a trapezoidal canal.

Example 1.1

Find the slope of the canal that has a height difference between two points of 30 cm in a length of 50 m.

Solution:

$$\text{Bottom slope } (\%) = \frac{30 \text{ cm}}{50 \text{ m}} \times 100 = \frac{0.3 \text{ m}}{50 \text{ m}} \times 100 = 0.6\%$$

1.4 PIPE SYSTEMS

Occasionally, it is more economical to build a pipe system than to line a canal. Pipe systems are generally substitutes for tertiary canals. The most common types of pipeline systems are open pipeline systems.

1.4.1 Open Pipeline Systems

Open pipeline systems are generally used in the rotational delivery among farm turnouts. They require the use of open stands made up of concrete or plastic to relieve the pressure (generally at 6 m of static pressure). They have overflow weirs in open

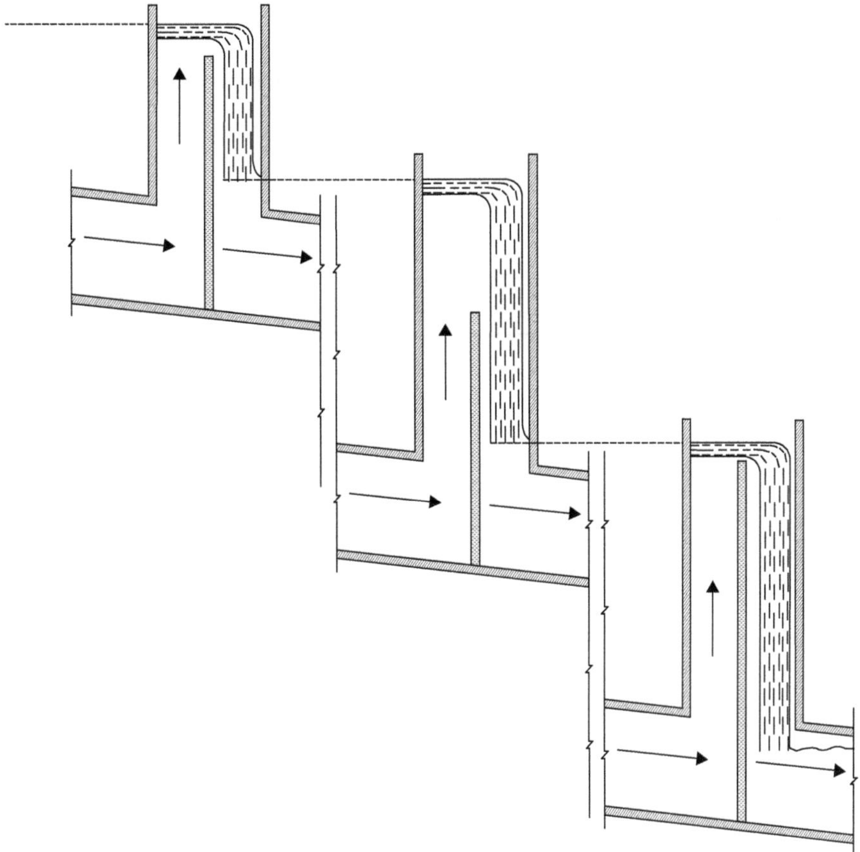

FIGURE 1.4 Open pipe system.

standpipes (Figure 1.4). Open standpipes, with an overflow weir, provide the passage of water to the downstream pipe section. The overflow weir retains the upstream water level to provide sufficient head for the upstream turnout, while also providing an overflow for all unused water. Reinforced concrete is the material that is used in open pipelines, which works with a maximum pressure of 6 m. Generally, all the flow that enters the system has to be used in one hydrant. Standpipes, such as the water hammer, are also utilized to release surges in pressure. In some cases where the water is pumped directly into the stand, the stand dampens the surges and protects the lines. Standpipes are also used as air vents, releasing trapped air from the pipes. The standpipes are also used to branch the pipe network. The flow rate is measured at the inlet to the pipeline and must exactly match the combined turnout deliveries.

1.4.2 Closed Pipeline Systems

Closed pipeline systems develop maximum static pressure. The advantage of using a closed pipe is that more pressure can be obtained at each farm turnout. This pressure may be required in some microirrigation or sprinkler irrigation systems.

Closed pipeline systems are also helpful for gated pipes, so greater lengths can be used. In some pipe networks, pressure is controlled at different farm turnouts by sizing the pipes at the design stage or by installing standpipes with standard heights over each hydrant.

1.4.3 Semi-Closed Pipelines

Semi-closed pipelines are used in a few irrigation districts, mainly in California. These systems have a float valve in each open stand, which controls the flow into the downstream section (Figure 1.5). The float valve is known as the Harris valve, which is a simple mechanical leveraged valve connected to a float that activates it with the rise and fall of the water surface on its downstream side within a standpipe. If there is a demand for flow downstream, the water level in the upstream stand drops slightly, causing the float to drop and the valve to open. The action is then repeated at each of the upstream stands, all the way to the source. Semi-closed systems can add

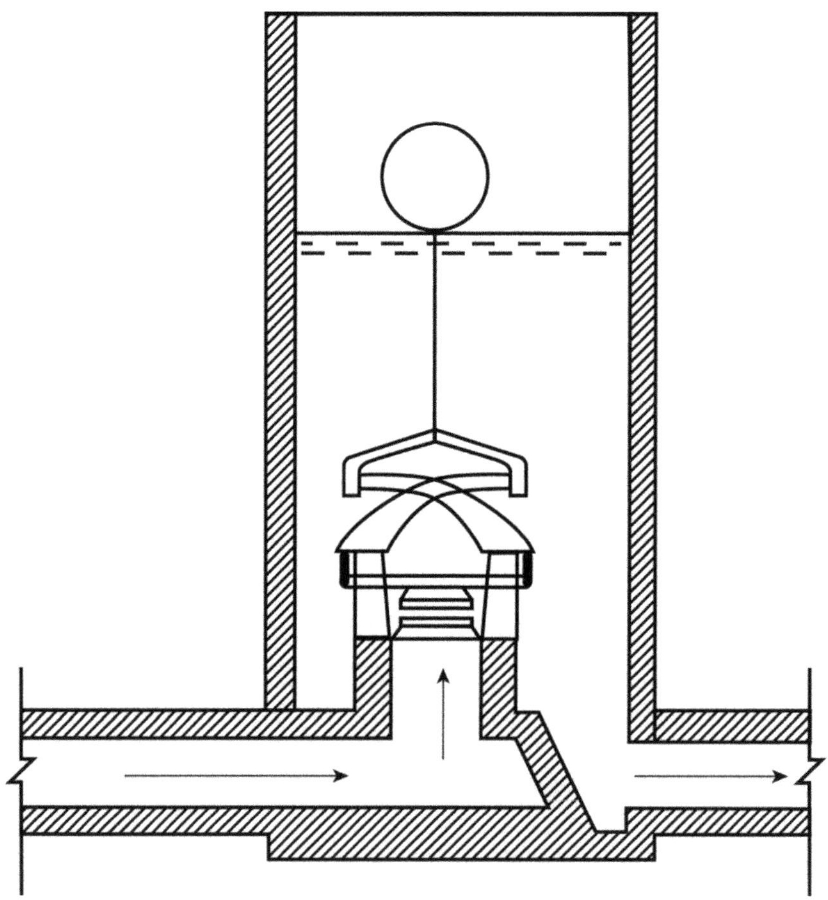

FIGURE 1.5 Float valve used in semi-closed pipe systems.

flexibility to the system by adding downstream control, as in the case of closed systems or hydraulic gates (e.g., AVIO gates) in canals. The advantage of the semi-closed system is that it can regulate a wide range of flow with small changes in the head. Therefore, a non-reinforced concrete pipeline can be used with this system. One of the problems with the semi-closed system is that it requires proper installation, and the water needs to be free of debris to avoid interference with the valve.

1.5 CANAL STRUCTURES

Canal structures are required in the canals to control the flow of irrigation water and deliver the required amount of water to different parts of the irrigation system. The main canal structures are classified as (1) erosion control structures, including (a) stilling basins, (b) drop and chutes, and (c) spillways; (2) cross structures, including (a) aqueducts, (b) culverts, and (c) inverted siphons and flumes; (3) division structures, including (a) division boxes, (b) turnout structures, and checks; (4) water measurement structures; and (5) water control structures. Some of the structures are used to measure and control water, and some others are used only to measure or control water.

1.5.1 Erosion Control Structures

Water flowing through the canals can reach very high velocities. The steeper the bottom slope, the higher the flow velocity. Water can detach soil particles from the sides and bottom of the canal, which are deposited on the bottom as the velocity slows down. The canal can be silted if velocity is not controlled with structures, and the banks may collapse. Some structures used to control erosion on canals include stilling basins (as shown in Figure 1.6), drop structures and chutes (as shown in Figure 1.7), spillways, and canal lining.

Stilling basins are used to protect canals from erosion, converting the energy of upstream fast-flowing water into turbulence, so the velocity is reduced downstream.

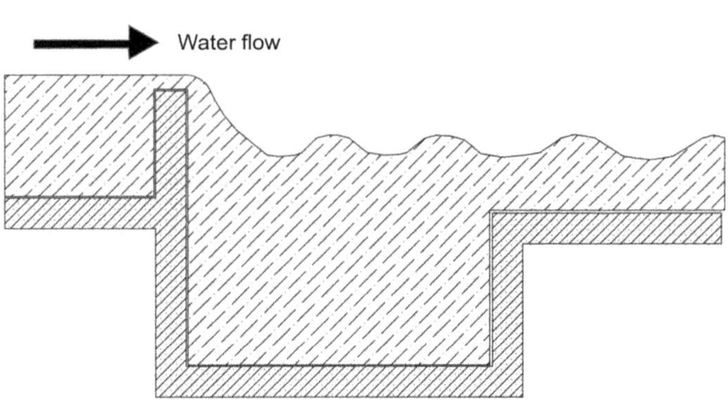

FIGURE 1.6 Stilling basin used to reduce velocities in canals.

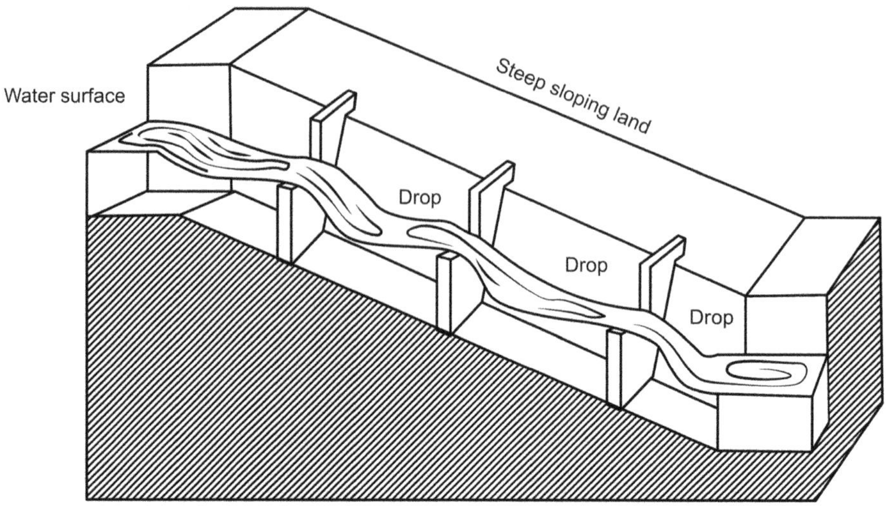

FIGURE 1.7 A series of drop structures to reduce water velocity.

These structures are needed in farm outlets when water is pumped to a canal or after some water to drop structures.

Drop structures or chutes are usually used in canals with a steep bottom slope to reduce the flow velocity and avoid canal erosion. As illustrated in Figure 1.7, a series of drop structures are constructed in the canal at different elevations. In this case, water can flow in a relatively flat section. If the canal has relatively big differences in elevation, chutes are usually used. Unlike drop structures that allow water to drop freely from a higher section to a lower one, the chute carries water through a steep, lined canal section.

Spillways are used to discharge excess water from a canal to a drainage canal to protect its banks. When water rises due to an intake structure being opened instead of being closed, water may pass the freeboard level, reach the crest of the canal embankment, and start overtopping it. When water overtops the canal, the bank of the canal can be destroyed. To avoid overtopping of the canal, an emergency outlet, called a spillway, is constructed in the canal. This structure consists of a small section of the canal bank reinforced with concrete, which is lowered below the crest. When the water rises, the spillway directs the excess water to a drainage canal. On some occasions, the spillway directs the water to some regulating reservoirs.

Besides being used to reduce infiltration and percolation losses, canal lining is used to protect the walls of the canal and avoid its collapse.

1.5.2 Crossing Structures

Crossing structures are often necessary to push water across roads, hillsides, natural depressions, and drainage ditches. They include aqueducts, culverts, and inverted siphons.

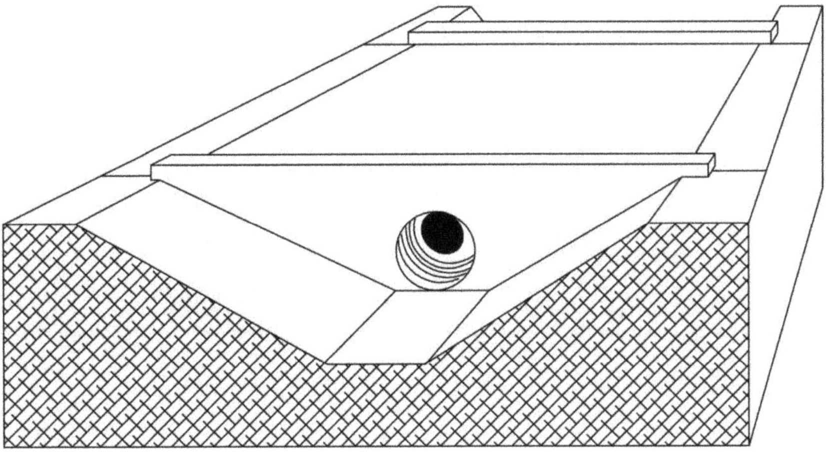

FIGURE 1.8 Culvert.

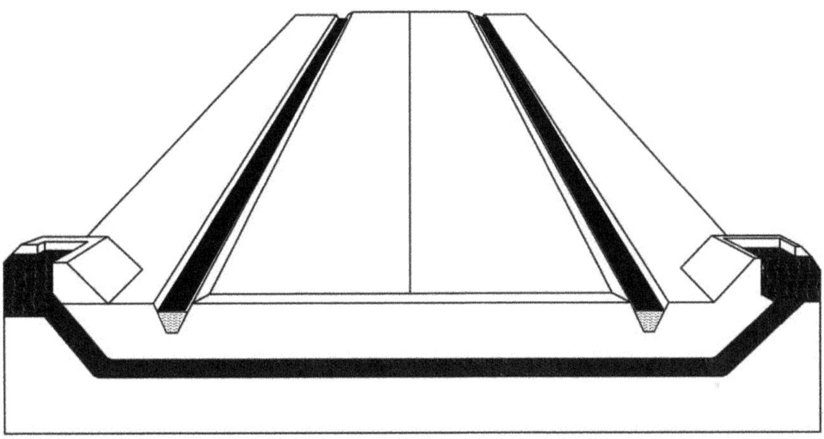

FIGURE 1.9 Inverted siphon.

a. **Aqueducts**: Aqueducts are used as self-supporting canals or pipe systems to carry water across drainage canals, gullies, and depressions.
b. **Culverts**: Culverts consist of a buried pipe with masonry or concrete head-walls at both the inlet and outlet, which carries water underneath a road, canal, or natural depressions (as shown in Figure 1.8).
c. **Inverted siphon**: Inverted siphon is needed to cross water underneath a road at the same level or below the canal bottom. As shown in Figure 1.9, it consists of an inlet and outlet, which are connected by a pipe.

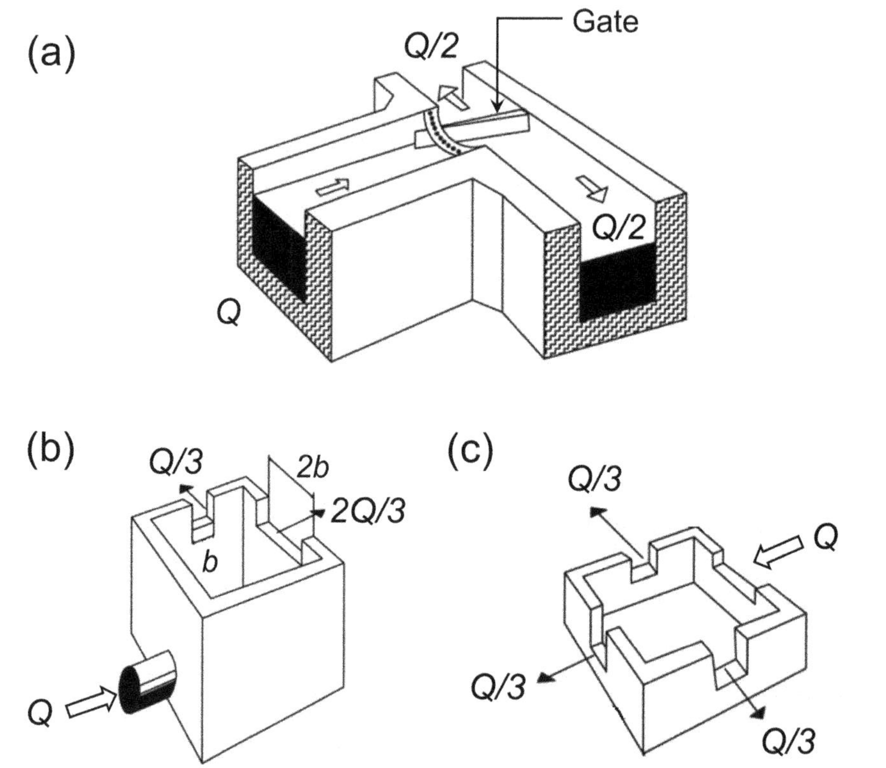

FIGURE 1.10 (a) two-way division box with evenly diverted flow, (b) two-way division box with unevenly diverted flow, and (c) three-way division box with evenly diverted flow.

1.5.3 Division Control Structures

Division boxes are required for easy and accurate distribution of water within the canals or from the canals to the farms. Division boxes are used to divide and distribute flow into two or more canals. There are gates in each box to help divert the water (Figure 1.10). Two-way division boxes can be used in a canal to divert its total flow evenly (Figure 1.10a) or unevenly, depending on the gate widths (as shown in Figure 1.10b). A three-way division box with flow evenly diverted is shown in Figure 1.10c.

Farm turnouts or off-takes are structures used to control the delivery of water from the canal to the farm (as shown in Figure 1.11). Some of those structures can also be used to measure water. There are several gates used to do this. Examples of these structures include sluice gates, double gates, and single baffle gates.

1.5.4 Water Measurement Structures

Water measurements are needed for better management and water conservation. By measuring the flow of irrigation water, farmers know how much water is applied in

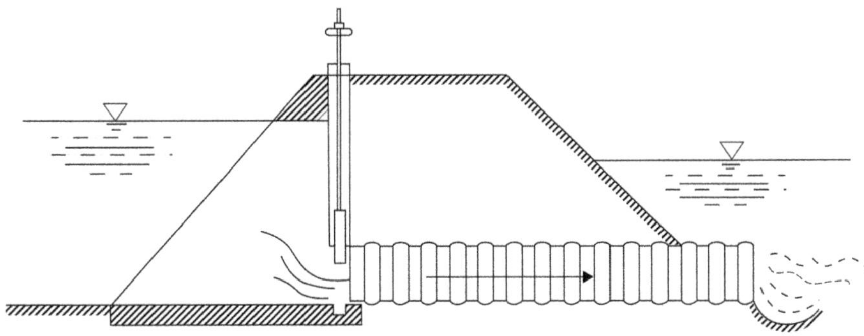

FIGURE 1.11 Farm turnout.

the field. Some irrigation districts or organizations charge for the water based on the volume delivered to the farmers. Water can be measured in secondary and tertiary canals with several structures, such as weirs and flumes. Water flows can be measured with scales or with pressure transducers fed by solar panels. There are special installation requirements for each structure, depending on site elevations and water velocity. Chapter 8 will cover these structures in more detail.

1.5.5 Water Level Control Structures

Several structures are used to divert or control the water levels on canals. For example, checks are structures that are used to raise water levels in the field ditch to divert water to the field. It includes permanent and portable checks, which are detailed in Chapter 3. Some of these water control structures are covered in Chapter 12.

1.6 CLASSIFICATION OF IRRIGATION SYSTEMS

Depending on the type of irrigation subsystems, irrigation systems can be classified as (1) source-conveyance-delivery-receiver systems, (2) source-conveyance-receiver systems, (3) source-delivery-receiver systems, and (4) source-receiver systems. A short description of these systems is in order.

1.6.1 Source-Conveyance-Delivery-Receiver System

All four components constitute an irrigation system. For example, canal irrigation, where the water is delivered from a dam, conveyed by main canals, then by canal branches, and finally by ditches of channels to the farm for irrigation. These are the most versatile and are usually designed for the irrigation of large agricultural lands.

1.6.2 Source-Conveyance-Receiver System

In this case, there is no delivery system. The water is transferred directly from the conveyance system to the farm or the receiver system. Examples are farmlands located along the rivers and canals, aqueducts, etc. Thus, water is received by the receiver system directly. Another example is the artesian well, where water is withdrawn from underground or an aquifer.

1.6.3 Source-Delivery-Receiver System

In this case, water is diverted directly from the source system to the receiver system. Examples are water withdrawn from a reservoir to the farm or water delivered from a well.

1.6.4 Source-Receiver System

The water is delivered to the farm directly from the source. This happens when water is pumped from a reservoir to the farm or from an aquifer.

1.7 DISCIPLINARY CONSIDERATIONS

Irrigation is an interdisciplinary field, and its operation and management are even more so. This is seen by the involvement of different disciplines in the understanding and development of different components.

1.7.1 Source System

The areas involved in planning, design, and development of irrigation systems include (1) climatology, (2) hydrology, (3) hydraulics, (4) geotechniques, (5) structures, and (6) geology. The disciplines (2)–(5) constitute engineering.

1.7.2 Conveyance System

The areas involved in the conveyance system include hydraulics, hydrology, geotechniques, and structures. This is primarily an engineering area.

1.7.3 Delivery System

The delivery system involves climatology, agricultural science, hydrology, hydraulics, geotechniques, and structures. This is primarily an engineering area.

1.7.4 Receiver System

The areas involved in the receiving system are agricultural sources, climatology, hydrology, hydraulics, remote sensing, hydrometry, etc. This combines engineering and agricultural areas.

1.8 SCALE CONSIDERATIONS

The operation and management of an irrigation system depend on the scale involved.

1.8.1 Space Scale

The scale reflects the spatial scale or size. This is primarily measured on the horizontal plane. The irrigation systems can be classified as small, medium, or large.

1.8.2 Time Scale

This scale is associated with the operation of the system. The time scale can be weekly, daily, monthly, seasonal, yearly, etc. The issues of maintenance and operation are associated with the temporal scale.

1.9 IRRIGATION ENTERPRISE

An irrigation enterprise embraces (1) all land within the area of jurisdiction, (2) the water to be used in the enterprise (including water released from the area), (3) the physical system of improvements to move and control water from the source or sources to the farm (including those necessary to care for unwanted water), (4) the people and equipment necessary to operate and manage the enterprise, (5) the people within the area, and (6) improvements and controls to provide for the use of water for other purposes (hydropower, navigation, flood control, etc.). The organizational structure of the enterprise is directly linked with the type of irrigation system. For example, if the system is a microirrigation system, its operation and management is quite different from that of a canal system or sprinkler system.

The enterprise is limited by (1) legal considerations, (2) economic limitations, (3) environmental considerations, and (4) ethical situations. Normally, it does not include activities within the farm areas except those directly affecting the operation of the system.

1.9.1 Objectives and Scope

First, the objectives of the operation and management of irrigation systems should be considered. The objectives to be achieved are (1) to deliver irrigation water equitably to all water users at the lowest possible cost commensurate with the type of service for which the water users are willing to pay and consistent with sound management practices and conservation of water; (2) to maintain the total irrigation system and drainage system, if present, to the degree necessary to preserve the capacity and condition of the facilities in a manner to avoid undue depreciation and to store and deliver irrigation water, as required by the water users; (3) to hold, to a minimum, the system seepage and operational waste; and (4) to be constantly alert to technical advancements and adopt new practices that will improve system operation efficiency.

Second, the operation and management of an irrigation enterprise include all of the various items related to the operation and management of all the facilities associated with the delivery of water to the farms. It must be emphasized that management implies more than just the business aspect but must also include the concept of maintaining a permanent agriculture. This embraces, by implication, any operation that has to do with the management of the land. This larger perspective of management must always be kept in mind by management and operational staff. An emphasis must be placed on business management and the operation and maintenance of physical facilities of the irrigation system if a successful and sustained enterprise is to be accomplished.

Third, many irrigation systems are extensive and complicated. It is anticipated that the construction of facilities for irrigation of agricultural lands will decrease in the future. Thus, there are many real unsolved problems of irrigation in the field of technical management of irrigated areas rather than in the design and construction. Social, political, ecological, and environmental factors will be more acute, as the population increases and the management of the irrigation enterprises will become more complicated. This will be more pronounced in the older projects, where all problems may not have been considered in depth during the planning, design, and construction stages.

Fourth, an understanding of the operation and management of irrigation systems is vital for the successful planning, design, and construction of such systems. Many of the difficulties in the operation and management of systems can be traced to the lack of proper consideration of these matters in the early stages of building the projects.

1.9.2 Aid for Operation and Management

The aids provided by the planner include (1) the concept of enterprise and (2) the goals and objectives; those provided by the designer include (1) design criteria, (2) designer operating instructions, and (3) maps and drawings; and those provided by the construction agency include (1) maps and final drawings, (2) information on sources of equipment, spare part lists, etc., and (3) miscellaneous information on geology, groundwater, etc., that was encountered and photography of the construction.

1.9.3 Use of System

When water is used for purposes other than irrigation in multipurpose projects, the design of the system and its operation and maintenance will vary considerably, depending on the number of other uses and priorities. Multiple-purpose water infrastructures provide a number of goods and services. For example, irrigation for agriculture through a network of distributed canals; hydropower generation; drinking water supply through reservoir storage and water distribution networks; water supply for industrial uses; flood control by regulating water discharge; transport and navigation; and water storage (OECD, 2017). Multiple-purpose water infrastructures account for about 30% of large dams globally. A general visualization of what a multiple-purpose project might include is shown in Figure 1.12.

1.9.4 Other Considerations

It must be taken into account that the operation and management of an irrigation system involve many considerations other than just the physical system of improvements. Some of these considerations are given as follows:

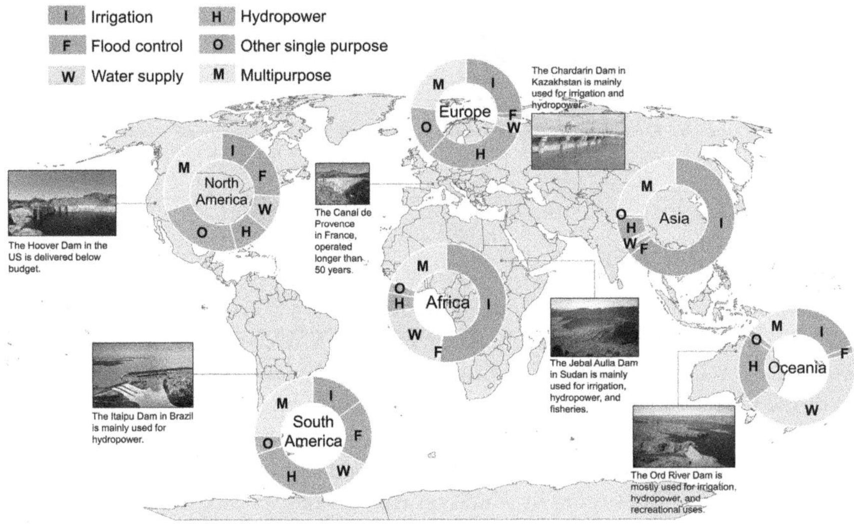

FIGURE 1.12 Examples of multiple-purpose water infrastructures in different regions. (Data from OECD (2018).)

a. The operation and management of an irrigation system must ensure reliable delivery of water to the receiver system. This means delivering water to the user when he needs it in the proper quantities and flow rates. The delivery of water must be accomplished with a minimum of waste. This is important to minimize the impact of damage and adverse consequences, such as conflicts when the supply is unreliable.
b. In a collective system, it is important to have equity in the distribution and avoid preference for specific users. For example, some users receive less water and flow rates, but not when water is needed.
c. The system should be flexible to adapt to changing water demands imposed by weather, management, or changing crop production conditions.

REFERENCES

OECD (2017). *Policy Perspective, Multi-Purpose Water Infrastructure: Recommendations to Maximize Economic Benefits*. OECD Publishing, Paris.

OECD (2018). *Strengthening the Role of Multi-Purpose Water Infrastructure: The Case of Shardara, Kazakhstan*. OECD Publishing, Paris.

2 Irrigation Systems
A Brief History

2.1 INTRODUCTION

Irrigation has been practiced on our globe for some 3,000–4,000 years. The exact date of the beginning will probably never be known. Technical writing was almost non-existent so long ago. However, from other writings, we find references to irrigation or the results of irrigation, although technical information is lacking.

Water is the key limiting factor for crop production and plant survival in most climates and parts of the world. In humid areas, rainfall amounts received vary between years, and the timing in which rainfall is received doesn't generally match crop water requirements for optimum production. In most arid areas of the world, irrigation is essentially necessary for crop production. Rainfall amounts are above the historical average during wet years and below it in dry years. In humid areas and dry years, irrigation is required to increase yield and quality per unit of area. A maximum potential yield can be attained when the crop is well irrigated, and a minimum yield is generally a consequence of dry-land farming. Crop production between these limits will depend on the irrigation applied, climate, water quality, crop response, stage of growth at which water stress was imposed, soil type, and water management.

Irrigation began as a necessity for human survival and expanded as growing populations and civilizations developed. Crop production was originally rain-fed or dry-land. Subsequently, agriculture was practiced along the river's edges, utilizing the receding floodwater. Then, technological advances allowed water to be diverted from rivers and later extracted from aquifers. Presently, irrigation must be justified economically, socially, and environmentally to be sustainable. Technological advances are incorporated to accomplish irrigation more efficiently and cost-effectively. These technological advances can imply implementing better planning or design of the system, using more efficient tools, and improving the operation and maintenance of irrigation schemes. By studying the first irrigation projects from ancient civilizations, we can study the evolution, progress, and fate of these irrigation projects and learn from their mistakes and successes in achieving sustainability.

This chapter provides a snapshot of irrigation practices of ancient civilizations using surface or groundwater, irrigation development in the 19th and 20th centuries, and the growth of irrigation systems worldwide.

DOI: 10.1201/9781003747444-3

2.2 IRRIGATION PRACTICES OF ANCIENT CIVILIZATIONS

2.2.1 Mesopotamia and Babylonia (Presently Iraq)

The Sumerians, who are considered as one of the first known civilizations, settled in an area known as Mesopotamia (now Iraq) around 7000 BC. Mesopotamia means "land between the rivers" and lies in an extensive valley with fertile soil between the Euphrates and the Tigris Rivers. Both rivers flow from the mountains of Turkey. The Euphrates also passes through Syria, and both rivers empty into the Persian Gulf.

As the Sumerian civilization flourished, several cities grew, closely confined to the banks of these rivers around 4000 BC. Sumerians drew water inland to the Valley by branching canals from the main river watercourses. Water was diverted to the canals through breaches cut in the levees of the river's upland. Initially, the operation of canals resulted in uncontrolled flooding. Later, these canals developed natural regimes despite their artificial origin. According to ancient records, several problems emerged between 2400 and 1700 BC, probably due to operation and maintenance problems. Over-irrigation created a rise in the groundwater table and salinization of areas, especially in the southern parts of the alluvial plains. Patches of saline ground were observed in several fields. An example of this, as explained by Jacobsen and Adams (1958), is that in 3500 BC, the proportion of wheat and barley was nearly equal in the southern part of the valley. About 1,000 years later, the less salt-tolerant wheat accounted for only one-sixth of the crop. In 2100 BC, wheat accounted for less than 2% of the planted area, and by 1700, it was abandoned. Concurrent with the shift from wheat to barley in the same southern area, there was a severe decline in productivity. Barley yields declined from 2,537 kg/ha in around 2400 BC to 1,460 kg/ha by 2100 BC. By around 1700 BC, the yield shrank to 897 kg/ha. The irrigation area was enlarged to an unprecedented scale around 226–664 AD. The artificial canals branched, and more feeding canals and weirs were built. Several brick sluice gates along the branches were also built. Many people were required to build the canals and maintain them free of weeds and silt. In spite of the outstanding engineering progress, several problems emerged, such as the accumulation of silt on canal beds, which required periodic maintenance. Social unrest, political authorities with military adventures, and intrigues caused the strong central authority to relinquish the complex maintenance system to local communities (Jacobsen and Adams, 1958). The lack of organization among these communities, who were ill-equipped to maintain the system, sharply reduced the flow and the head of water in branch canals, causing a temporary abandonment of the area in the 11th and 12th centuries.

2.2.2 Egypt (Nile River)

One of the first civilizations settled in the fertile floodplains of the Nile River around 7000 BC. During that time, sufficient rainfall occurred in the upper Sahara. Annual flooding in the area near the river spread rich mineral-coated silt, which made the land fertile. After the water receded, Egyptians did planting, and the crops were irrigated with ditches and canals. Even though climate change around 2500 BC dried the Sahara, irrigation allowed cities to continue thriving and growing, forming a great empire. Advances in engineering skills allowed Egyptians to build a network of irrigation canals and sluice gates on the sides of the river. They grew crops such as wheat to make bread, barley to make beer, and flax to weave linen for clothes.

2.2.3 China

The first Chinese settlements started on the banks of the Yellow and Yangtze Rivers around 7000 BC. The rivers provided fertile soil but caused frequent devastating floods. This allowed the people to grow rice in the wet areas. Chinese irrigation probably spread from Babylonia and probably began before 2200 BC as communities grew larger and the first Chinese dynasties appeared. The government's success was often measured by the wisdom of their monarchies and the progress in controlling the waters of its rivers. A good example was Yu the Great, who restored flooded lands during the Xia Dynasty around 2200–1700 BC (Li, 1914). Notably, Chinese irrigation works included the Dujiangyan built in 256 BC by the State of Qin. The Dujiangyan irrigation system was located near the Min River, which is the longest tributary of the Yangtze River. It included three main constructions, i.e., Fish Mouth Levee, Flying Sand Weir, and Bottle-Neck Channel. These three main constructions worked together, with the result that the area became the most productive agricultural region in China. This irrigation system is working properly to date, irrigating an approximate area of 500,000 ha of fields. Another notable irrigation work is the Grand Canal, also known as the longest ancient canal, which was built in stages over several centuries. This canal connects the cities of Beijing and Hangzhou and extends over 1,770 km, linking the Yellow and Yangtze Rivers (Needham, 1986).

2.2.4 India

The Indus Valley Civilization (2500–1900 BC) was known for its sophisticated canal systems and hydraulic structures. The Rigveda, an ancient Indian scripture, mentions well-style irrigation methods, including kupa and avata wells that were perennially full of water (Puri, 1968). The water, drawn using a varatra (rope strap) and a cakra (wheel) to lift kosa (pails), was then distributed through surmi susira (broad channels) and subsequently into khanitrima (diverting channels) for field irrigation.

Indian scholar Pāṇini documented the tapping of several rivers, such as the Sindhu and Sarayu, for irrigation purposes around 400 BC (Gopal, 2008). During the Maurya Empire era (300 BC), the state charged farmers for irrigation services from rivers, indicating a more organized approach to water management. The Yogasutra, written by Patanjali around 400 AD, further illustrates ancient irrigation practices (White, 2014). In Tamil Nadu, a significant advancement was the construction of the Grand Anicut (canal) across the Kaveri River during the 300 AD, a design still in use today (FAO, 2014).

The medieval era witnessed an expansion of irrigation systems, particularly under Sultanate rulers. Firoz Shah Tughlaq (14th century) built extensive canal systems in the Indo-Gangetic doab and regions west of the Yamuna River (Siddiqui, 1986). These canals were later expanded upon by Mughal rulers, providing significant water resources for agriculture and settlements in northern India.

In the colonial era, the British initiated the modernization of irrigation in India by constructing large dams, canals, and channels. By 1800, about 800,000 ha were irrigated, a number that increased eightfold by 1940, spanning regions like Uttar Pradesh, Bihar, Punjab, Assam, and Orissa (FAO, 2014). The Ganges Canal, reaching 350 miles from Haridwar to Kanpur, was a notable achievement. This period saw a dramatic increase in the land under irrigation, partly in response to major famines in the 19th century.

2.2.5 Americas

On the coast of Peru at the Viru Valley, Andean societies based their growth on irrigation systems and construction of terraces between 400 BC and the Time of Christ. By 800 AD, the population of the Viru Valley reached its peak, and from 1200 AD, it dramatically decreased. People were forced to relocate from the densely settled valley bottoms to the upper narrows of the valleys. Historians attribute the relocation to several reasons, including the increasing salinity and rising water tables (Willey, 1953).

About 300 BC, the Teotihuacans relocated to what is now Central Mexico. They irrigated their fields by diverting water from shallow spring-fed streams into simple irrigation canals that flowed into their fields.

Around 1300–1400 AD, a culture that developed an outstanding hydraulic scheme by draining land and manipulating water resources was the Aztecs (Michael Coe referenced by Sabloff, 1989). The Aztecs reclaimed swamps and constructed agricultural plots called "chinampas" and built an immense dike to keep salty water from the eastern and northern parts of the lake system to mix and diminish the water quality of the west and south lakes (Michael Coe referenced by Sabloff, 1989). The Aztecs built their city on two islands in the middle of the lake. They also created chinampas in the shallower depths of the lake by inserting stakes and planting willow trees in small plots that were filled with weeds and mud. They took the silt from the bottom of the canals and poured it on the chinampas. The Aztecs produced two crops per year, such as corn, beans, and squash. They were able to irrigate their crops with sub-irrigation from canals. They were also able to control salinity by rotating crops and alternating planting years. They also used the channel system to distribute their products.

In the Americas, the Hohokam and the Anasazi subsisted in arid and semi-arid lands as a consequence of their irrigation infrastructure. The Hohokam constructed a complex irrigation system in the desert lowlands of Arizona's Salt–Gila River basin, Arizona, between 300 BC and 1 AD. They built more than 300 miles of main canals and 700 miles of distribution canals in the Salt River basin. It is believed that the irrigation network was destroyed by flooding. Poor maintenance caused their canals to silt, and waterlogging and poor drainage produced salinity around 1400 AD (Mays, 2007).

In the high deserts of the Colorado Plateau, within the San Juan Basin of northwestern New Mexico, the Anasazi had their homeland. The Anasazi developed an interesting method of collecting and diverting runoff from Chaco Canyon. They collected water from a side canyon at the top of the upper mesa and diverted it into canals by an earthen masonry dam near the mouth of the side canyon. The canals averaged 4.5 m in width and 1.4 m in depth, and some were lined with stone slabs, while others were bordered with masonry walls. The canals ended at the masonry head gate where water was diverted to the field via small ditches. The irrigation system operated from 600 to 1200 AD. It is believed that this civilization collapsed mainly because of deforestation and the clearing of vegetation, which resulted in erosion of the arroyos in lower elevations of the water supply levels to that of the irrigated fields (Mays, 2007).

2.3 IRRIGATION FROM GROUNDWATER

About 75% of the water used for all purposes (excluding hydropower and navigation) is from surface water sources (FAO, UN, 2021). However, in Asia and the Near East, (1) extensive dry areas, (2) dense population, and (3) dominance of agriculture led to the use of groundwater, which was important to the development of the area. These conditions led to the development of the art of constructing wells and infiltration galleries. Accounts of well water and well construction are found in ancient literature, particularly the Biblical record in Genesis.

Very large diameter wells were excavated in the Near East, some of them with ramps to admit donkeys and a depth of about 50 m. The churn drill was invented in China in 211 BC by the Qin dynasty, resulting in deep wells of small diameters (Geng, 1997). The principle of the churn drill is the same today; the machinery has been greatly improved. An account of an ancient Chinese well 1,500 m deep drilled with tools is available.

Kanats (long infiltration galleries) were used (and are still in use) to collect water from alluvial fans and soft sedimentary rocks. These were formed with a series of wells, or vertical shafts, connected by tunnels. Tunnels had flat slopes and reached the surface at the point of the use of water. This system probably originated in what is now Iran some 2,500 years ago. The practice spread rapidly east to Afghanistan and west to Egypt (later, it spread west to Morocco and east to India). One system, located on the left or west flank of Nile Valley in Egypt, built about 500 BC, provided irrigation for 4,700 km^2 (Olmstead, 1948). The most extensive systems are in the alluvial fans from the Elburz Mountains in Iran.

Today, drilling technology has advanced, more efficient pumps have been developed, and more is known about irrigation technology. Groundwater is used more as a source of irrigation water, although proportionally much less than the use of surface water, yet it is important to sustain irrigated agriculture. Even today, many areas of the world have ancient irrigation systems in use, and they seem to be serving their purpose well.

2.4 IRRIGATION DEVELOPMENTS IN THE 19TH AND 20TH CENTURIES

Several technological inventions, such as the use of new materials and irrigation techniques, have improved productivity per unit of water applied and per unit of area, have reduced production costs, and have increased the sustainability of the systems in the last century. Advances in irrigation, such as drainage, measurement of water flow, and on-farm irrigation management, have been developed in several diverse areas.

John Johnston was the first person to lay tile to drain wet soils on a farm field in Geneva, New York, in 1835. Excess water from irrigation in soils with poor drainage and high water tables can now be artificially drained with tile drainage. Drainage can improve root aeration and dispose of excess water and salts, creating a better environment for plants to grow. The depth and spacing between drain tiles are important factors that are being continuously improved with mathematical calculations and computer simulations based on soil properties.

The use of pumps or gravity-fed systems permitted the development of drip and sprinkler irrigation, which were rapidly adopted and improved during the 20th century. The first experiments with drip irrigation started in Germany in 1860. They used subsurface irrigation from drainage tiles of short clay pipes with open joints. The lines were buried at 0.8 m depth and spaced every 5 m. The system worked for 20 years. This system was later modified in Germany in 1920 by using perforated plastic pipes to irrigate (Keller and Karmelli, 1974). Similar experiments were conducted in the Soviet Union and France in 1923. Then irrigation was extended throughout Europe, South Africa, Israel, Mexico, and the United States. Drip irrigation changed the concept of using the soil as water storage by the concept of meeting evapotranspiration water demand with more frequent irrigation (Howell et al., 1980).

Advances in flow measurements have allowed for better irrigation management. The Parshall Measuring Flume was invented by Ralph L. Parshall in 1922 at the Hydraulics Laboratory of the Lory Student Center at Colorado State University in Fort Collins. The flume is still being used worldwide to this day.

In sprinkler irrigation, several advances allowed its expansion. One invention was the impact sprinkler head by Orton Englehardt in Glendora, California, in 1933 and commercialized by Rain Bird (ASABE, 2008). Another invention that was patented by Frank Zybach on July 22, 1952, a tenant farmer and inventor living near Strasburg, Colorado, was the "Self-Propelled Sprinkling Irrigating Apparatus". The center pivot consists of several segments of pipe joined together and supported by trusses and towers, with the sprinkler positioned along the pipe and moving continuously around the pivot. Water is fed from the pivot at the center of the circle. Zybach formed a partnership with A. E. Trowbridge, an entrepreneur and businessman, in 1953 to manufacture center pivots in Columbus, Nebraska. In 1954, they sold the manufacturing rights to Valley Manufacturing, starting a worldwide industry that included several manufacturers. There were several companies that manufactured the center pivots in Nebraska, and they have since been improved (ASABE, 2008). The center pivot was the first automatic system that produced high irrigation efficiency and uniformity. According to the ASABE, center pivots have become the most widely used form of sprinkler irrigation. By 1992, over 110,000 machines were being used to irrigate 14 million acres in the United States.

In surface irrigation, the use of laser equipment to grade the land was developed by agricultural engineers, James Fouss and Norman Fausey of the U.S. Department of Agriculture (USDA)'s Agricultural Research Service (ARS) at Ohio State University in the mid-1960s (ASABE, 2008). Since then, this technology has continuously improved and has helped increase irrigation efficiency.

2.5 GROWTH OF IRRIGATION SYSTEMS

The practice of irrigation is almost worldwide. Irrigation systems are expensive, must be considered as relatively permanent, and are quite complicated. Irrigation systems cannot be operated and maintained without a considerable number of technically trained individuals. Water is in limited supply in the world and must be optimally used. Irrigation is the only use that competes with many other uses. Management must be practiced to maintain a permanent irrigated culture.

Early irrigation systems were direct diversions from streams and served small areas. In some areas, these were single wells. Later, cooperative schemes were formed where, perhaps, two or more irrigators banded together to build slightly more elaborate systems. Still later, various types of organizations were formed to provide irrigation water to large tracts of land. Some of these included storage facilities. As these systems grew in extent and cost, they progressed from single-purpose projects for irrigation to multipurpose projects, which are common today. The multipurpose projects are usually large. These projects serve many other purposes, such as production of hydropower, navigation, flood control, sediment control, recreation, fishing, and municipal and industrial supplies.

Governments throughout the world became involved with different agencies and provided capital and expertise to construct bigger dams to expand irrigation. In the United States, agriculture expanded rapidly between 1900 and 1940 thanks to the ability of generating and selling hydroelectric power, making water affordable for irrigation purposes. The private sector has played an important role in expanding irrigation from large aquifers such as the Ogallala in the Great Plains, in the southern states, and in Florida (CAST, 1998).

The total geographical area, cultivated area, and irrigated area of the world by continent are shown in Table 2.1. Cultivated areas account for only about 12% of the global land surface, yet they provide more than 80% and 60% of worldwide calorie and protein requirements, respectively (Ritchie and Roser, 2019). Based on estimates from the Food and Agriculture Organization (FAO), further expansion of cultivated

TABLE 2.1
Summary Statistics of Irrigated Areas by Continent in 2019

Regions	Total Geographical Area (Mha)	Area Cultivated (Mha)	Area Irrigated (Mha)	Area Irrigated (Percent)	Population[a] (Millions)
America	4,065.1	368.2	56.0	15.2	1,022.8
Asia	3,199.9	590.1	239.6	40.6	4,641.1
Africa	3,035.0	275.7	16.4	6.0	1,340.6
Europe	2,341.2	288.8	26.3	9.1	747.6
Oceania	849.6	33.3	3.3	9.9	42.7
World	13,497.3	1,556.1	341.6	22.0	7,794.8

Source: Data from FAO (2021).
[a] Population data were in 2020.

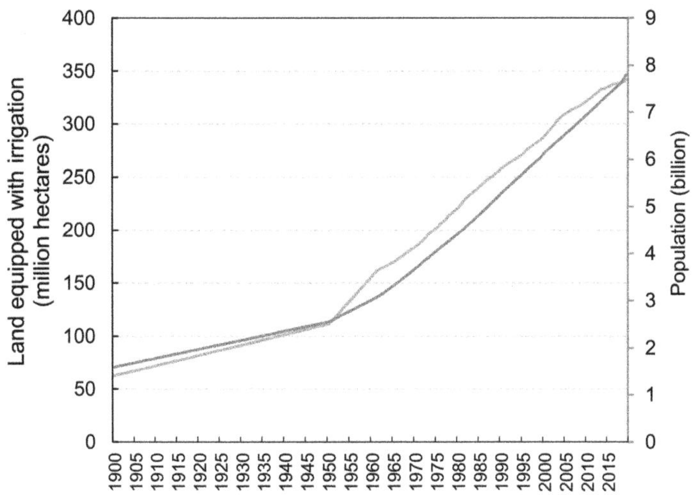

FIGURE 2.1 World population and world irrigated land. (Data from FAO (2021).)

areas is very limited, with a projected net increase of 0.7 M km^2 by 2050 compared with the 2010 level (Alexandratos and Bruinsma, 2012). Therefore, increasing productivity on existing cultivated land is considered the primary solution to feeding the growing population. With proper design and management, modern irrigation methods can significantly increase crop yields and therefore play an important role in future food security. As shown in Figure 2.1, at the beginning of the 1900s, about 60 Mha of the cultivated land was irrigated, but it rose to more than 110 Mha in 1950. A more rapid increase was observed after the 1950s, and it increased to 341.6 Mha in 2019, accounting for about 22% of the cultivated area (Table 2.1). There is a trend indicating the relation between global irrigated area and population, as shown in Figure 2.1. This trend will be less apparent with globalization and as surpluses move from high-production areas to highly populated demand areas.

The largest irrigated areas are in India and China, which together account for about 42% of the global irrigated cultivated area (Table 2.2). Domestic food production in these two countries highly relies on irrigation, with relatively large proportions of irrigated land. Other countries, such as Egypt, Uzbekistan, Bangladesh, Pakistan, Iran, and Iraq, also have high proportions of irrigated cultivated areas. Figure 2.2 shows the spatial distribution of global irrigated cultivated areas (Siebert et al., 2013).

The types of irrigation systems have also changed with time. For example, surface irrigation accounted for 98% of the irrigated land in the United States in the 1950s, and this type of irrigation system has declined to 32% in 2017 (ICID, 2021). Meanwhile, the sprinkler irrigation system has gradually increased from 2% in 1950 to 60% in 2017. Microirrigation, such as drip or trickle irrigation, began in the 1960s and accounted for 8% of the irrigated land in the United States in 2017. Table 2.3 lists 12 leading countries equipped with sprinkler and microirrigation methods.

TABLE 2.2
Summary Statistics of the Largest Countries Irrigated Area Wise

Country	Total Geographical Area (Mha)	Area Cultivated (Mha)	Area Irrigated (Mha)	Area Irrigated (Percent)	Population[a] (Millions)
1. China	960.0	135.7	74.2	54.7	1,439.3
2. India	328.7	169.3	70.4	41.6	1,380.0
3. United States	983.2	160.4	26.9	16.8	331.0
4. Pakistan	79.6	31.3	20.2	64.5	220.9
5. Iran	174.5	17.5	9.6	54.7	84.0
6. Brazil	851.6	63.5	7.0	10.9	212.6
7. Mexico	196.4	22.1	6.8	30.8	128.9
8. Indonesia	191.7	51.3	6.7	13.1	273.5
9. Thailand	51.3	21.3	6.4	30.1	69.8
10. Bangladesh	14.8	8.8	5.6	63.1	164.7
11. Turkey	78.5	23.2	5.2	22.5	84.3
12. Vietnam	33.1	11.7	4.6	39.2	97.3

Source: Data from FAO (2021).
[a] Population data were from 2020.

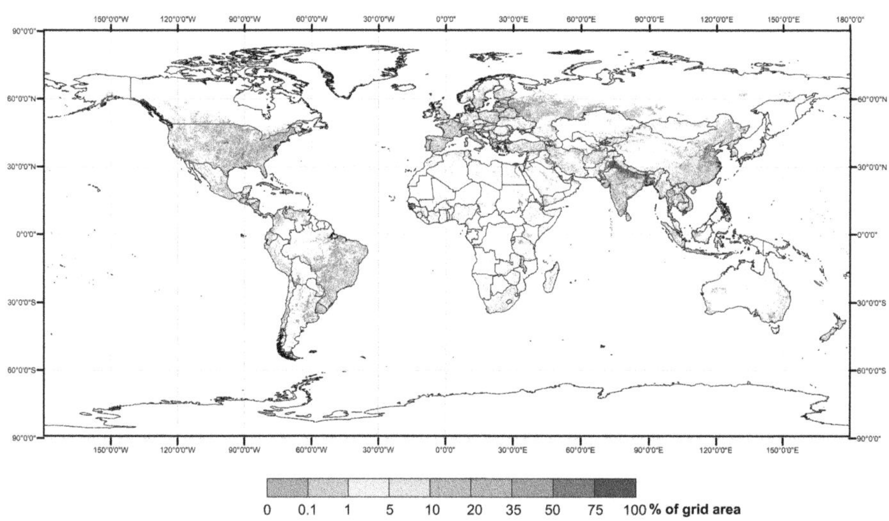

FIGURE 2.2 Global distribution of irrigated cultivated area in the percentage of grid area (%).

2.6 OPERATION AND MANAGEMENT

The operation and management of an irrigation system must be an integral part of the development of irrigation projects. That is, the original concept of the project should include consideration of operation and management. Also, continued consideration

TABLE 2.3
The 12 Leading Countries Equipped with Sprinkler and Microirrigation Methods

Country	Sprinkler Irrigation (Mha)	Microirrigation (Mha)	Sprinkler + Microirrigation		Year
			Total (Mha)	% of Total Irrigated Area	
United States	14.12	1.98	16.10	68.6	2017 and 2013
China	3.73	5.27	9.00	13.7	2017
India	3.04	1.90	4.94	8.0	2010
Brazil	3.86	0.62	4.48	77.3	2013
Spain	0.89	1.79	2.68	73.7	2015
Turkey	1.43	1.12	2.55	38.3	2020
Russia	2.50	0.05	2.55	56.6	2012
Iran	1.00	1.02	2.02	23.9	2019
France	1.38	0.10	1.48	57.0	2017
Italy	0.96	0.42	1.38	57.1	2013
South Africa	0.92	0.37	1.29	77.0	2017
Australia	0.82	0.22	1.04	48.2	2017

Source: Data from ICID (2021).

should be given during (1) reconnaissance studies, (2) feasibility studies, (3) design, (4) construction, and (5) the transition period from construction to operation.

Thus, operation and management problems do not just begin after construction entirely. These problems are greatly alleviated if proper planning is achieved throughout the development of the original concept. Design criteria, designs, operating instructions, estimated operation, management, and research costs included in economic studies to arrive at annual costs, and similar information will aid the operation and management.

REFERENCES

Alexandratos, N. and Bruinsma, J. (2012). *World Agriculture: Towards 2030/2050 – The 2012 Revision*. ESA Working Paper No. 12–03. FAO, Rome.

ASABE (2008). ASABE historic agricultural engineering landmarks. https://asabe.org/about-us/about-asabe/history/asabe-historic-landmarks

CAST (1998). Effective Use of Water in Irrigated Agriculture. Council for Agriculture and Science and Technology. Report No. 113. ISSN 0194-4088.

FAO (2014). *AQUASTAT Country Profile-India*. Food and Agriculture Organization of the United Nations (FAO), Rome.

FAO (2021). *FAOSTAT Database*. Food and Agriculture Organization of the United Nations (FAO), Rome. https://www.fao.org/faostat/en/#data

FAO, UN (2021). *Progress on Level of Water Stress. Global Status and Acceleration Needs for SDG Indicator 6.4.2*. Food and Agriculture Organization of the United Nations and United Nations Water, Rome.

Geng, R. (1997). Advanced geodrilling techniques in China. In: H. Guo, V. Singhroy, and T.G. Farr, Eds., *New Technology for Geosciences: Proceedings of the 30th International Geological Congress*. VSP, Utrecht. pp. 264.

Gopal, L. (2008). *History of Agriculture in India, up to c. 1200 A.D.* CSC and Concept Pub. Co. for the Project of History of Indian Science, Philosophy, and Culture, New Delhi, pp. 762.

Howell, T.A., Stevenson, D.S., Aljibury, F.K., Gitlin, H.M., Wu, I.P., Warrick, A.W. and Raats, P.A.C. (1980). Design and operation of trickle (drip) systems. In: M.E. Jensen, Ed., *Design and Operation of Farm Irrigation Systems*. 1st ed. ASAE Monograph No. 3. The American Society of Agricultural Engineers, St Joseph, Mich, pp. 663–713.

ICID (2021). *Annual Report 2020–21. Agricultural Water Management for Sustainable Rural Development*. International Commission on Irrigation and Drainage, New Delhi.

Jacobsen, T. and Adams, R.M. (1958). Salt and silt in ancient mesopotamian agriculture. *Science, New Series*, 128(3334), 1251–1258.

Keller, J. and Karmelli, D. (1974). *Trickle Irrigation Design*. 1st ed. Rain Bird Sprinkler Mfg. Corp., Glendora, pp. 133.

Li, U.B. (1914). *Outlines of Chinese History*. Shanghai, Commercial Press, edited by Professor Whiteside, of Soochow, China. https://archive.org/details/outlinesofchines00liun/page/n3/mode/2up

Mays, L.W. (2007). Water sustainability of ancient civilizations in Mesoamerica and the American Southwest. *Water Science and Technology: Water Supply*, 7(1), 229–236.

Needham, J. (1986). *Science and Civilization in China: Volume 4, Physics and Physical Technology, Part 3, Civil Engineering and Nautics*. Cambridge University Press, Cambridge.

Olmstead, A.T. (1948). *History of the Persian Empire*. The University of Chicago Press, Chicago, IL, pp. 224.

Puri, B.N. (1968). Irrigation and agricultural economy in Ancient India. *Annals of the Bhandarkar Oriental Research Institute*, 48/49, Golden Jubilee Volume 1917–1967, 383–390.

Ritchie, H. and Roser, M. (2019). How much of the world's land would we need in order to feed the global population with the average diet of a given country? https://ourworldindata.org/agricultural-land-by-global-diets

Sabloff, J. (1989). *The Cities of Ancient Mexico Reconstructing a Lost World*. Thames and Hudson, New York, 224 pages.

Siddiqui, H.I. (1986). Water works and irrigation system in India during Pre-Mughal Times. *Journal of the Economic and Social History of the Orient*, 29(1), 52–77.

Siebert, S., Henrich, V., Frenken, K., and Burke, J. (2013). *Global Map of Irrigation Areas Version 5*. Rheinische Friedrich-Wilhelms-University, B., Germany. Food and Agriculture Organization of the United Nations, Rome.

United Nations Population Division (2019). World Population Prospects 2019. https://www.un.org/development/desa/pd/news/world-population-prospects-2019-0

Willey, G.R. (1953). *Prehistoric Settlement Pattern in the Viru Valley, Peru*. Smithsonian Institute, Bureau of American Ethnology, Bulleting 155, Washington, DC, 453 p.

White, D. (2014). *The Yoga Sutra of Patanjali: A Biography*. Princeton University Press, Princeton, NJ, pp. 139.

3 On-Farm Irrigation Practices

3.1 INTRODUCTION

Irrigation is the process of applying water to the soil to provide moisture for plant growth. Some knowledge of the methods of application and their characteristics is therefore useful for the operation and management of the irrigation system. The irrigation organization (e.g., a district or a government department) makes decisions about the operation of the conveyance system, and the end-water user (i.e., the farmer) makes on-farm irrigation decisions. Understanding the operation and management of on-farm irrigation practices can help improve the service to the end user, efficiency in the operation, and water conservation. The on-farm irrigation application may be accomplished in three ways: by surface means, by sprinkling, or through microirrigation. In this chapter, we will discuss these three on-farm irrigation practices.

3.2 SURFACE IRRIGATION

Surface irrigation is the oldest and most widely used irrigation method. It is used in deep soil with medium to fine textures where the interval between irrigations can be spaced. Surface irrigation requires the application of big volumes of water, and generally, water is rotated from one field to another. Water is supplied and delivered to the soil, where it infiltrates as water advances through the soil. The excess is removed in some surface systems. This is contrasted to sprinkler and microirrigation, in which water is distributed through pressurized pipes. Surface irrigation systems include flooding in basins or checks, border irrigation, furrow irrigation, and wild flooding. Water is conveyed to the soil surface, and it infiltrates. Wild flooding allows water to flow freely over the field surface and is without any regulation of the application or uniformity. Therefore, wild flooding is not introduced here.

3.2.1 Flooding in Basins or Checks

A basin consists of a small area surrounded by dykes to control runoff (Figure 3.1). The field needs to be level and subdivided into several small areas or basins. The size of a basin is usually selected to fit the machinery being used, and its size depends on the available stream size and soil type (Table 3.1). The width of the basin primarily depends on the topographic slope, as shown in Table 3.2. For steep slopes, the basin

DOI: 10.1201/9781003747444-4

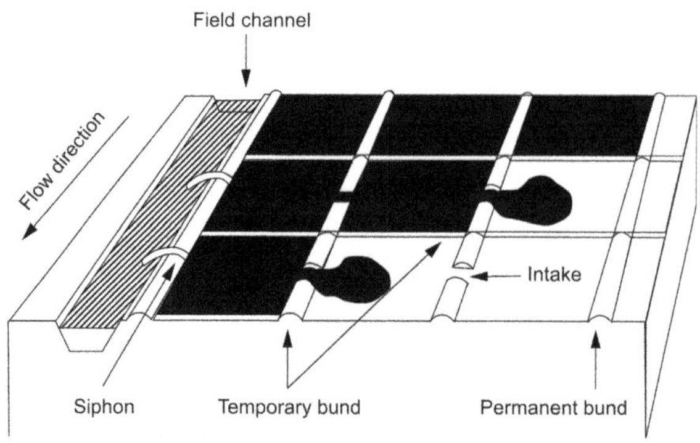

FIGURE 3.1 Basin flood irrigation.

TABLE 3.1
Maximum Basin Areas (m²) for Various Soil Types and Available Stream Sizes

Flow Rate (L/s)	Sand	Sandy Loam	Clay Loam	Clay
5	35	100	200	350
10	65	200	400	650
15	100	300	600	1,000
30	200	600	1,200	2,000
60	400	1,200	2,400	4,000
90	600	1,800	3,600	6,000

Source: After Brouwer et al. (1985).

TABLE 3.2
Approximate Values for the Maximum Basin Width for Different Land Slopes

	Maximum Width (m)	
Slope (%)	Average	Range
0.2	45	35–55
0.3	37	30–45
0.4	32	25–40
0.5	28	20–35
0.6	25	20–30
0.8	22	15–30
1.0	20	15–20
1.2	17	10–20
1.5	13	10–20
2.0	10	5–15
3.0	7	5–10
4.0	5	3–8

Source: After Brouwer et al. (1985).

FIGURE 3.2 Zero level basin irrigation (a rice plantation in South Carolina, United States).

width should be narrow. Otherwise, too much earth movement is required to obtain level basins. Other factors that affect basin width include the depth of fertile soil, basin construction method, and agricultural practices. Medium- to fine-textured soils are more suitable for basin irrigation than coarse-textured soils, since they have relatively high water-holding capacity. Basin irrigation is commonly used for rice paddies or for crops that are unaffected by ponded water resting for periods of more than 24 hours. Basins are used to flood a small area as fast as possible and prevent water from flowing to adjacent fields. With laser technology, it is now possible to increase the size of the basins reported in Table 3.1 and increase uniformity as the zero level basin, as shown in Figure 3.2.

The flow of water in a field ditch can be obstructed with a check structure to raise the water level and allow higher discharge. The check structures can be temporary (Figure 3.3) or permanent (Figures 3.4 and 3.5). In Figure 3.3, the citrus is irrigated with border irrigation methods, and water is diverted to the border by cutting into the field ditch. Permanent check structures are illustrated in Figure 3.4, and a wooden check structure in a rice plantation in South Carolina, United States, is shown in Figure 3.5.

3.2.2 Border Irrigation

Border irrigation consists of long graded strips of land separated by earth dykes with free drainage at the lower end (Figure 3.6). Borders can adapt to the topography of

FIGURE 3.3 Temporary check used to irrigate citrus with border irrigation (flood irrigation in citrus orchards in Harlingen Irrigation District in Texas).

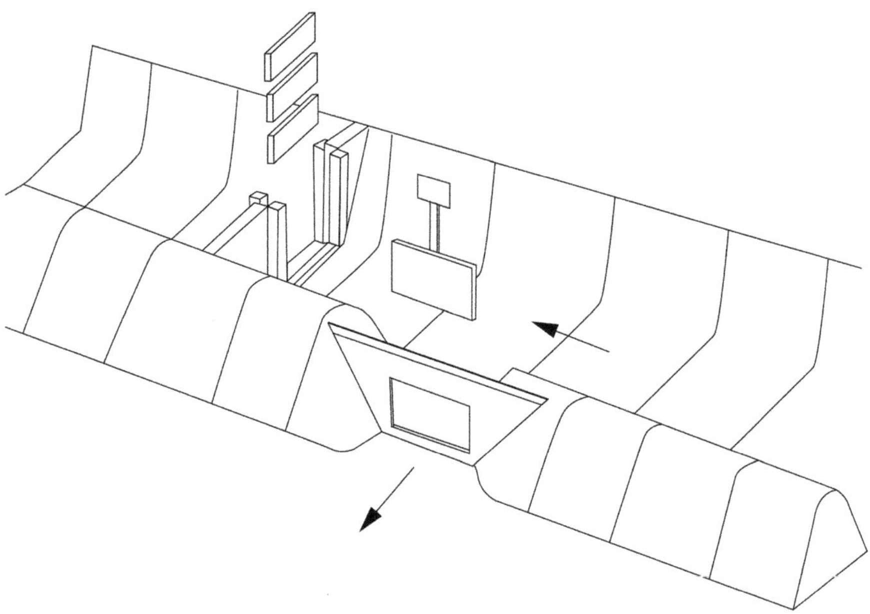

FIGURE 3.4 Permanent check structures.

FIGURE 3.5 Permanent check structures in a rice plantation in South Carolina.

the land, can have rectangular shapes (Figure 3.6), or can follow the contour of the land (Figure 3.7) on steeper (slope) land and generally have longer lengths than have the basins. Border irrigation is typically more advantageous for larger, mechanized farms due to its design, which facilitates long, uninterrupted field lengths for easier machinery operations. Borders can vary in size, often ranging from 3 to 30 m in width and extending up to 800 m or more in length, depending on various factors. However, this method may be less effective for small-scale farms that depend on manual labor or animal-powered cultivation.

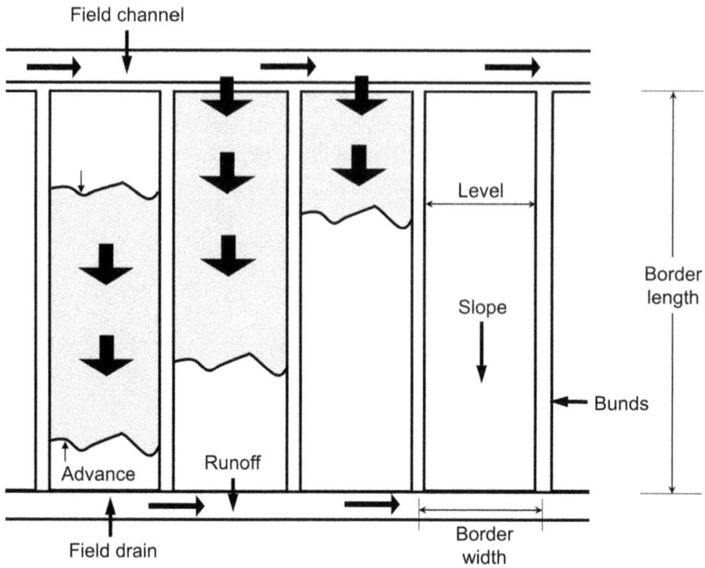

FIGURE 3.6 An example of a border irrigation system.

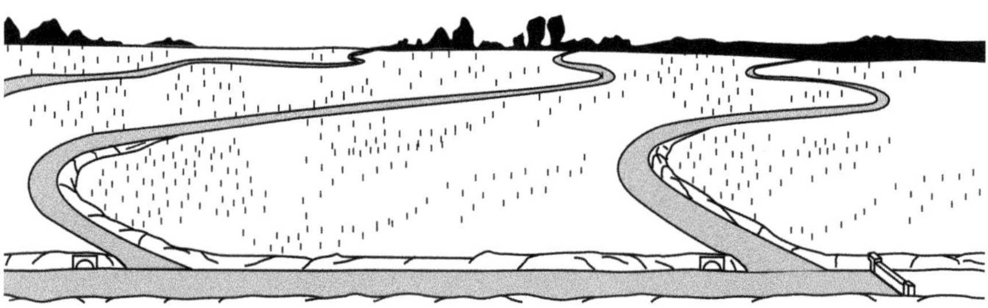

FIGURE 3.7 An example of a contour border irrigation system. (Adapted from USDA-NRCS, 2012.)

Regarding terrain suitability, border slopes should be consistently uniform. A minimal slope of 0.05% is necessary for effective drainage, while a maximum slope of 2% is recommended to minimize soil erosion risks. As for soil requirements, border irrigation is best implemented in areas with deep, homogeneous loam or clay soils that have moderate infiltration rates. In contrast, heavy, clay soils may pose challenges for border irrigation due to the prolonged time required for water infiltration. In such cases, basin irrigation is often a more suitable alternative. When it comes to crop suitability, border irrigation is ideal for closely grown crops, such as pasture or alfalfa, which can thrive under this irrigation method (Brouwer et al., 1985).

The width of a border will depend on the machinery being used. Widths of 15–20 m are commonly used on flat soils; on slopes of 0.3%–0.4%, they should not exceed 10–12 m; and on slopes of 0.5% or greater, the width should not exceed

TABLE 3.3
Suggested Maximum Typical Flow Rates, Border Widths, and Lengths for Different Soil Types

Soil Type	Border Slope (%)	Unit Flow Per Meter Width (L/s)	Width (m)	Length (m)
Sand (infiltration rate greater than 25 mm/h)	0.2–0.4	10–15	12–30	60–90
	0.4–0.6	8–10	9–12	60–90
	0.6–1.0	5–8	6–9	75
Sandy Loam (infiltration rate 10–25 mm/h)	0.2–0.4	5–7	12–30	90–250
	0.4–0.6	4–6	6–12	90–180
	0.6–1.0	2–4	6	90
Clay (infiltration rate less than 10 mm/h)	0.2–0.4	3–4	12–30	180–300
	0.4–0.6	2–3	6–12	90–180
	0.6–1.0	1–2	6	90

Source: After Brouwer et al. (1985).

6–9 m. Water can be applied using gated pipes or flexible pipes, or can be applied from earth ditches using siphon tubes, spikes, or by opening the ditch. Water mainly moves vertically downward as the sheet of water advances between the two dykes. Some typical flow rates for different soils are shown in Table 3.3.

The most common irrigation method for citrus and other fruit trees on the farm is flood irrigation with graded borders (Figure 3.8). To irrigate efficiently with flood irrigation, it is recommended to level the land to the appropriate grade before establishing the orchard and control water applications with valves or structures (Figure 3.9). Fruit orchards that are bordered and properly graded do not produce runoff. To distribute water faster and more efficiently, it is recommended to install alfalfa or orchard valves at different locations in the orchard using gated or flexible pipes. Permanent borders can be built every two rows, with an irrigation valve between each pair (Figure 3.8). Temporary borders may be single or double rows, depending on the grower's preferences. For better control and faster irrigation, it is recommended to build one border per row of trees. The border edge is about 30 cm high. To reduce the irrigated area, temporary borders can be placed along one side of the rows of young trees. This method, called strip flooding or narrow-border flood, allows faster water advancement.

3.2.3 Furrow Irrigation

In partial flooding, the entire irrigated area is partially flooded. The furrows may be small furrows or channels that carry water down the slope between crop rows (Figure 3.10). Water infiltrates, as it moves down the furrow. The crop is generally grown on top of the ridge (Figure 3.10). Furrow irrigation is suitable for most crops and crops that cannot stand under water for periods of 12–24 hours. Suitable crops for furrow irrigation include the following: (1) row crops like maize, sunflower,

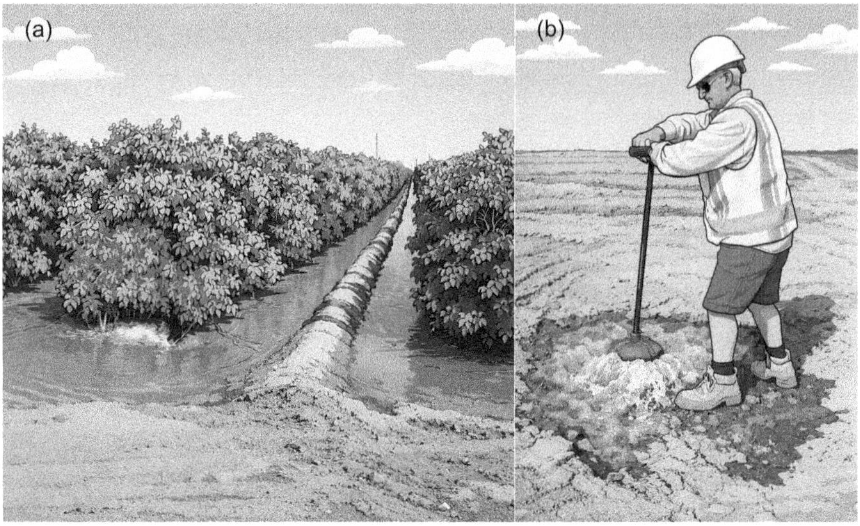

FIGURE 3.8 Border irrigation using an alfalfa (orchard) valve. (a) Citrus trees with borders and (b) an alfalfa (orchard) valve used for border irrigation. Each valve covers one border with two rows of citrus trees. (Adapted from Enciso et al., 2008.)

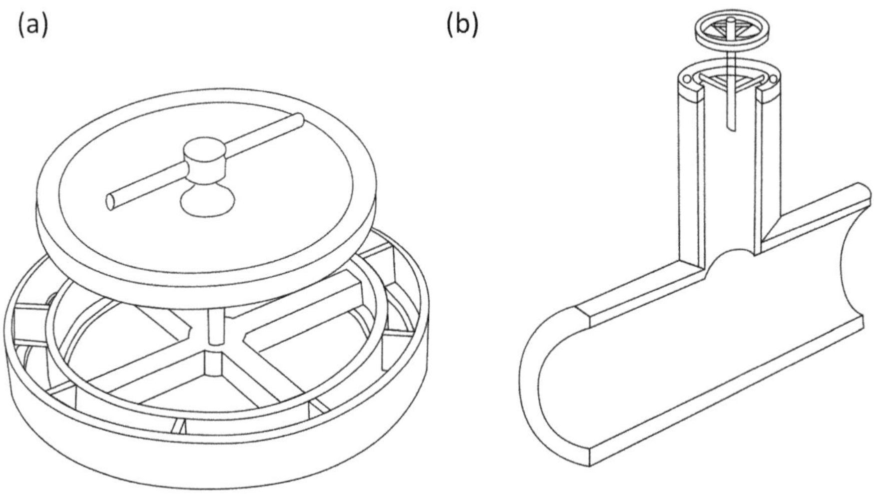

FIGURE 3.9 Alfalfa valve used in surface irrigation. (a) valve assembly, and (b) cross-sectional view showing the riser and pipeline connection.

sugarcane, and soybean; (2) crops sensitive to flooding, such as tomatoes, various vegetables, potatoes, and beans; (3) broadcast crops like wheat; and (4) fruit trees, including citrus and grape (Figure 3.11).

There are several subdivisions of these furrows, such as contour, corrugation, and zig-zag shapes. On sloped terrain, it's recommended that furrows align with

FIGURE 3.10 Illustration of ridges and furrows in a furrow irrigation.

FIGURE 3.11 Small furrows used to irrigate grapes.

the land's contours. This requires precise contour mapping before furrow creation. Corrugation irrigation is suited for densely planted crops. It involves the use of small streams of water flowing through V-shaped furrows or corrugations. These corrugations, typically smaller than the furrows used for row crops, allow water to seep into the soil and then spread laterally, reaching the spaces between the corrugations. This method is effective in ensuring adequate irrigation even with fluctuating water supplies throughout the season, making it a reliable and efficient technique for crop irrigation in areas with varying water availability (Stanley, 1954). For tree irrigation, a single furrow next to the tree line is sufficient at the beginning, but as trees grow, multiple furrows may be needed for an adequate water supply. A unique zig-zag design is sometimes implemented to enhance water distribution (Figure 3.12).

Using flexible plastic pipes and gated pipes, furrow irrigation can conserve water and reduce labor costs. This system involves laying out flexible plastic pipes along the designated furrow lines in the field (Figure 3.13). These pipes, often equipped with gates or openings at regular intervals, allow for the controlled release of water directly into the furrows (Figure 3.14). The use of gated pipes promotes more uniform water distribution, thereby reducing waste and improving water use efficiency. This method is especially useful in water-limited regions because it allows farmers to precisely manage the amount of water delivered to each furrow, thus optimizing irrigation and promoting better crop growth. Additionally, the flexibility and portability of these pipes make them ideal for varying field sizes and shapes, enhancing

FIGURE 3.12 Zig-zag furrows for tree irrigation.

FIGURE 3.13 Furrow irrigation using flexible plastic pipes.

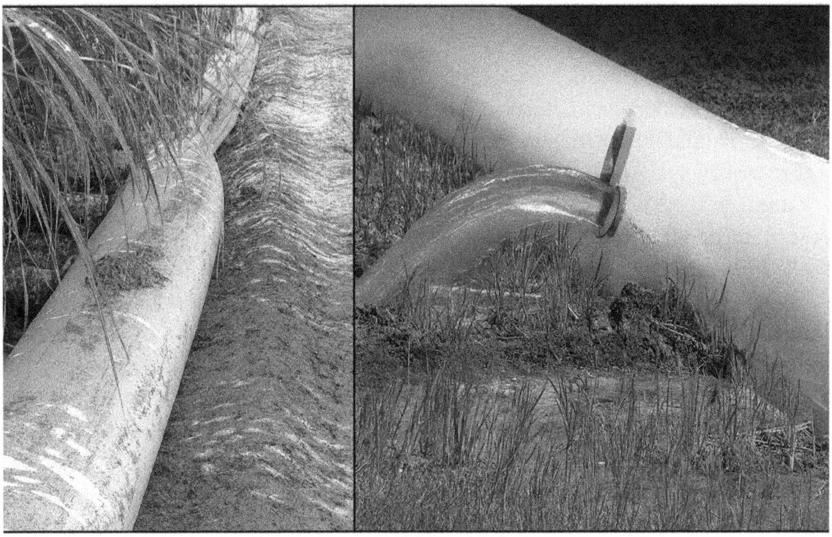

FIGURE 3.14 Furrow irrigation using a gated pipe.

TABLE 3.4
Practical Lengths in Meters of Sloping Furrows for Different Soil Types, Slopes, and Depths of Water to be Applied

Furrow Slope (%)	Maximum Flow Rate Per Furrow (L/s)	Average Depth of Water Applied (mm)					
		75	150	50	100	50	75
		Clay (m)		Loam (m)		Sand (m)	
0.05	12.0	300	400	120	270	60	90
0.1	6.0	340	440	180	340	90	120
0.2	3.0	370	470	220	370	120	190
0.3	2.0	400	500	280	400	150	220
0.5	1.2	400	500	280	370	120	190
1.0	0.6	280	400	250	300	90	150
1.5	0.4	250	340	220	280	80	120
2.0	0.3	220	270	180	250	60	90

Source: Adapted from Booher (1974).

their adaptability to different agricultural settings. This system not only conserves water but also reduces labor costs and improves efficiency, making it a valuable tool in modern farming practices.

The length of furrow is generally adapted to the field shape, but it should not be too long to improve irrigation efficiency and avoid excessive percolation at the upstream end of the row. The maximum recommended lengths and stream sizes for different desired irrigation depths and soil types are shown in Table 3.4.

There are several management strategies for furrow irrigation to irrigate more rapidly and conserve water. These management strategies are aimed at reducing either percolation or runoff losses. The strategies that reduce percolation losses are (1) irrigation of wide space alternate furrows; (2) irrigation of furrows compacted by tractor wheels and surge irrigation; (3) irrigation that reduces runoff losses through runoff recovery systems; (4) irrigation of a portion of the furrow length; and (5) the use of dams at the lower end of the furrow.

Alternate or wide-spaced furrows provide some additional storage capacity for rainfall. This strategy is best suited for fine-textured soils and can reduce water use by 33% in some clay loam soils and 20% in Richfield clay loam soils (Enciso, 1991).

Irrigation of wheel compacted furrows has reduced the intake by 33% and deep percolation by 50% in ¼ mile furrows without reducing yield (Musick et al., 1985). The objective of compacting furrows is to help water advance as fast as possible to the end of the furrow to achieve good irrigation uniformity. The compaction of furrows is generally done with the wheels of the tractors. Packers and smotherers have also been used to help water advance more quickly to the lower end of the furrow. The effect of compaction varies with soil texture, soil moisture, and the speed at which the tractor is driven.

Surge irrigation consists of applying a limited water delivery rate to a larger irrigated area than would be possible with a continuous flow. This is done by alternating the surges to a set of furrows on each side of a controller valve assembly. The mechanism by which surge irrigation reduces infiltration is the consolidation of the furrow perimeter during furrow interruptions, the filling of cracks with sediment load during flow interruptions, the filtration of sediment when the water supply is interrupted, and the greater detachment of sediment caused by the rapid advance of the surged stream front (Kemper et al., 1988). The infiltration reduction varies with the soil and is difficult to predict.

An irrigation strategy developed to reduce runoff is the limited irrigation dryland (LID) system proposed by Stewart (1983). The LID system consists of dividing the furrow into three sections: one of full irrigation, one of limited irrigation, and one of dryland. The first half of the furrow is fully irrigated. The runoff from the full irrigation areas is received in the next quarter of the section, which is considered the water limiting area. If the tailwater reaches the lower fourth of the area (dryland area), the runoff is contained by a dam. The dam also retains moisture and prevents rainfall runoff. In each section, different seed densities and fertilization doses are used to increase the profitability of the crop.

The recovery runoff system consists of a small reservoir that collects runoff in the low corner of the field, which is then pumped back through a pipeline to the head end to be used for irrigation of the same field.

3.2.4 Summary of Surface Irrigation

The selection of an irrigation method depends on development costs, field geometry, the amount and skill level of labor required, soil type, the type of crops being grown, availability of water resources, climate, and other factors. It is subject to choosing

TABLE 3.5
Comparison of Surface Irrigation Methods

Criteria	Basin Irrigation	Border Irrigation	Furrow Irrigation
Development costs	High	Moderate to high	Low
Most appropriate field geometry	Variable	Rectangular	Rectangular
Amount and skill of labor input required	Low labor and moderate skill required	Moderate labor and high skill required	High labor and high skill required
Land leveling and smoothing	Extensive land leveling is required initially, but smoothing is less critical if done periodically	Moderate initial investment and regular smoothing is critical	Minimal required but needed for high efficiency. Smoothing needed regularly
Soils	Moderate- to fine-textured soils	Moderate- to fine-textured soils	Coarse- to moderate-textured soils
Crops	Solid-stand crops	Row/solid-stand crops	Row crops
Water supply	High discharge, short duration, infrequent supply	Moderately high discharge	Low discharge, long duration, frequent supply
Climate	All	All, but better in low to moderate rainfall	All, but better in low rainfall
Principal risk	Scalding	Scalding	Erosion
Efficiency and uniformity	High	High with blocked ends	Relatively low to moderate

Source: Adapted from USDA-NRCS (2012).

one type of surface irrigation method over another. A general comparison of these three surface irrigation methods is summarized in Table 3.5.

3.3 SPRINKLER IRRIGATION

In sprinkler irrigation, the water is applied above the soil surface as a spray. Water is pushed under pressure through some nozzles contained in the sprinklers. The pressure can be supplied mechanically by a water pump or gravitationally by the difference in elevation between two points. The sprinkler system can vary in design to apply water to the soil at rates equal to or less than the infiltration rate. The component of the sprinkler system is the water pump, the mainline, and laterals with their sprinklers. Sprinkler systems provide a high degree of water control and can be used on steep or undulating land (with low application rates). Sprinkler irrigation generally utilizes smaller flow rates, which is more effective than surface irrigation methods. In canal systems, a sprinkler irrigation system should be located in water canals that convey water continuously, because they usually apply smaller irrigation depths at shorter irrigation intervals, requiring a constant water supply. The higher irrigation frequency can be advantageous on soils of high permeability or low water-holding capacity. Sprinkler irrigation can also be used for fertilizer application, frost protection, and temperature control, providing a comprehensive solution for farm irrigation and management. There are many types of sprinkler systems, but all may be grouped into three general classes according to their movement on the field: (1) continuous movement, (2) periodic movement, and (3) stationary or fixed systems. Continuous movement irrigation systems usually have a higher uniformity of application than periodic movement and stationary or fixed irrigation systems, and their travel speed can be adjusted to reduce runoff.

3.3.1 Continuous Movement

The sprinkler operates while moving on the field either linearly or circularly, and it can be classified as a central pivot, lateral movement, or traveling gun. On canal systems, these require continuous water supplies; therefore, they should only be installed in canals that have free water demand or continuous water flow.

3.3.1.1 Center Pivot

The lateral moves continuously around a pivot to irrigate a circular area (Figure 3.15). Water is supplied from the pivot and applied through the sprinkler heads (Figure 3.16). Center pivots are commonly about 300 to 400 m long, although systems may range from about 35 m for single-span units to 800 m for large installations. The most common center pivot lateral is 400 m (1,312 ft) long, uses a 168 mm (6-5/8 in.) diameter pipe, and covers a circle of 50 ha (124 acres) inscribed within a "quarter section" of land with an area of 65 ha (161 acres). Water is applied at an instantaneous rate of 25–100 mm/h.

The selection of a pump should consider the pressure required to offset friction losses in the supply line to the center point, plus the pressure required by the

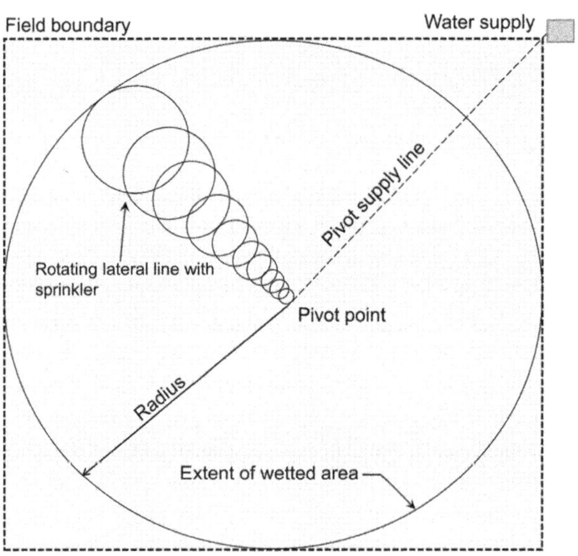

FIGURE 3.15 An illustration of the center pivot system.

FIGURE 3.16 Illustration of a center pivot system: (a) view from the pivot point, (b) a sprinkler, and (c) view from the end gun.

center pivot machine. The pressure and flow needed for the center pivot will depend on the sprinkler package. Medium pressure impacts require pressure from 40 to 50 psi, and high pressure impacts from 55 to 65 psi. Other sprinkler packages are MESA (Medium Elevation Sprinkler Application), LEPA (Low Energy Precision Application), and LESA (Low Energy Spray Application). The LESA system may require between 15 and 30 psi, and the LEPA system may require only 15 psi. These systems apply low rates (25–75 mm) per rotation, and each rotation takes one to three days to complete the circle.

3.3.1.2 Lateral Linear System

The lateral moves continuously in a straight line with a guided system, irrigating rectangular fields (Figure 3.17a and b). This system is mainly installed on level fields. There are two types of lateral linear systems. One kind contains an attached pump that takes water from a canal. A second type takes water with a flexible hose from a hydrant, irrigating as it moves and pulling the flexible hose. Generally, a rod is placed between two hydrants to stop the machine. It needs to be connected and disconnected from one hydrant to another to irrigate the next position, requiring a lot of labor. The rate of water applied, and the total depth of water depends on the speed of the linear system and the sprinkler package selected. The pump needs to offset the friction losses of the main line and provide the pressure and flow rate required by the machine.

FIGURE 3.17 Linear system: (a) water supply from a canal and (b) water supply from a hose. (Adapted from USDA-NRCS, 2016.)

3.3.1.3 Traveling Gun

The traveling gun is a single high-capacity self-propelled sprinkler in which water is supplied through a flexible hose. There are two types of traveling guns: cable-tow traveling guns and hose-reel traveling guns. In both systems, the guns are mounted on carts or trailers, which move slowly and continuously through the field as the guns operate. The rate of water application and total depth applied depends on the flow rate supplied to the gun, the diameter of coverage, and the speed at which the gun travels.

3.3.1.3.1 *Cable-Tow Traveling Guns*

Cable-tow systems automatically tow a large gun through the field by winding up a cable. The gun is mounted on a moving vehicle, which also contains a cable reel and winch. The cable is stretched across the field in the desired direction of travel, and its end is firmly anchored to the end of the tow path. As water flows to the gun, an impeller drive unit or water piston is used to power the winch. The cable-tow traveling gun pulls itself across the field by winding up the cable. As the traveler moves along its path, the sprinkler wets a strip of land. The travel speed is adjustable from a few feet per minute (fpm) to 10 fpm or more. Water is delivered to the gun through a flexible hose that is also towed along with the system. A tow path approximately 10 ft wide is required for this type of traveling gun, because the flexible hose loops behind the cart. The most common gun type sprinkler discharge is 32 L/s, wets a diameter of 120 m, and can irrigate up to 32 ha. The typical hose length is 400 m long. This system requires high pressures (80–120 psi) to operate the sprinklers, which uses high energy, thus making it expensive to operate.

3.3.1.3.2 *Hose-Reel Traveling Guns*

Hose-reel systems require less labor than cable-tow systems because they are easier to set up for operation. A hose-reel traveling gun uses a large reel unit to pull the sprinkler, as it winds up the hose (Figure 3.18). The sprinkler irrigates a strip of land, as it is pulled. The hose is semi-rigid and does not collapse on the reel, so that water can be continuously pumped through it during operation. Hose-reel traveling guns have a separate pump unit. The hose and gun are laid out in the desired direction of travel. The winding speed can be controlled according to the depth that needs to be applied, and it can travel unattended for a distance equal to the length of the hose.

FIGURE 3.18 Hose-reel traveling gun. (Adapted from USDA-NRCS, 2016.)

Hose-reel systems are used for the same crops as cable-tow gun systems, and they have the same discharge, wetting diameters, and pressure characteristics.

3.3.2 Periodic Movement

On these systems, the mainline is generally fixed, and the laterals are moved from one position to another. There are laterals that can be moved by hand or portable self-moved laterals. They are mounted above the soil surface on wheels and the sprinklers are mounted on the aluminum lateral pipe. The main lines and sub mains are generally buried and the laterals are fed by a hydrant. The laterals can be moved manually or mechanically. There are four types of self-propelled sprinkler systems, classified by the method of movement: (1) Side-wheel roll, (2) hand moved, (3) end tow, and (4) gun sprinklers. These systems are also called set systems, because sprinklers need to be set in a fixed position during a time period to apply a given irrigation depth.

3.3.2.1 Side-Wheel-Roll Sprinkler Systems

The side roll consists of rigidly coupled aluminum pipes mounted on large wheels, with the pipe acting as an axel (Figure 3.19). Sprinklers are kept in an upright position by means of a weighted swivel coupling on each sprinkler. The lateral pipe is typically rotated by a chain drive system powered by a small gasoline engine of about 8 Horse Power (HP) and transmission located near the center of the lateral. The lateral pipe can have a length of up to 490 m. The most popular side-wheel-roll is 400 m, with 35 sprinklers, a 100 mm (4 in.) diameter pipe, and an operating pressure of about 4 atm (60 psi) at the inlet. The side roll irrigates a strip of land and then moves about 35 m every 3–12 hours to irrigate another set. The lateral pipe is mounted only 3–4 ft above the soil surface; therefore, it is only adaptable to short crops—mainly short forage crops and pastures.

FIGURE 3.19 Side-wheel-roll sprinkler systems. (Adapted from USDA-NRCS, 2016.)

FIGURE 3.20 Hand moved portable system.

3.3.2.2 Hand-Moved Lateral Systems

Portable, hand-moved sprinkler systems consist of connecting several aluminum pipe sections using quick couplings (Figure 3.20). Each pipe section is 20, 30, or 40 ft long, and it contains a short riser pipe supporting a sprinkler head mounted generally at the coupling. Buried PVC mainlines and manifolds are sometimes installed, and lateral connections are made through permanent hydrant valves that bring the water to the lateral. These systems are sometimes called semi-permanent because the main lines are permanently installed and only the laterals are portable. Portable, hand-moved sprinkler systems are manually moved from zone to zone. Then, the lateral line is disassembled and moved to the next position to be irrigated. It requires a lot of labor to move the sprinklers between positions, and its initial cost is low. Because the pipes must be manually moved, these systems are not adaptable to tall crops, such as corn or other crops, which may make its movement difficult.

3.3.3 Stationary or Fixed Systems

These are very rare because of their high cost. The laterals are placed permanently and do not need to be moved from one position to another (Figure 3.21). The laterals are placed after planting and at the end of the irrigation season and are controlled by valves. The main line is generally buried.

3.4 MICROIRRIGATION

Microirrigation, also referred to as drip or trickle irrigation, is a technique that applies water directly to the crop root zone in small amounts, either as drops, tiny streams of droplets (microsprinkler), or a miniature spray (micro-sprayers), using emitters placed along a water delivery line (Figure 3.22). The device that applies water to the soil is called an emitter. Emitters reduce the pressure in the distribution network through a small orifice or a long, narrow flow path, allowing water to be delivered in small quantities at low pressure. Emitters wet a limited area of the soil surface, allowing water to move horizontally and vertically.

FIGURE 3.21 Stationary or fixed systems.

FIGURE 3.22 Microsprinkler irrigation systems.

Most microirrigation systems include a pump unit, a filter, chemigation units, a mainline lateral, and accessories, such as pressure regulators, connections, and pressure relief valves (Figure 3.23). The pump unit takes water from the source and supplies the pressure required by the system. The main advantages of microirrigation are the high uniformity (more than 90%), better fertilization and chemigation control, and labor and water savings. Microirrigation systems can be used in areas with limited water supplies, where the amount of water available per unit area is low due to low canal or irrigation well capacities, or in undulating terrain. Microirrigation systems make the most of the limited water supply. The disadvantages are its high initial cost, high maintenance requirements, and the requirement of higher management skills.

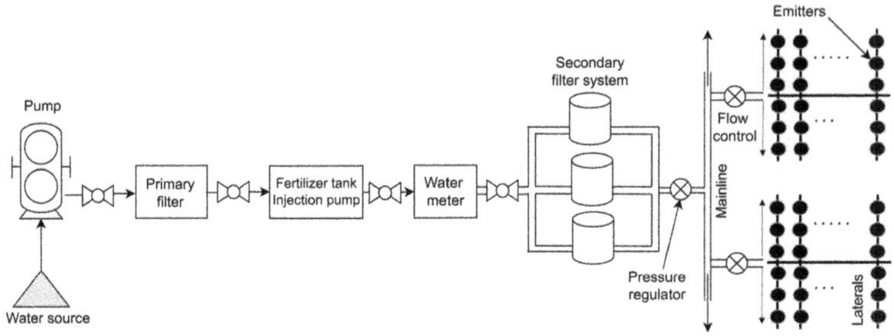

FIGURE 3.23 Typical drip irrigation system layout.

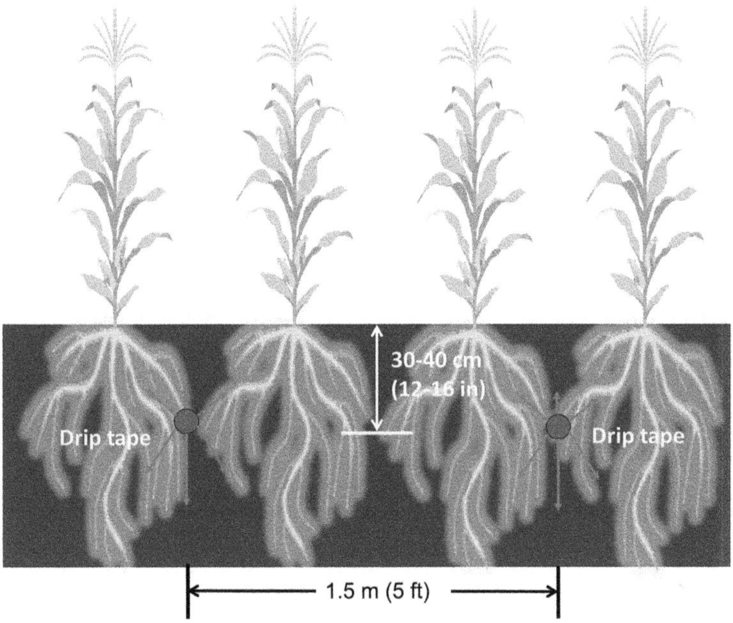

FIGURE 3.24 Typical spacing and depth of subsurface drip irrigation systems.

The systems generally operate under pressure lower than 30 psi. Most of the pressure is required to backflush the filters.

3.4.1 Sub-irrigation

Sub-irrigation can be broadly defined as regulating the groundwater table by artificially adding water below the soil surface. The water level should be maintained at a level that ensures the best balance of water and air in the root zone. Water is supplied through tiles of perforated pipes laid at a shallow depth below the ground surface. The surrounding soil is moistened by gravity and diffusion. The typical spacing of subsurface drip is about 1.5 m (5 ft) and the typical depth is 30–40 cm (12–16 in.) (Figure 3.24). This method reduces evaporation and soil surface cooling, places

FIGURE 3.25 Subsurface drip irrigation systems.

moisture in or near the root zone (where needed), and may be used to advantage when only small water flows are available. Soil types are important considerations in the use of this method. An example of a subsurface drip irrigation system is shown in Figure 3.25.

REFERENCES

Booher, L.J. (1974). Surface Irrigation. FAO Agricultural Development. Paper No. 95. United Nations Food and Agriculture Organization. Rome, Italy.

Brouwer, C., Prins, K., Kay, M., and Heibloem, M. (1985). *Irrigation Water Management: Irrigation Methods*. Training Manual No. 5. FAO, Rome.

Enciso, J. (1991). An Infiltration Model for Managing Furrow Irrigation with Limited Water. Ph.D. Dissertation, The University of Nebraska – Lincoln.

Enciso, J., Sauls, J.E., Wiedenfeld, R.P., and Nelson, S.D. (2008). *Impacts of Irrigation on Citrus in the Lower Rio Grande Valley*. B-6205 05-08; AgriLife Extension-Texas A&M System, College Station, TX.

Kemper, W.D., Trout, T.J., Humphcrys, A.S., and Bullock, M.S. (1988). Mechanisms by which surge irrigation reduces furrow infiltration rates in a silty loam soil. *Transaction of the ASAE*, 31(3), 821–829.

Musick, J.T., Pringle, F.B., and Johnson, P.N. (1985). Furrow compaction for controlling excessive irrigation water-intake. *Transaction of the ASAE*, 28, 502–506.

Stanley, W.R. (1954). *Corrugation Irrigation*. U.S. Department of Agriculture, Soil Conservation Service, Washington, DC.

Stewart, B.A. (1983). Automated seeding rate selector. *Agronomy* Journal, 75(1), 152–153.

USDA-NRCS (2012). Surface irrigation. In: C. Prestwich, J. Busch, J. Hester, B. Sandoval, L. Nelson, C. Stoner, and M. Tweffort, Eds., *National Engineering Handbook*. Part 623: Irrigation-Chapter 4. USDA-NRCS, Washington, DC, pp. 4–1–4–63.

USDA-NRCS (2016). Sprinkler irrigation. In: C. Prestwich, J. Walker, P. Robinson, and J. Busch, Eds., finalized under the guidance of R. Sampson, *National Engineering Handbook*. Part 623: Irrigation-Chapter 11. USDA-NRCS, Washington, DC, pp. 11–1–11–191.

4 Organization for Irrigation Operation and Management

4.1 INTRODUCTION

Irrigation projects play a crucial role in providing water to agricultural lands for crop production. The projects involve a comprehensive range of engineering works, including the acquisition, conveyance, and delivery of water, as well as the development and maintenance of crop-growing areas. Additionally, some irrigation projects may incorporate both drainage and supply works. At specific points in the system, the water may also be used for other purposes, making irrigation projects an integral component of many water supply systems.

Irrigation projects vary in size from small single-farm units to extensive areas of several hundred thousand acres. For a small farm project, the work needed to obtain and convey water may be as simple as a low-diversion weir or an inexpensive pumping plant, together with shallow ditches and a few minor control structures. For a large project, work may include many farms, and all features of the irrigation system are larger, more complicated, and more expensive. When water is used for purposes other than irrigation, the system becomes even more complicated. As irrigation systems become increasingly complicated, so does management. Many functions are added, increased responsibilities result, and, overall, operation and maintenance acquire major proportions.

Subsequent to an adequate determination of irrigability and drainability of soils and of an adequate water supply, intelligent management is essential to the establishment and improvement of irrigation and drainage systems. Implied here is that the area proposed for development is irrigable and drainable, and that an adequate water supply is available to satisfy needs. It must be emphasized that water supplies in the world are exhaustible, and therefore, the available water should be used where irrigation can be accomplished with the greatest benefit. This should be the foremost consideration in planning and establishing, or improving, irrigation and drainage systems. In a larger sense, management implies more than just business aspects; it must also include the idea of a permanent and sustainable agriculture. This larger view must always be kept in mind by the management and operating staff. This chapter discusses different organizations for the operation and control of irrigation systems.

DOI: 10.1201/9781003747444-5

4.2 TYPES OF ORGANIZATIONS

The legal and administrative aspects of organizations that manage irrigation enterprises are so diverse that it is not practical to present anything more than a brief summary of the most important aspects. In irrigation enterprise management, one must become intimately acquainted with the legal and administrative instruments applicable to the local area. Irrigation practices and procedures should be followed and improved when desired.

The type of organization formed to deliver irrigation water, flood control and/or drain land is vitally important. Legal powers and the relationship between the organization and water users are governed by the kind of organization. Good irrigation projects fail if the wrong organization is adopted. If the organization does not fit the type of operation called for, changes must be made. The most successful enterprises are those owned and operated by water users. These are operated on a non-profit basis. Many private associations formed to make a profit through the conveyance and delivery of irrigation water have become insolvent and have been taken over by water users.

In general, three types of organizations can be distinguished: (1) private, which include individual projects, mutual, and commercial companies; (2) quasi-public under public laws, which prescribe procedures and public agencies participating in the organization and management without assuming direct financial responsibility; and (3) public, which are (a) organized under public laws, (b) administered by public agencies, and (c) financed by public funds.

In general, most organizations are constituted according to the existing legal framework. In most instances, the laws governing the overall organization specify: (1) how it is to be formed (by petition, elections, etc.); (2) who is responsible for what; (3) who may serve in the various management positions (qualifications); and (4) just what can and cannot be done legally.

4.2.1 Private Organizations

Private organizations can vary from individuals to groups of individuals to private institutions. Individuals need to operate and manage their irrigation systems in a sustainable manner, especially financially. Usually, individually owned irrigation systems are small, irrigating small areas.

These organizations are mainly controlled by farmers or private institutions that have a board of directors, which hires the district manager. The manager has some responsibilities, such as hiring personnel, training personnel, and operating and maintaining the irrigation district. The operating and maintenance costs of the system will be paid completely by the marketed water and based on the volumes of water delivered. The main characteristic is that these are organized by water users for their benefit. Some governments promote, encourage, and assist these organizations by financing irrigation water improvements through the provision of low interest loans and grants.

4.2.2 Quasi-Public Organizations or Mixed Control Organizations

These organizations are controlled by government officials and by farmers. Generally, the dam and main canal are controlled by government officials, while the secondary and tertiary canals are controlled by farmers' associations or water user organizations. Each tertiary canal can have its own farmers' association. These organizations should have a management plan indicating the area covered by the district and the possible water allocation, which is generally provided "as available" according to the amounts stored by the dam. Farmers' organizations have the power to collect water fees and control the operation and maintenance of the irrigation system. In several countries, governments are moving to decentralization and market-oriented approaches to managing water. The responsibility for operating and maintaining the irrigation districts has been transferred to water users' organizations. The governments have maintained just the responsibility of some main canals.

4.2.3 Public Organizations

These organizations are controlled by government officials, and the water rights are allocated to different farmers' organizations through local, state, or federal governments. These organizations are responsible for allocating water rights to various farmers' organizations, and the rules and regulations governing the allocation of these rights are determined by the government. One advantage of public organizations is that they can access government funding and resources, and technical skills which can be particularly important for large-scale irrigation projects. However, public organizations may have limited flexibility. Additionally, the allocation of water rights can be a contentious issue, particularly in areas where water resources are scarce or subject to competing demands.

4.2.4 Irrigation Organizations in the United States

According to the U.S. Department of Agriculture, there were 2,677 irrigation organizations to manage surface water and groundwater for irrigation in the United States in 2019 (USDA-NASS, 2020). The two primary functions of these organizations are water delivery and groundwater management. 98.3% (2,543) of these organizations are involved in water delivery, and 19.5% in the regulation or influence of groundwater use for irrigation (Table 4.1). The Pacific, Southeastern, and High Plains regions in the United States rely more on groundwater use for irrigation. In addition, these organizations have secondary functions, including municipal or residential water supply, electricity generation, recreation or wildlife management, agricultural drainage management, flood retention, and others. Table 4.2 lists the proportions of secondary functions of irrigation organizations in different regions in the United States.

The proportions of different organization types in different regions are shown in Table 4.3. Irrigation districts account for a large proportion of water delivery organizations, particularly in the High Plains (86.5%) and Southeastern regions (87.5%).

TABLE 4.1
The Total Number and Proportion of Primary Functions of Irrigation Organizations in Different Regions in the United States

Geographic Area	Number	Irrigation Water Delivery (%)	Groundwater Management (%)
Eastern Rockies	898	98.3	19.5
High Plains	177	54.2	48.0
Northwestern	424	97.4	16.7
Pacific	245	95.5	64.5
Southeastern	13	61.5	92.3
Southwestern	920	98.8	25.4
United States	2,677	95.0	27.5

Source: Data from USDA-NASS (2020).

Note that they are not equal to 100% due to the overlap in classification. Organizations can have more than one primary function. The geographic area covers 24 U.S. states where a majority of irrigation occurs. Eastern Rockies include the states of Colorado, Montana, and Wyoming; High Plains include Kansas, Nebraska, North Dakota, Oklahoma, South Dakota, and Texas; Northwestern region includes Idaho, Oregon, and Washington; Pacific region includes California and Nevada; Southeastern region includes Arkansas, Florida, Georgia, Louisiana, Mississippi, North Carolina, South Carolina; and Southwestern region includes Arizona, New Mexico, and Utah.

TABLE 4.2
Proportions of Secondary Functions of Irrigation Organizations in Different Regions in the United States

Geographic Area	Municipal or Residential Water (%)	Electricity Generation (%)	Recreation or Wildlife Management (%)	Agricultural Drainage Management (%)	Flood Retention-Main River Channel (%)	Other (%)
Eastern Rockies	11.2	2.0	6.8	10.0	2.9	7.5
High Plains	17.5	2.3	14.1	9.0	7.3	11.3
Northwestern	13.7	5.9	3.5	9.0	4.2	11.8
Pacific	18.8	6.9	6.1	14.3	9.8	24.5
Southeastern	7.7	—	23.1	61.5	7.7	38.5
Southwestern	13.8	2.5	5.3	11.3	4.6	7.9
United States	13.6	3.2	6.3	10.9	4.6	10.3

Source: Data from USDA-NASS (2020).

— indicates zero. Note that they are not equal to 100% due to the overlap in classification. Organizations can have more than one secondary function.

TABLE 4.3
Types of Irrigation Organizations in Different Regions in the United States

Geographic Area	Incorporated (%)	Unincorporated (%)	Irrigation Districts (%)	U.S. Bureau of Indian Affairs Organizations (%)	Other (%)
Eastern Rockies	11.2	62.7	20.2	1.7	4.2
High Plains	—	8.3	86.5	2.1	3.1
Northwestern	7.5	43.3	43.1	0.5	5.6
Pacific	—	7.7	75.6	2.6	14.1
Southeastern	—	12.5	87.5	—	—
Southwestern	23.7	43.5	26.4	1.5	5.0
United States	13.6	45.4	33.9	1.5	5.5

Source: Data from USDA-NASS (2020).
— indicates zero. Incorporated refers to a legal entity where shareholders jointly own a water conveyance system. This entity is established under legal requirements and is owned by its users, who usually pay for water usage. Unincorporated refers to an informal collaboration between ditch users, typically involving a partnership or informal group of two or more farmers who operate an irrigation water supply system for their personal needs. It is common for such partnerships to operate without any official formal organization.

4.3 CHARACTERISTICS OF INCORPORATED AND UNINCORPORATED ORGANIZATIONS

4.3.1 Procedure

Incorporated irrigation organizations are established under the provisions of state laws governing corporations. Articles of incorporation are required to be filled with state officials, and the organization must have a formal organizational structure, including a board of directors, officers, and bylaws. Membership is completely voluntary. There may be non-members or holdouts within areas served. Unincorporated organizations are formed through written or verbal agreements between members when there is no formal or legal procedure. Membership is voluntary, and there is no legal requirement to file articles of incorporation.

4.3.2 Powers

Incorporated irrigation organizations may act under the provisions of state law to issue stocks, incur indebtedness, issue bonds or notes, mortgage company property, acquire water rights and other property, develop or acquire water supplies for members, levy assessment against capital stock, collect tolls for water, and condemn the right-of-way in some states. These powers enable incorporated irrigation organizations to operate more efficiently and effectively by providing them with access to additional resources and the ability to make legally binding decisions. Unincorporated irrigation organizations, on the other hand, may act as a single natural person and may condemn right-of-way in some states. They may also incur indebtedness as tenants in common with unlimited liability. However, these organizations do not have the legal standing of an incorporated entity, and cannot, therefore, issue stocks or bonds, mortgage property, or acquire water rights or other property.

4.3.3 Capital and Financing

Acquiring capital for construction, operation, and maintenance is a significant challenge for unincorporated irrigation organizations. As there is no legal entity, the members rely on their own vocabulary to acquire funds. Members are also jointly and individually liable for any debts incurred on behalf of the Company. This joint liability means that the credit standing of the organization is usually low, making it difficult to obtain financing from external sources. One of the most common financing methods used by incorporated organizations is bond issues. These bonds are secured by physical facilities and water rights owned by the Company. However, the bonds are usually not secured for life on member land, although this may be specified in the articles of incorporation. For new construction, there is no good security for bond issues; hence, it is generally used for improvements.

4.3.4 Revenue

To function effectively, irrigation organizations require revenue to control construction, operation, and maintenance costs. Incorporated irrigation organizations may assess against stocks, which become a lien upon the stocks. If a member fails to pay the assessment, the organization can sell the stocks to recover the amount owed. Alternatively, the organization can withhold water delivery from delinquent members until they pay the assessment. Some mutual companies may require liens on land to enforce the payment of assessments. Assessments are usually made annually or as provided in the bylaws. Toll charges may also be made on the volume of water delivered to cover operation and maintenance costs. Toll charges may be levied for various reasons, including the cost of pumping water, maintaining canals, and the infrastructure. Unincorporated irrigation organizations depend on the payment of a prorated share of expenses by its members. This share is based on the amount of water used by each member. However, the process of collecting these payments can be challenging and sometimes may require court action to enforce payment.

4.3.5 Water Rights

Incorporated irrigation organizations have more flexibility in water rights ownership, depending on the circumstances of the company's origin. The landowner may retain water rights and transfer them to the company in exchange for stocks, or the company may have acquired the right to water for a given service area and sold stock-retaining water rights, making stock appurtenant or floating. The water rights of incorporated irrigation organizations may be subject to the appropriate system, which means that the beneficial user is the only possible owner of water. The system operates under the doctrine of "first in time, first in right", meaning that the earliest user who puts the water to a beneficial use gains the right to continue that use. For unincorporated irrigation organizations, landowners typically own the water rights associated with their land. The irrigation water is considered part of the land and cannot be sold separately. If a landowner sells his land, the water rights associated with the land are typically transferred to the new owner.

4.3.6 Management

Incorporated irrigation organizations are managed by a Board of Directors and Officers as specified by corporation laws. The Board of Directors is responsible for overseeing the overall operation and direction of the organization. Incorporated irrigation organizations generally have certain salaried officials, especially a superintendent and secretary-treasurer. The superintendent is responsible for overseeing the construction, operation, and maintenance of the irrigation system, while the secretary-treasurer is responsible for keeping financial records and managing the company's finances. Unincorporated irrigation organizations, on the other hand, are usually managed by a loose organization with informal management and no salaried officials. The lack of formal management in unincorporated irrigation organizations could make it difficult to make strategic decisions and ensure that all members are on the same page.

4.3.7 Supervision

Incorporated irrigation organizations are subject to public supervision, indicating that they are regulated by state corporation laws. State corporation laws govern the formation, operation, and dissolution of the corporation. Unincorporated irrigation organizations have no public supervision. There are no prescribed rules for membership, governance, or operation.

4.4 IRRIGATION DISTRICTS

Irrigation districts are entities established to deliver untreated water for irrigation to farmers and landowners within a defined geographic area and to provide for the drainage of lands. An irrigation district operating may contract with municipalities, political subdivisions, water supply corporations, or water users to deliver untreated water. The first irrigation district, Turlock Irrigation District, was established in California in 1887, as needed for irrigation water for agriculture became evident (Turlock Irrigation District, 2011). Since then, irrigation districts have become a common method of providing water for agriculture throughout the United States.

An irrigation district may cooperate with the United States under the federal reclamation laws for the purpose of (1) construction of irrigation and drainage facilities necessary to maintain the irrigability of land; (2) purchase, extension, operation, or maintenance of constructed facilities; or (3) assumption, as principal or guarantor of indebtedness to the United States on account of district lands. As shown in Table 4.3, irrigation districts play an important role in water delivery for irrigation in the United States.

4.4.1 Characteristics

Irrigation districts are public or quasi-municipal corporations established under state law to provide water at cost to lands within their defined geographic boundaries. They exist solely to provide water to lands within their boundaries and operate on a cost-recovery basis. This means that they do not seek to generate a profit from the sale of water, but rather to cover their operational and maintenance costs.

The procedures for organizing an irrigation district are described as follows. A group of landowners within the proposed district must petition for its creation, providing detailed information about the district's boundaries, water supply, and other pertinent data required by State law. County courts have jurisdiction over the process, including hearing objections, receiving counter-petitions, ordering feasibility studies if necessary, and ultimately either ordering the formation of the district or calling for an election to decide its formation. The creation of an irrigation district is a major decision with potential long-term effects on the local community. As such, the process is designed to be transparent and democratic, with opportunities for public input and review.

4.4.2 Powers

The powers of irrigation districts are vast and varied, enabling it to fulfill its mandate of providing water for irrigation to lands within its boundaries. One of the most important powers of an irrigation district is the ability to acquire water supplies and water rights. This power enables the district to ensure that it has sufficient water to meet the needs of farmers and landowners. Additionally, an irrigation district has the power to acquire real property, including land, materials, and things considered necessary to accomplish its objective.

An irrigation district is also authorized to exercise eminent domain, which allows the district to acquire property for public use, even if the owners are unwilling to sell or are asking for an exorbitant price. Another key power of an irrigation district is the ability to issue bonds. These bonds allow the district to acquire property to raise funds to finance the construction of works, improvements, and facilities necessary for the irrigation of land within their boundaries. The district can also assume indebtedness, which allows it to take on the responsibility for repaying debts incurred by other entities.

To finance its activities, an irrigation district can assess a special tax on lands within its boundaries. This tax becomes a lien on the land and must be paid by the landowner. The district also has the power to construct works and improvements necessary for the irrigation of land in the district. This includes the construction of pumping plants, improvement works, and other facilities necessary to provide water for irrigation. Additionally, the district has the power to construct drainage ditches or other facilities for drainage to ensure that the land in the district is properly drained. The district can also construct levees to protect the land in the district from overflow.

The powers granted to an irrigation district are essential for its ability to provide water for irrigation to lands within its boundaries. From acquiring water supplies and water rights to issuing bonds and assessing special taxes, these powers enable the irrigation district to finance and construct the necessary works and improvements to fulfill its mandate. While some of these powers may be controversial, they are necessary for an irrigation district to fulfill its important role in supporting agriculture and ensuring the sustainable use of water resources.

4.4.3 Capital and Finance

One of the main challenges for irrigation districts is the capital and financing required to operate and maintain their facilities. There are many ways of financing and raising capital for irrigation enterprises, and historically, some methods have been more successful than others. One method of financing irrigation districts that has not been successful is the formation of districts to improve, enlarge, or take over existing facilities. This is because such districts do not have adequate security to attract investors. Bonds have been a chief source of capital for irrigation districts, although ID (irrigation districts) bonds have not always been attractive to investors. The security for bonding typically consists of the power to levy special tax assessments, which are used to repay bondholders. Many districts have defaulted on bond payments, requiring refinancing.

Another trend in financing irrigation districts is the use of public–private partnerships (PPPs) to finance and operate irrigation facilities. PPPs involve collaboration between the public sector (i.e., the irrigation district) and private entities, such as investors or companies, to finance and operate the facilities. This approach can provide irrigation districts with access to additional capital, expertise, and technology.

In addition to bonding and PPPs, irrigation districts can also raise capital through grants and loans from government agencies or private organizations. However, such funding sources often come with specific requirements and restrictions, such as the need to match funds or comply with certain regulations. Overall, the financing and capital needs of irrigation districts are significant, and the methods of financing and raising capital can be complex and challenging. However, with careful planning, management, and collaboration with private and public entities, irrigation districts can successfully finance and maintain their facilities, providing vital water for agriculture and other uses.

4.4.4 Revenues

Revenues are collected in several ways, including assessments on land, tolls on water used, and tax levies. The primary way in which irrigation districts collect revenues is through assessments of the land. This means that each landowner within the district is charged a fee based on the amount of land they own. The fee can be a fixed amount per acre or can be based on the cost of providing water. Another way in which irrigation districts can collect revenue is through tax levies. The district can levy taxes against the assessed value of the property within the district. The method of assessment can be a uniform rate per acre, ad valorem, or benefit basis. Ad valorem means that the tax is based on the value of the property, while benefit basis means that the tax is based on the benefits received by the property from the irrigation system.

The collection of revenues is usually done by county officials along with other tax collections. The irrigation district may deny water service to any person who refuses to pay the required charges and assessments for water service, or who fails or refuses to pay any taxes levied against their property 6 months after those taxes become delinquent.

4.4.5 Water Right Ownership

The irrigation district may have appropriate water in the same way as does an individual. Some districts do not own water rights, but only own the physical facilities necessary to provide water to landowners. Where a water right is held by the ID, it is usually not appurtenant (conjoined) to specific lands. This means that the district has the right to use the water, but it is not tied to any specific property. This allows the district to allocate water based on need and to ensure that water is distributed equitably among landowners. In contrast, when individuals hold water rights, the water is usually appurtenant (conjoined) to the land. This means that water rights are tied to a specific property, and the landowner has the right to use the water for irrigation. This type of water rights ownership is more common in smaller irrigation systems and allows landowners to have more control over their water supply. In several districts where water comes from a reservoir, the water right is provided "as available", and the actual amount of water released from the reservoir may vary from year to year. This means that the amount of water available for irrigation is dependent on the availability of water in the reservoir. This type of water rights ownership requires careful management to ensure that water is distributed equitably among landowners.

4.4.6 Management

Effective management is essential for the success of irrigation districts. In most irrigation districts, an elected board of directors serves as the governing body. This board is responsible for overseeing the operations of the district, including setting policies, managing finances, and ensuring that water is distributed equitably among landowners. The board is usually composed of members who are themselves landowners in the district, elected by other landowners.

The board of directors typically employs a salaried manager to oversee the day-to-day operations of the district. The manager is responsible for implementing the policies set by the board, managing staff, and ensuring that the district's operations run smoothly. The manager often works closely with the board of directors to ensure that the district's goals are met.

Irrigation districts may have a complex organizational structure, depending on their size and scope of their operations. The structure typically includes operations staff responsible for the physical management of the district's infrastructure, maintenance staff responsible for repairing and maintaining the infrastructure, and clerical staff responsible for administrative tasks, such as accounting, record keeping, and customer service. The county treasurer usually acts as the ID treasurer.

This means that the county treasurer is responsible for collecting and disbursing funds for the district. The country treasurer may also be responsible for ensuring that the district's finances are managed effectively and that the district is following financial regulations.

4.4.7 Public Supervision

Public supervision is an important aspect of ensuring that irrigation districts operate effectively and transparently. County officials, such as commissioners, play a significant role in overseeing irrigation districts. They are responsible for ensuring that the district's operations are in compliance with state and federal laws and regulations. Country officials also play a role in the initial feasibility studies for irrigation district projects and may approve bond issues for funding. County officials may certify bonds to ensure that they meet state requirements. The certification process helps ensure that bondholders are protected, and that the district's financial management is transparent and accountable. County officials are responsible for overseeing the management of irrigation districts. They may specify accounting procedures and other operational requirements by statute to ensure that the district is operating efficiently and effectively. By providing close supervision, county officials help ensure that irrigation districts operate sustainably and equitably for the benefit of all stakeholders.

4.5 GROUNDWATER CONSERVATION DISTRICTS

Groundwater conservation districts (GCDs) are established to manage the use of groundwater resources and, in many situations, to control their use and conservation locally. These districts are typically created by state or local governments, or by organized individuals through a petition to a controlling court or state/federal government. While their responsibilities may vary, these districts generally focus on a set of core objectives designed to ensure sustainable management and protection of groundwater resources. These objectives typically include monitoring and reporting groundwater conditions, managing groundwater quality (e.g., protecting the area from saltwater intrusion), collecting groundwater pumping data, balancing water extraction with rainfall recharge, charging fees for pumping groundwater, permitting the development of new wells, controlling soil subsidence caused by withdrawn water, and protecting regions from water transfers to other areas. These districts can be promoted by county, state, or federal governments or formed by organized individuals through a petition to a commissioner's court, state, or federal government.

4.5.1 Powers

To manage groundwater resources effectively, GCDs are authorized with different powers to enforce rules and regulations. One of the primary powers of GCDs is the development and adoption of comprehensive management plans that aim to ensure the efficient use of groundwater and prevent waste. GCDs also have the power to require permits for drilling, equipping, completing, or altering the size of a well. This ensures that the wells are constructed and maintained in a manner that minimizes the risk of contamination and maximizes the efficient use of groundwater resources. Another power of GCDs is conducting surveys of groundwater resources. This includes monitoring static water levels, maintaining well inventories, assessing water quality, pump capacities, and volumes of water extracted. These surveys help the districts to better understand the available groundwater resources and their usage patterns. GCDs also have the power to monitor water pumpage, which helps them regulate the usage and prevent over-extraction of groundwater. Additionally, districts may regulate the spacing of wells or the production of wells to ensure efficient utilization of groundwater resources. Finally, GCDs may also require that unused or abandoned wells be capped or plugged, which aims to prevent any contamination of the groundwater resources and ensure that the unused wells do not act as conduits for water loss through leakage.

4.5.2 Capital and Finance

The size and scope of a GCD's staff depend on several factors, including the district's goals, available funding, and the contributions of local taxpayers. In some cases, a district may have only one part-time employee, while others may have a team (Porter et al., 1999). The financing of GCDs also varies from state to state and country to country. For example, in Texas, the Texas Water Code, Chapter 52 allows groundwater districts to levy property taxes to pay for the maintenance and operating expenses according to the assessed value of the farms and properties within the district. Most districts in Texas are funded through these ad valorem taxes, and the tax rates are set by local elections. In some other places, GCDs may be financed through user fees assessed, based on the volume of water pumped or the volume permitted. This approach is typically used in areas where there is a high demand for water resources and where the districts are responsible for ensuring equitable distribution and allocation of groundwater.

Regardless of the funding model used, it is crucial for GCDs to have a clear and transparent system of financial management. This includes establishing a budget, tracking expenses, and maintaining accurate records of revenue and expenditure. Effective financial management enables districts to make informed decisions about how to allocate resources and prioritize their activities, ultimately ensuring the sustainable management and conservation of groundwater resources for the benefit of both present and future generations.

4.5.3 Revenues

The main source of revenue for most GCDs is through water fees, which are assessed based on the volume of water pumped by individual users. Districts can also generate revenue through permit fees, application fees, and water analysis. Permit fees are charged when a user applies for a permit to drill, equip, or alter the size of a well. The fee amount varies, depending on the district and the type of permit being requested. The revenue generated from these fees could be used to cover the costs associated with processing the permit application and conducting necessary inspections.

Application fees are another potential revenue source for GCDs. These fees are charged when a user applies for a variance or exemption to a district rule or regulation. The fee amount varies depending on the district and the type of application being submitted. The revenue generated from these fees can be used to cover the costs associated with reviewing the application and conducting necessary investigations.

Water analysis fees are another source of revenue for some GCDs. These fees are charged when a user requests a water quality analysis of his or her well. The fee amount varies depending on the district and the type of analysis being conducted. The revenue generated from these fees can be used to cover the costs associated with conducting the analysis and providing the results to the user.

In addition to these sources of revenue, some GCDs may receive grants or other forms of funding from government agencies or private organizations. These funds can be used to cover a wide range of costs, such as conducting research or implementing conservation programs.

4.5.4 Water Ownership

Water ownership in GCDs varies from county to county. In some counties, water is held in trust by the state and allocated to the water users through permits or "water rights". A water right specifies the volume, frequency, and time period during which water can be taken. In states like Texas, the law distinguishes between surface water and groundwater. All surface water is considered state water, with the exception of diffused water or water captured from stormwater runoff that is not in a watercourse. Groundwater in Texas is associated with the English common law doctrine of "rule of capture", which states that landowners have the right to capture the groundwater beneath their property and use it as they see fit (Texas water code § 36.002). This law is interpreted as that once water is captured by a well and delivered to the surface, it belongs to the landowner. Landowners can even sell all the water that they captured on their property. In 1951, the first GCD (High Plains Underground Water Conservation District) was established in Texas. Currently, there are 98 GCDs in Texas, covering nearly 70% of the state area. These districts are responsible for protecting, preserving, conserving, and preventing groundwater from pollution within their boundaries. These have prevented moving water from one geological formation to another and have protected the aquifers from withdrawing at a rate that allows poorer quality water to encroach into the groundwater reservoir. GCDs have also played a crucial role in safeguarding groundwater reservoirs from contamination by saltwater, other harmful substances, and well runoff. Their efforts further extend to prevent the percolation of pollutants through the strata above the water table, thus ensuring the long-term sustainability and quality of groundwater resources (Fipps, 2002).

4.6 DRAINAGE CONSERVATION DISTRICTS

Drainage conservation districts (DCDs) are entities that are formed to manage and maintain water drainage systems in a particular area. The districts are responsible for the establishment and maintenance of canals, drains, ditches, laterals, and levees.

The district will also designate the streams and bayous necessary to be cleaned, deepened, and straightened, based on water quality, flood risk, and ecological importance, with the goal of improving water quality, reducing flood risk, and enhancing fish and wildlife habitat.

4.6.1 Powers

Once established, DCDs are granted certain powers to enable them to fulfill their purpose. In general, DCDs have the power to supervise the construction, maintenance, and repair of canals, drains, ditches, levees, culverts, and bridges within the district. The district may need to estimate the cost of proposed improvements to the drainage infrastructure. This includes the development of detailed cost estimates for the construction of new canals, drains, ditches, and levees, as well as estimates of the annual cost of maintaining the improvements. The engineer assigned to the project should develop a map of the district with its boundaries. In addition, the engineer needs to ascertain and procure proper and necessary outlets for the proposed canals, drains, and ditches necessary to drain the district.

Another important aspect of DCD's authority is the preparation of maps and profiles of canals, drains, ditches, and levees within the district, as well as their outlets extending beyond district boundaries. This includes mapping the existing drainage infrastructure, identifying areas that require improvement or repair, and ensuring that the drainage system functions efficiently and effectively. DCDs also have the power to enter into contracts and agreements to facilitate the implementation of drainage infrastructure projects within the district, which may include agreements with landowners, contractors, and other parties involved in the construction or maintenance of drainage infrastructure. Additionally, DCDs may have the power to acquire, purchase, or lease property for the purpose of constructing or maintaining drainage infrastructure.

4.6.2 Capital and Finance

The financing and capital of DCDs vary from state to state and county to county. In Texas, the Texas Water Code, Chapter 56 authorizes districts to issue bonds to pay for drainage improvements. The state also authorizes the DCDs to assess and collect annual taxes by levying property taxes on an equal or uniform basis or rate on each acre of land in the district and striving to be fair and equitable to all landowners. The tax revenue collected is used to pay the interest and principal on the bonds issued by the district. Taxes collected by bonds shall be placed in the interest and sinking fund. The districts are also authorized to have a tax assessed and collected on district property sufficient to maintain, repair, and preserve district improvements and to pay any legal debts, demands, or obligations of the district. However, the tax may not exceed one-half of 1% of the total assessed valuation of the district for that year.

The district may also receive grants or other funds from state or federal agencies to fund drainage improvement projects. They may use these funds to finance capital projects, such as building canals, drains, ditches, levees, culverts, and bridges. The district may also use these funds to conduct studies and surveys to assess the district's drainage needs.

To manage the district's finances, the district board appoints a treasurer and auditor. The treasurer is responsible for the safekeeping and investment of the district's funds, and the auditor audits the district's financial records to ensure compliance with state laws and regulations.

4.7 COMMERCIAL COMPANIES

4.7.1 Purpose and Chief Values

Commercial companies are organizations that construct and operate irrigation systems for profit. These companies are established with the chief values of developing new lands and selling them for profit. They are involved in the construction and maintenance of irrigation systems and may provide water services to farmers and other clients on an annual basis.

4.7.2 Types

There are three types of commercial companies. The first type is the construction and development company, which is temporary in nature. This type of company is typically established to develop and sell new lands. The company may also sell water and facilities to the buyers at the completion of the project. In some cases, the company may revert to a mutual company. The second type of commercial company is the private company, which is permanent in nature. This type of company sells water to farmers and other clients at an annual rate. Private companies typically have a specific service area. They are not subject to public regulation, but their rates may be negotiated between the company and its customers. The third type of commercial company is the public utility company (permanent). This type of company provides water services to farmers and other clients and is subject to public regulation. The rates charged by public utility companies are determined by public utility commissions or other regulatory bodies.

Commercial companies are generally incorporated for legal actions and financial reasons. Incorporation provides legal protection to the owners of the company and may aid in enforcing collections from customers who fail to pay their bills. It also allows the company to raise capital by issuing stocks or bonds.

4.7.3 Financing

Commercial companies require significant funds to undertake various tasks assigned to them. There are different ways through which these companies can raise funds to finance their activities. One way to raise funds is through tax levies within the district. The tax levies are assessed on all landowners within the district, which are usually based on the assessed value of the land. Another way is through the sale of bonds. General obligation bonds or revenue bonds may be sold to raise funds for specific projects. The approval of these bonds is subjected to proper procedures and regulations. The proceeds from the sale of these bonds are used to finance specific projects, and the repayment of the bonds is guaranteed by the district's tax revenues.

Commercial companies also have various powers, including the power of eminent domain, the power to conduct studies, investigations, and surveys, and the power to conduct demonstration projects within the district with the cooperation of other interested agencies. In short, this legislation would provide the districts created with a full set of tools for use in solving present and future natural resource problems.

4.8 IRRIGATION SYSTEMS OPERATED BY GOVERNMENTS

Government-operated irrigation systems are managed by governmental departments to ensure efficient allocation and fair distribution of water resources. They are designed to address the varied needs of stakeholders, including the agricultural sector, local communities, and the environment. These departments are responsible for executing water-related policies and legislation, overseeing the construction and maintenance of irrigation infrastructure, and managing the financial aspects of water distribution. Such systems are particularly prevalent in extensive gravity flow networks, exemplified by the expansive irrigation canals found in countries like India, Pakistan, Brazil, and China. The organizational structure of these large-scale gravity flow systems is often complex and multi-layered, with high-level system organizations providing the framework for water transport (Figure 4.1). This is achieved through a combination of natural waterways and engineered channels, including rivers, canals, reservoirs, and diversion structures.

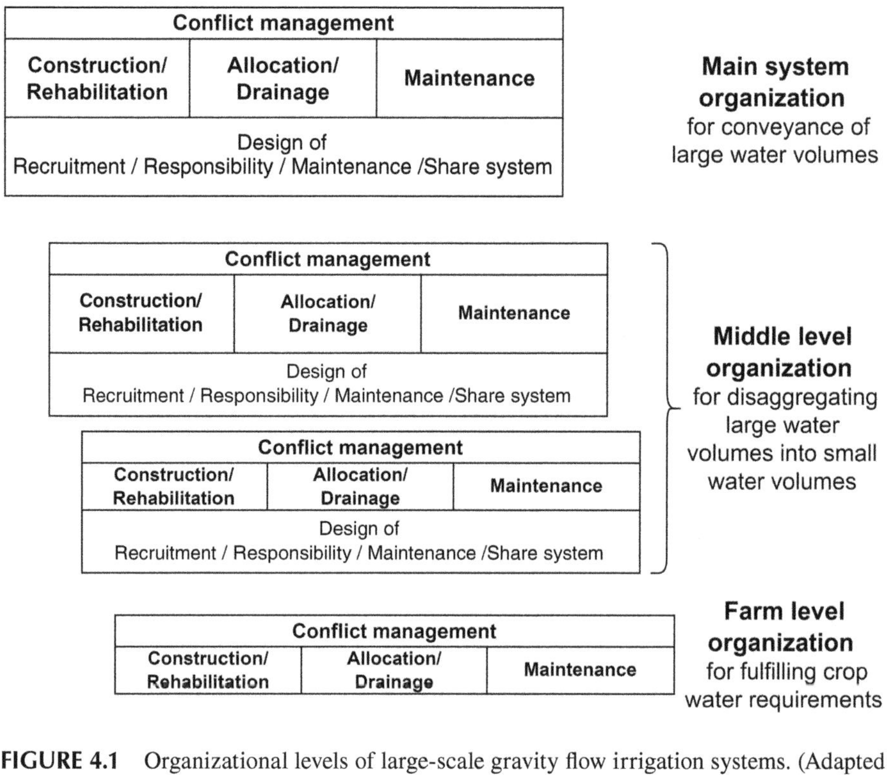

FIGURE 4.1 Organizational levels of large-scale gravity flow irrigation systems. (Adapted from Freeman et al., 2021.)

These main system organizations operate under the premise that strategic movement of water toward designated agricultural areas will foster the development of localized water control, as this becomes a necessity for the cultivation and maintenance of these lands. Collaborative efforts between these main system organizations, middle level organizations, and local farm level originations are essential. Such partnerships aim to ensure the irrigation systems function optimally, meeting the agronomic demands of the nation. The irrigation canal systems owned and managed by governments or water resources departments in China and India will be discussed as examples.

4.8.1 China

4.8.1.1 Power

In China, water resources management is primarily overseen by the Ministry of Water Resources (MWR) and its provincial affiliates, establishing a hierarchical framework for water governance across different administrative levels. The MWR formulates national policies, plans, and standards for water resources management and oversees their implementation across the country, since the first Water Law was enacted in 1988. This includes the development of water resources from major rivers, lakes, and projects that span multiple provinces. Local governments manage water projects within their jurisdiction. For each of the seven major river basins, the National River Basin Commission (NRBC) was established to manage water resources via delegation from the MWR. These commissions have the authority to control water extractions from the mainstream of these basins, leaving the management of withdrawals from tributaries to local Water Resource Bureaus (WRBs).

Provincial, prefectural, and county governments all have WRBs linked vertically to the MWR in Beijing. Provincial WRBs operate under the provincial governments and are responsible for implementing national water policies within their administrative boundaries. They manage large-scale irrigation projects that span multiple prefectures within the province and coordinate with lower levels of government to ensure the alignment of local projects with national standards and goals. Provinces have the autonomy to execute their own water management strategies, sometimes at the expense of neighboring provinces and in conflict with broader plans.

Prefectural and county WRBs focus on the management of water resources, including irrigation, within their specific administrative boundaries. They oversee the development and maintenance of medium-sized irrigation projects and ensure that local practices comply with both provincial and national policies.

Irrigation districts are specific management units created to administer water resources for irrigation over areas that may span multiple counties or prefectures within a province. They are tasked with the operational aspects of irrigation, including water distribution, maintenance of irrigation infrastructure, and coordination of water use among agricultural users. An irrigation district reports to the WRB that covers its entire administrative boundaries, which means its administrative alignment can vary depending on whether it spans multiple counties or prefectures. For instance, if an irrigation district spans multiple prefectures, it falls under the jurisdiction of the provincial WRB. Conversely, if it covers areas within multiple counties but all within the same prefecture, it is governed by the prefecture's WRB. This arrangement underscores the complexities of implementing uniform water management policies across various levels.

In the mid-1990s, the World Bank initiated a significant push toward the reform of irrigation management, marking the beginning of substantial changes within China's surface irrigation institutions. This reform transitioned management responsibilities from village collectives to water user associations (WUAs), which are participatory organizations founded on farmer involvement. In response to these developments, in 2002, China's central government released guidelines aimed at improving the water infrastructure management systems across the nation, advocating for the creation and support of WUAs. Although WUAs have emerged as a key element of these reforms, they are among various strategies adopted by local administrations. Notably, some village committees have chosen to directly engage individual farmers for local water management tasks (Wang et al., 2023). Research conducted in the Ningxia and Henan Provinces within the Yellow River Basin indicated that, before 2004, a greater number of villages (30%) showed a preference for contractual management methods as opposed to setting up WUAs (22%) (Huang et al., 2008).

4.8.1.2 Financing

The financing of large canal irrigation systems in China is multifaceted, involving various funding sources to support their construction, operation, and maintenance. A major portion of the financing comes from the central government, which allocates funds through various ministries, including the MWR. Provincial and municipal governments also allocate budgetary resources for irrigation projects within their jurisdictions. These funds may cover a portion of the construction costs, operational expenses, and local initiatives to improve irrigation efficiency. In addition, at the sub-provincial level, irrigation districts and WUAs collect fees from farmers and other users based on the volume of water used or the area irrigated. These fees contribute to the operational and maintenance costs of the irrigation systems.

Historically, national funding sources have been focused on new investments, while the maintenance of the system has been largely dependent on local governments (Lohmar et al., 2003). However, the revenue from user fees, often insufficient due to the water pricing structures (Nyberg and Rozelle, 1999), has not always adequately supported the operational and maintenance demands. As a result, many irrigation infrastructures have been dysfunctional due to low investment or lack of regular maintenance (Wang et al., 2020). Similar challenges were faced by large-scale irrigation projects in sub-Saharan Africa during the early 20th century (Higginbottom et al., 2021).

In recent years, investment strategies for irrigation systems in China have emphasized enhancing the maintenance of existing infrastructure over the initiation of new projects. To leverage additional capital and improve the efficiency of irrigation systems, China has increasingly adopted PPP models. These partnerships represent a collaborative effort between governmental bodies and private sector stakeholders to fund, construct, and manage extensive canal irrigation projects. Additionally, for projects requiring significant financial inputs, local governments or designated entities may issue bonds to raise funds from investors. These bonds can provide a substantial source of financing for large-scale irrigation projects, with the promise of future repayment based on revenue generated from water user fees or governmental support.

4.8.2 India

4.8.2.1 Power

In India, agricultural lands are predominantly privately owned, often fragmented into small holdings. Most irrigation schemes are financially sponsored and owned by state governments, which follow an irrigation policy aimed at providing water for at least one major crop annually to as many scattered landholdings as possible (Chaude et al., 2023). The organizational structure of canal irrigation in India is hierarchical, involving multiple levels of governance—from the central government to local communities. This structure ensures effective management, distribution, and maintenance of canal irrigation systems across the country. In 1974, the Government of India initiated a Command Area Development (CAD) program to improve the efficiency and productivity of irrigation systems. The program focuses on optimizing the use of water resources, enhancing agricultural output, and ensuring equitable water distribution to farmers.

The key components of the CAD program include the following: (1) Modernization and maintenance: Upgrading and maintaining irrigation systems for efficient operation; (2) Drainage system: Developing and maintaining main drainage systems beyond individual farm blocks; (3) Field-level infrastructure: Constructing field channels, field drains, and field access roads within each outlet command; (4) Land leveling: Leveling land on an outlet command basis to suit the type of irrigation being implemented; (5) Consolidation of land holdings: Redrawing field boundaries and consolidating land holdings to improve irrigation efficiency; (6) Water distribution: Enforcing a rotational water delivery system to ensure equitable, timely, and adequate water supply to individual fields; (7) Groundwater development: Developing groundwater resources (e.g., tube wells, bore wells) and promoting conjunctive use with surface water; (8) Cropping Patterns: Adopting and enforcing suitable cropping patterns to optimize water use; (9) Input supply: Providing seeds, fertilizers, pesticides, and credit services to farmers; (10) Market infrastructure: Developing regulated markets, processing facilities, warehouses, and link roads; (11) Support for small farmers: Implementing special programs for small and marginal farmers, as well as agricultural laborers; (12) Diversification: Promoting diversification into activities such as animal husbandry, farm forestry, poultry, and pisciculture; (13) Soil conservation: Implementing soil conservation measures and afforestation where necessary; and (14) Extension services: Conducting farmer training and extension programs through the training and visit system.

On-farm development (OFD) is a critical component of the CAD program, focusing on improving the efficiency of irrigation water use at the field level. OFD works primarily involve the construction of field channels, land leveling/shaping, and realignment of field boundaries. These activities ensure that irrigation water reaches every field within the command area of an outlet, thereby maximizing agricultural productivity. Previously, the focus was on delivering canal water to 40–60 ha blocks at government expense. However, the emphasis has shifted to smaller 5–8 ha blocks, significantly increasing the scope of OFD works required. This shift underscores the need for active farmer involvement to ensure accountability and meet the urgent demand for on-farm development.

4.8.2.2 Financing

In India, the operation and maintenance of irrigation systems are funded from non-plan funds, which are part of the regular government budget. However, a significant portion of these funds is allocated for staff salaries, leaving minimal resources for actual maintenance work, such as repairs and infrastructure upgrades. This misallocation of funds has led to the widespread deterioration of irrigation systems, including canals, pipelines, and drainage networks. The National Water Policy (2012) stipulates that irrigation water rates should, at a minimum, cover the full operation and maintenance costs and ideally contribute to the recovery of capital expenditures. However, in practice, water rates in India are heavily subsidized and set far below the actual cost of operation and maintenance. Compounding this issue is the poor recovery of water charges, with collection rates often below 50% due to weak enforcement mechanisms and farmers reluctance to pay.

Addressing these challenges will require coordinated policy reforms, more sustainable revenue mechanisms, and greater involvement from both farmers and private stakeholders.

REFERENCES

Chaube, U.C., Pandey, A., and Singh, V.P. (2023). *Canal Irrigation Systems in India: Operation, Maintenance, and Management.* Water Science and Technology Library, Volume 126. Springer Nature, Cham.

Fipps, G. (2002). *Managing Texas Groundwater Resources Through Groundwater Conservation Districts. Texas AgriLife Extension Service.* The Texas A&M University System, College Station, Texas. B-1612.

Freeman, D.M., Bhandarkar, V., Shinn, E., Wilkins-Wells, J., and Wilkins-Wells, P. (2021). *Local Organizations for Social Development: Concepts and Cases of Irrigation Organization.* Routledge, New York.

Higginbottom, T.P., Adhikari, R., Dimova, R., Redicker, S., and Foster, T. (2021). Performance of large-scale irrigation projects in sub-Saharan Africa. *Nature Sustainability*, 4(6), 501–508.

Huang, Q., Rozelle, S., Msangi, S., Wang, J., and Huang, J., 2008. Water management reform and the choice of contractual form in China. *Environment and Development Economics*, 13, 171–200.

Lohmar, B., Wang, J., Rozelle, S., Huang, J., and Dawe, D. (2003). *China's Agricultural Water Policy Reforms: Increasing Investment, Resolving Conflicts, and Revising Incentives.* Agriculture Information Bulletin Number 782. Market and Trade Economics Division, Economic Research Service, U.S. Department of Agriculture, Washington, DC.

National Water Policy (2012). Government of Inda Ministry of Water Resources.

Nyberg, A. and Rozelle, S. (1999). *Accelerating China's Rural Transformation.* World Bank, Washington, DC.

Porter, D., Persyn, R., and Enciso, J. (1999). Groundwater Conservation Districts: Succes Stories. Texas AgriLife Extension Service. The Texas A&M University System, College Station, TX. B-6087.

Turlock Irrigation District (2011). Proposed Study Plan. Don Pedro Project. FERC No. 2299. Turlock Irrigation District and Modesto Irrigation District. https://tuolumnerivertac.com/Documents/Don%20Pedro%20PSP%20110725.pdf

USDA-NASS (2020). *Irrigation Organizations*. U.S. Department of Agriculture, National Agricultural Statistics Service, Washington, DC.

Wang, J., Huang, Q., Huang, J., and Scottt, R. (2023). *Managing Water on China's Farms: Institutions, Policies and the Transformation of Irrigation under Scarcity*. Nikki Levy, San Diego, CA.

Wang, J., Zhu, Y., Sun, T., Huang, J., Zhang, L., Guan, B., and Huang, Q. (2020). Forty years of irrigation development and reform in China. *Australian Journal of Agricultural and Resource Economics*, 64(1), 126–149.

5 Administration

5.1 INTRODUCTION

Administration is the management and organization of a system, project, or program. In the context of irrigation systems, it involves coordinating activities and resources to ensure that water is delivered efficiently to crops and that the irrigation system is properly maintained and effectively managed. To ensure efficient and effective use of water resources, an organization for the administration, operation, maintenance, and care of an irrigation and/or drainage system is necessary. A well-administered irrigation system will increase crop yield, improve water management, and reduce water wastage. Additionally, it will ensure that water is distributed equitably among users, reducing conflicts and promoting social cohesion. An effective administration system will ensure that there is a clear delineation of roles and responsibilities for each stakeholder involved in the system. This includes the farmers, water users, irrigation officials, and other stakeholders. An effective administrative system also provides mechanisms to resolve disputes and conflicts that may arise.

However, managing an irrigation system can be a challenging and complex task. It requires careful planning, regular maintenance, and continuous monitoring to ensure effective and efficient operation. Additionally, there are often a number of different stakeholders involved in irrigation system management, including farmers, government agencies, and water districts. This chapter discusses the administration structures in different irrigation organizations and examines the roles and responsibilities of different stakeholders.

5.2 ADMINISTRATION STRUCTURES IN IRRIGATION SYSTEMS

5.2.1 At the Project Level

The governing body of an irrigation system varies, depending on the size and complexity of the system. For smaller cooperative or federal projects, the governing body is generally a board of directors elected by the people being served. The board of directors is responsible for managing the day-to-day affairs of the irrigation system. They ensure that water is distributed equitably, maintenance is regularly carried out, and conflicts are resolved timely. For example, small local groups in the Far East and Asia are communally managed by the farmers themselves with little or no external government assistance (Sagardoy et al., 1986). Although they often have excellent organizational features, most indigenous schemes can be improved at relatively little cost through better agricultural extension and assistance on various aspects of system design. Small schemes constructed by the government often perform poorly, as operational responsibility is handed over almost immediately to the farmers without

DOI: 10.1201/9781003747444-6

sufficient time being devoted to technical or management training. A rural engineering service with junior field staff based at local offices of Farmers' Service Centres should be established to provide specialist advice on operation and maintenance. Large tubewells often prove to be much less efficient owing to problems associated with their management, and the choice between operation by a publicly owned operator and a supervised farmers' group may be difficult to make.

On larger projects, particularly those that are composed of several smaller irrigation systems that are now served by common or related storage and conveyance systems, the governing body may be a board consisting of representatives of individual areas to be served. This board has many advantages, including better representation of the interests of different groups of farmers, more effective communication, and more efficient decision-making process. Large projects generally require a professional staff and central water management activities, such as water distribution and maintenance, assessment and collection of water charges, and assistance and extension to farmers on water management. Other services, including agricultural services, commercial services, and basic infrastructure and social services, may also be provided. In new settlement schemes, it is important to include commercial services within the project organization, and the government may need to ensure that commercial services are central to the project organization. The inclusion of responsibility for basic infrastructure, social services, and commercial services within the irrigation organization is usually necessary where there are no other institutions in the area capable of performing these tasks.

5.2.2 Types of Administration Structures

There are three main types of administration structures: (1) community-led administration, (2) government-led administration, and (3) hybrid administration. Each structure has its own advantages and limitations, and the best choice depends on factors such as the system's scale, the availability of technical expertise and resources, and the level of community participation and ownership.

In community-led administrative structures, the responsibility for managing and operating the irrigation system rests with the community. This structure is often seen in small-scale irrigation systems, where the community members come together to form an association to manage the system. In this structure, the community members are responsible for maintaining the system and ensuring that the water is distributed fairly among the users. The advantage of community-led administration is that it promotes community participation and ownership, which can lead to better management of the system. Additionally, it can lead to a more equitable distribution of water among the users. However, the disadvantage is that it may lead to a lack of technical expertise and resources to manage the system effectively. Examples of this type include Irrigation Districts and Water Districts in the United States, Comunidades de Regantes in Spain, Subak Irrigation Associations in Indonesia, Irrigation Associations in Afghanistan, and Asociaciones de Canalistas in Chile. These groups are referred to as Associations of Irrigation Water Users or, more briefly, Irrigation Associations (IA) (Sagardoy et al., 1986).

In government-led administration structures, the government takes the lead in managing and operating the irrigation system. The government may own and manage the system entirely, or it may contract out the management to private entities. In this structure, the government oversees system maintenance and ensures fair distribution of water among users. The advantage of government-led administration is that it ensures a high level of technical expertise and resources to manage the system. Additionally, government-led administration can provide a stable source of funding for the system. However, the disadvantage is that it may lead to bureaucracy, delay in decision-making, and a lack of accountability to the community. Government-led administration structures are found in many countries, such as Spain, Turkey, Bolivia, Iraq, most of Ecuador, and Kenya. They are known as Public Irrigation Schemes (PIS) (Sagardoy et al., 1986).

A hybrid administration structure combines elements of both government-led and community-led administrative structures. In this structure, the government works in partnership with the community to manage the system. The government may provide technical assistance and resources, while the community is responsible for the day-to-day management and operation of the system. The advantage of hybrid administration is that it combines the strengths of both government-led and community-led administrative structures. This structure can lead to better management of the system and more equitable distribution of water among the users. However, the disadvantage is that it may lead to conflicts between the government and the community over decision-making and accountability. In several Asian countries, irrigation schemes where government officials manage the primary irrigation system, while farmers' associations control the tertiary canals, are increasing. These groups are also referred to as "Irrigation Schemes with Mixed Control" (Sagardoy et al., 1986).

5.3 ADMINISTRATION BY IRRIGATION ASSOCIATIONS

The board of directors or the governing body is responsible for the overall administration of the irrigation system. They ensure that the system is operated and maintained efficiently, effectively, and sustainably. They also oversee the distribution of water, monitor water usage, and ensure that water is distributed equitably among users. A careful study of the brochure "Guidelines for New Directors of Irrigation Districts", published by the Extension Service of the University of Nebraska, gives an excellent understanding of the qualifications and duties of members of the board of directors (Axthelm, 1968). Much of the information is appropriate for anyone serving on a governing body of an irrigation enterprise, whether it be constituted as a private, quasi-public, or public organization. The organization is subdivided into several bodies according to its complexity. A typical organizational chart of irrigation districts in the United States is shown in Figure 5.1.

Some of the bodies of the organization are as follows:

a. **General assembly**: This is composed of the farmers of the association, and their main function is to select the board of directors and approve or disapprove the management plans.

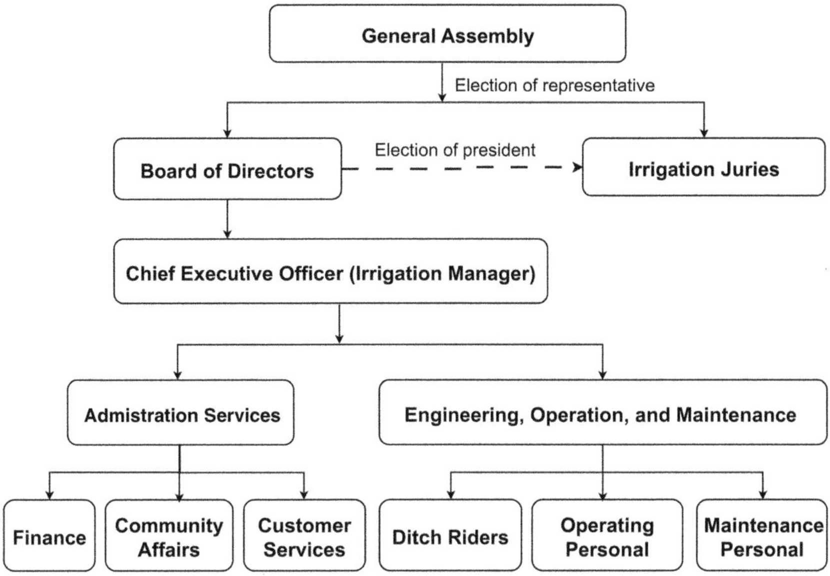

FIGURE 5.1 A typical irrigation district organizational chart.

b. **Board of directors**: They supervise the management plan, construction improvements, administration of the district, and training of the personnel. They are also responsible for selecting the general manager.
c. **Chief executive officer (irrigation manager)**: Regardless of the type of governing board, the most important single individual in the operation and management of the irrigation venture will be a well-qualified individual for overall supervision of the project operation and maintenance. The irrigation manager should have technical capabilities and knowledge in hydraulics, statistics, topography, agriculture, economics, and business management. This individual should be experienced in this type of business venture, and be one who respects his position, is respected by the people being served, and whose judgment will be respected and sought. The manager should also have social skills to understand the users and to deal with their operation problems since most of them are human-related. The balance of the organization will vary with project needs and can only be established after the operation and maintenance practices and procedures for the individual projects are analyzed or established.
d. **Operating personnel**: These personnel will be needed to regulate water supply and delivery facilities; a reservoir superintendent may have to be permanently assigned to duty at water storage facilities if they are remote; and an operation manager must be assigned to patrol the waterways as well as make the necessary daily adjustments in water control facilities.

The irrigation district may be divided into sections or into small user organizations according to the size of the district, distribution of the canals, topography, pumping stations, drains, and access to roads. In large districts, an irrigation manager may have one or more operation managers who supervise a group of ditch riders.

e. **Ditch riders and Zanjero**: Zanjeros, a Spanish term used in the Southwestern United States, designate the personnel to control water deliveries and patrol waterways on irrigation projects in the United States. They are important for the successful operation of an irrigation project and, undoubtedly, have been since the practice of irrigation began. A ditch rider has the responsibility of delivering water to a group of water users and operating the canals, as well as providing information to the manager about the water volumes and times delivered to each water user or farmer. The farmer communicates with the water user group and the administration department. In addition to distributing the project's water and maintaining the integrity of its conveyance system for the principal commodity, the ditch rider plays a crucial role in public relations by fostering communication and positive relationships with stakeholders, including farmers, landowners, and regulatory agencies. A well-qualified and responsible ditch rider, also known as a zanjero, is an invaluable asset to the successful operation and administration of any irrigation project. As a general criterion, a single ditch rider can usually serve 5,000–8,000 acres (2,000–3,000 ha) of land depending, of course, on the area covered and the number of facilities that must be controlled.

f. **Maintenance personnel**: Maintenance organizations vary and must be planned for the work that must be accomplished. On smaller projects, only a nucleus of a crew is required during the operating season. Accordingly, the operating staff can be utilized for maintenance work after the irrigation season is over. In some areas, water is delivered year-round. Under these circumstances, operating staff and maintenance staff must be maintained at all times. There are some distinct advantages and economics to be realized in the control approach to project operation and maintenance. Fewer administrative personnel are required for the administration of several smaller and individual project systems. For equipment maintenance, less outlay is required, since much of the equipment necessary for the maintenance of an irrigation and drainage system can only be utilized for specific jobs and is idle much of the time if retained for small acreage. Of course, the organization necessary to operate a system in which the water is used for various purposes other than irrigation must vary to fit the particular circumstances.

g. **Administration and office staff**: They are in charge of collecting water fees, maintaining water use records, and other administrative tasks. Office assistance is also necessary in administering the project and in maintaining constant communication with the ditch riders, water masters, and water users. On smaller projects, one or two persons in the headquarters office may be all that is necessary; however, on larger projects, a large staff is required to keep up with daily activities. Some of the larger projects now use automatic billing and electronic equipment for office records.

h. **Irrigation juries**: They are responsible for determining punishments for users who fail to respect rules and regulations.

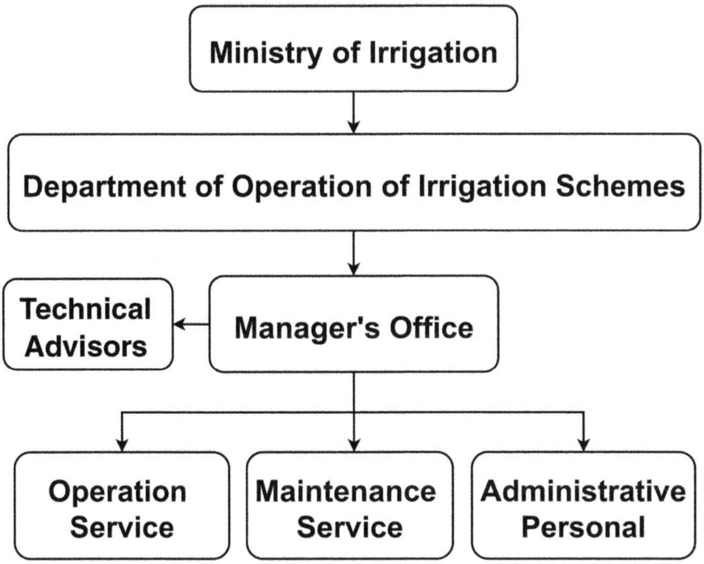

FIGURE 5.2 A typical administration in public irrigation schemes.

5.4 ADMINISTRATION IN PUBLIC IRRIGATION SCHEMES

PIS are government-controlled organizations that manage irrigation projects. In some countries, PIS are established on a permanent basis, while in others, they are established as a transitional measure to transfer responsibility to farmers. PIS are closely connected to the organization used during the construction stage of the irrigation project, and they often employ most of the executive and administrative personnel who worked on the construction. This can lead to insufficient attention to agricultural problems.

PIS has a clear line of command, starting from the national level and going down to the lowest units. As shown in Figure 5.2, the Manager's Office is in charge of implementing the plan approved by the central office, while the Executive Bodies are responsible for operational, maintenance, and administrative units. PIS offers good possibilities for rationalizing water distribution, but this potential is often not realized. It also keeps a considerable amount of technical data, making it useful for future operation and maintenance of the irrigation system. PIS is essential when using sophisticated water distribution techniques that farmers' organizations cannot manage.

5.5 CASE STUDIES

5.5.1 China

China's irrigation administration is characterized by a complex and hierarchical system (Figure 5.3). At the national level, the Ministry of Water Resources (MWR) serves as the central authority, overseeing large-scale irrigation projects that span multiple provinces, such as those along the Yellow River and Yangtze River basins.

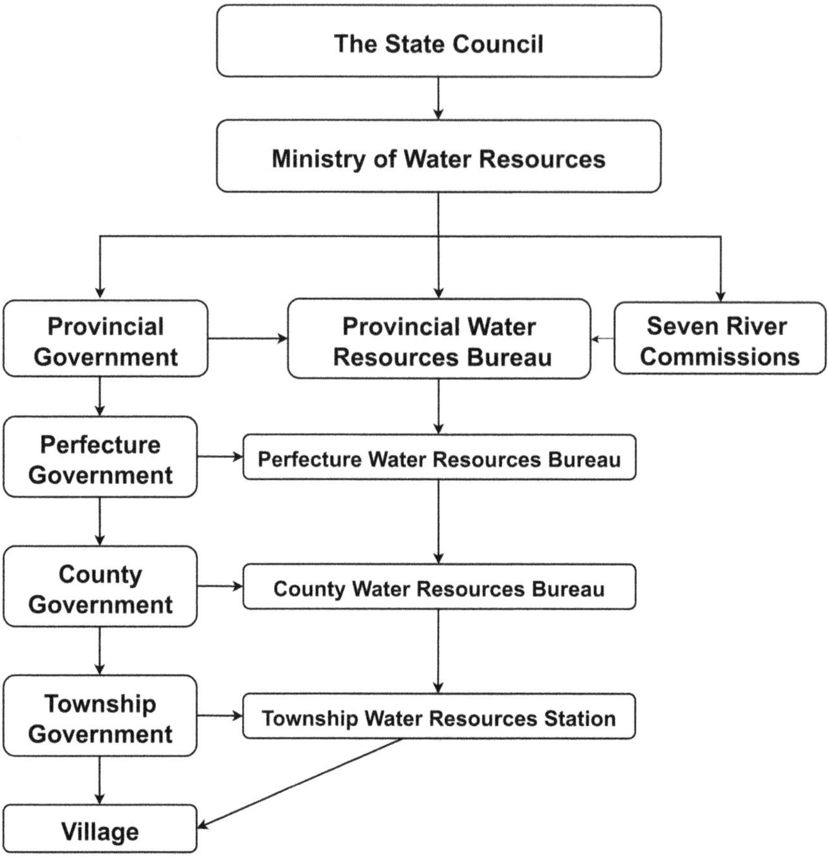

FIGURE 5.3 Administration structure in China. (Adapted from Lohmar et al., 2003.)

The national government is responsible for investing in and developing water resources from large rivers, lakes, and projects that cover more than one province. Local governments manage irrigation systems within their administrative districts, forming a multi-tiered governance structure.

Outside of the central government, a network of subnational water management institutions plays a significant role in shaping water policy. Provincial, prefectural, and county governments each have their own Water Resources Bureaus (WRBs), which are linked vertically to the MWR in Beijing. Formally, the subnational offices are charged with implementing the rules and policies advanced by the national authorities. In reality, however, the heads of local WRBs are appointed by and report to the leaders of their own jurisdictions (such as provincial governors or county magistrates). These horizontal ties frequently dominate the vertical ones. As a result, WRBs often design and execute water policies based on the specific needs of their jurisdictions, leading to significant heterogeneity in water policies across regions. This decentralized approach allows for flexibility but can also result in inconsistencies and challenges in coordinating national water management goals. One example is the Yellow River Basin, where irrigation districts manage main and branch canals. The Yellow

River Conservation Commission oversees the operation of main canals and branch canals, which transport water from the Yellow River and its reservoirs to lower-level distribution networks (He et al., 2020). County governments, township governments, and village committees are responsible for managing tertiary canals and field-level distribution. This localized approach allows communities to adapt water distribution to their specific agricultural needs and conditions. This decentralized approach has improved water delivery efficiency in some areas but has also created challenges in coordination and equitable distribution. For instance, upstream regions often have better access to water, leaving downstream areas vulnerable to shortages.

5.5.2 India

Canal irrigation systems in India play an important role in supporting the country's agricultural sector. The administration of these canal systems is primarily managed by state-level Irrigation Departments, which oversee the operation, maintenance, and revenue collection of irrigation infrastructure (Figure 5.4). The administration system is characterized by a mix of centralized control and localized responsibilities, leading to both efficiencies and challenges.

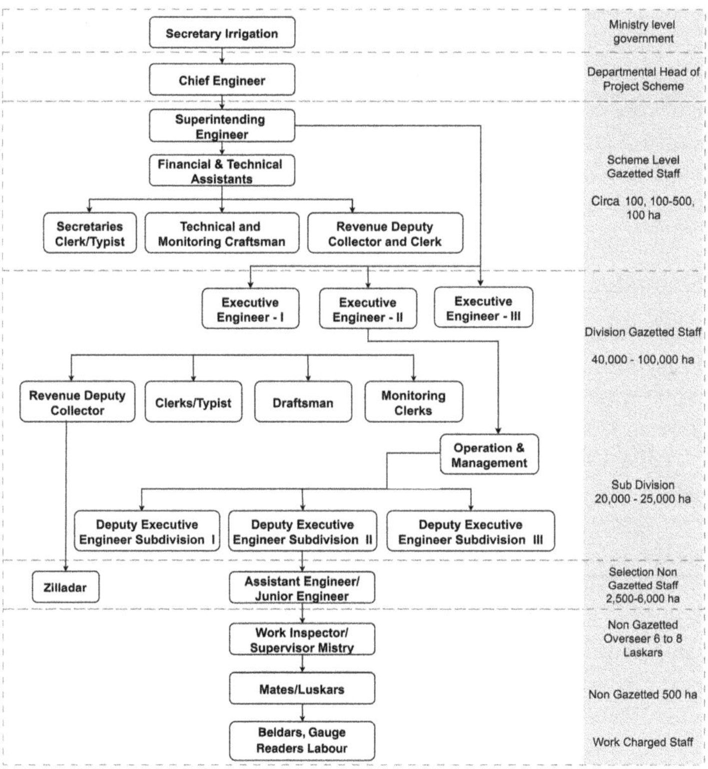

FIGURE 5.4 Organizational chart of a typical major irrigation scheme in India. (Adapted from Chaube et al., 2023.)

The operation and maintenance of canal systems are primarily handled by civil or irrigation engineers trained in specialized Water and Land Management Institutes. These engineers are responsible for maintaining and ensuring the efficient performance of irrigation infrastructure, including canals, reservoirs, and distribution networks. Despite their technical expertise, the system faces several issues, such as aging infrastructure, siltation, and inadequate maintenance due to limited funding and resources. These challenges often lead to water losses and reduced system efficiency, undermining the overall effectiveness of the irrigation network.

The canal system is revenue-oriented, with a primary focus on collecting water charges from farmers. While this approach ensures a steady income for maintaining infrastructure, it often prioritizes revenue generation over service delivery. This has led to inefficiencies in water distribution, as water may not be allocated based on actual agricultural needs. Additionally, the lack of accountability to farmers means that delays or inadequate water delivery are rarely addressed adequately, leaving farmers with limited recourse.

One of the most significant shortcomings of the system is that farmers' water rights are not formally recognized. If a canal system fails to deliver water on time or provides an inadequate quantity, farmers have limited options. At most, they can claim a remission in water charges, but this does not compensate for potential crop losses or income reductions. This lack of legal recognition undermines farmers' ability to demand reliable and equitable water distribution, further exacerbating inequities in the system. This disconnect between canal administration and on-farm water management creates gaps in service delivery and reduces the overall effectiveness of the irrigation system. Reforms that prioritize farmers' participation, equitable water distribution, and sustainable infrastructure management are essential for improving the efficiency and reliability of India's canal irrigation systems.

5.5.3 Australia

Irrigation administration in Australia has evolved considerably in recent decades, largely in response to the need for efficient water management, sustainability, and farmer empowerment. Historically, irrigation systems in Australia were managed by state governments, with little involvement from farmers or local communities. However, by the 1970s, it became evident that centralized management was inefficient, leading to rising costs, poor system operation, and inadequate maintenance. Farmers, recognizing these inefficiencies, began advocating for greater control over irrigation management. This movement, supported by government policies and industry organizations, laid the foundation for Irrigation Management Transfer (IMT). The Water Reform Framework (1994) was an improvement policy that mandated greater responsibility for irrigation management to be transferred to water users (Council of Australian Governments, 1994). This led to the emergence of various Irrigation Water Providers (IWPs), which operate under state-specific legislation such as the Water Management Act 2000 in New South Wales and the Irrigation Act 1994 in South Australia.

Today, the core principle of Australia's irrigation administration is the decentralization of management responsibilities from government agencies to local entities, often owned and operated by farmers. This shift has been facilitated by the formation of Water User Associations (WUAs) and Water Services Committees (WSCs), which empower farmers to participate in planning, decision-making, and maintenance activities (Poddar et al., 2011). This approach has not only improved the efficiency of water delivery but has also fostered a sense of ownership and accountability among farmers.

Irrigation management in Australia is characterized by various legal and corporate structures, including private companies, statutory trusts, and cooperatives. For instance, Murray Irrigation Limited and Murrumbidgee Irrigation are private companies owned by irrigators, while entities like Renmark Irrigation Trust operate as statutory trusts (Poddar et al., 2011). These organizations manage large-scale irrigation systems, often covering hundreds of thousands of hectares and serving thousands of farmers. This diversity in organizational forms allows for flexibility and adaptability to local conditions and needs.

Water markets play a crucial role in Australia's irrigation administration, particularly in the Murray-Darling Basin (MDB). In the MDB, water ownership has been separated from the land, allowing non-landholders, such as financial investors and environmental water holders, directly own water rights via water markets (Grafton and Horne, 2014). This separation has created a flexible and dynamic system where water entitlements can be traded, enabling farmers to better manage water scarcity and allocate resources more efficiently. The ability to trade water entitlements has provided farmers with significant flexibility, particularly in times of drought or variable water availability. Farmers can purchase additional water during periods of scarcity to sustain their crops or sell surplus water during times of abundance, generating additional income. This market-based approach has not only improved the efficiency of water allocation but has also encouraged more sustainable water use practices. By responding to market signals, farmers can optimize their water use to maximize productivity and profitability, while also contributing to the overall resilience of the irrigation system.

However, challenges, such as the lack of transparency in trade restrictions and allocation announcements, as well as the need for robust accounting of water extraction and use, have been reported, highlighting the necessity for further reforms to optimize the functionality of water market (Seidl et al., 2020). These issues have created significant barriers to the efficient operation of water markets, particularly in the MDB, where water trading is a critical mechanism for managing water scarcity and resource allocation (Seidl et al., 2020). Farmers and other market participants often face difficulties in accessing clear and timely information about water allocations, trade rules, and restrictions, which can hinder their ability to make informed decisions.

For instance, delays in allocation announcements or sudden changes in trade regulations can disrupt agricultural operations and reduce confidence in the market's reliability. Another challenge is the need for robust accounting of water extraction and use. Accurate and reliable data on water extraction and use are fundamental to the effective functioning of water markets. However, inconsistencies in water accounting practices, combined with limited monitoring and enforcement, have raised concerns about the accuracy of water use reporting. In some cases, over-extraction or unauthorized use of water has gone undetected, undermining the integrity of the market and leading to inequities in water distribution.

To address these challenges, there is a pressing need for enhanced monitoring systems, standardized accounting practices, and stricter enforcement of water use regulations. Using advanced technologies, such as real-time monitoring and remote sensing, can substantially improve the accuracy and reliability of water accounting. These measures would ensure that water allocations are used as intended, fostering greater trust and efficiency in the market.

REFERENCES

Axthelm, D. (1968). EC68–774 Guidelines for New Directors of Irrigation Districts. Historical Materials from University of Nebraska-Lincoln Extension. 3873. https://digitalcommons.unl.edu/extensionhist/3873

Chaube, U.C., Pandey, A., and Singh, V.P. (2023). *Canal Irrigation Systems in India: Operation, Maintenance, and Management*. Water Science and Technology Library. Volume 126. Springer Nature, Cham.

Council of Australian Governments (1994). The Council of Australia Government's Water Reform Framework. Extract from Council of Australian Governments: Hobart 25 February 1994 Communique.

Grafton, Q.R. and Horne, T. (2014). Water markets in the Murray-Darling Basin. *Agricultural Water Management*, 145, 61–71.

He, C.S., Harden, C.P., and Liu, Y.X. (2020). Comparison of water resources management between China and the United States. *Geography and Sustainability*, 1(2), 98–108.

Lohmar, B., Wang, J., Rozelle, S., Huang, J., and Dawe, D. (2003). *China's Agricultural Water Policy Reforms: Increasing Investment, Resolving Conflicts, and Revising Incentives*. Agriculture Information Bulletin Number 782. U.S. Department of Agriculture, Market and Trade Economics Division, Economic Research Service, Washington, DC.

Poddar, R., Qureshi, M.E., and Syme, G. (2011). Comparing irrigation management reforms in Australia and India: A special reference to participatory irrigation management. *Irrigation and Drainage*, 60(2), 139–150.

Sagardoy, J.A., Bottrall, A., and Uittenbogaard, G.O. (1986). *Organization, Operation and Maintenance of Irrigation Schemes*. FAO Irrigation and Drainage Paper 40. Food and Agriculture Organization of the United Nations, Rome.

Seidl, C., Wheeler, S.A., and Zuo, A. (2020). Treating water markets like stock markets: Key water market reform lessons in the Murray-Darling Basin. *Journal of Hydrology*, 581, 124399. https://doi.org/10.1016/j.jhydrol.2019.124399

Section 2

Operation of Irrigation Systems

6 Water Delivery and Irrigation Scheduling

NOTATION

Q	discharge in the channel (m^3/s, cfs)
FC	field capacity (m/m, in./in.)
PWP	permanent wilting point (m/m, in./in.)
f_d	allowable depletion (fraction)
AWC	available water capacity (mm)
AD	allowable depletion (mm)
ET_o	reference evapotranspiration (mm/day)
K_c	crop coefficient
ET_c	crop evapotranspiration (mm/day)
TAW	total available water (mm)
R_d	root zone depth (cm)
$f_{d\max}$	maximum allowable fraction depletion
U_f	upward flow of water from the groundwater (mm/day)
SWD	soil water deficit or soil water depletion (mm)
d_e	effective irrigation depth (mm)
P_e	effective precipitation (mm)

6.1 INTRODUCTION

The objective of irrigation systems is to deliver water to each individual field in the quantity the field should receive and at the time its use is needed. The success of the operation of an irrigation system is measured by the extent to which this objective is achieved. To achieve increased crop yield and higher efficiency, it is important to implement appropriate water delivery methods, establish proper distribution control, and comply with relevant legal regulations. This chapter provides an analysis of different water delivery methods, distribution control mechanisms, water delivery scheduling, and rotation schedules. Additionally, it discusses various factors that influence the selection of a water delivery system and examines the legal perspectives of water delivery.

6.2 METHODS OF WATER DELIVERY

There are four principal methods for the delivery of water: (1) continuous supply, (2) supply on rotation, (3) supply-on-demand, and (4) arranged supply. It seldom happens that any one of these four methods is strictly followed to deliver irrigation water to a project. Depending on the prevailing conditions, it is common to modify

DOI: 10.1201/9781003747444-8

a particular method or combine two or all four methods for application at various times or locations.

6.2.1 Continuous Delivery

In the continuous delivery method, water users (farmers) have a continuous right to access water when needed. Each user receives his share of water as a continuous flow in proportion to the area of his land. This method is most suitable for areas with appropriate water rights where users can demand their legal allotment at any time, regardless of their actual need. This approach typically results in minimum design capacity but requires farmers to store water on their property. This method may waste water and time and may contribute to waterlogging when farm tracts are small (and flow rates are small) or on sandy soil. If extreme conditions make rotation or demand systems impractical because of the resulting low farm irrigation efficiencies, then the continuous delivery method should not be used.

The actual use of water is normally intermittent and depends on the local conditions of each farm and the crops growing thereon. Suppose that in extreme cases, all crops are of the same kind in an area; then all farms will desire their full flow of water. This means that laterals or sub-laterals (the smaller distribution channels that deliver water directly to the farms) must have a capacity equal to the total water rights under them. The main canal can be of somewhat low capacity because it generally doesn't deliver water to all farms at the same time under normal conditions. If the main canal is undersized relative to the laterals, then during these peak demand periods, the total water required may not be met. This situation underlines the importance of designing canal systems with a clear understanding of how water will actually be delivered and used, ensuring that capacity is adequate when high simultaneous demand occurs.

The operation of canals under continuous-flow delivery is usually easier because there is less fluctuation in use. The water surface can be maintained at proper levels above the outlets to deliver the required quantities of water to each farm with minimal regulation. However, when water usage is more irregular, such as at the beginning and end of the season, regulation of canals and turnouts will be prompted in connection with demand.

The continuous delivery method is well-suited to canals with good water rights for the duration of flow in streams and a consistent water supply. It is also suitable for farms with pressurized irrigation systems, such as sprinklers or drip irrigation, which require a small but continuous flow rate. Although the delivery of water may be continuous from the turnouts, the supply on the farm may be rotated between different fields. This method is preferable where soil, topography, or crop conditions reduce the size of the irrigation head, which refers to the flow of water distributed at a single irrigation upstream source location, and it is also often called the irrigation stream size, or where irrigation heads are relatively large in proportion to the size of the farm.

Example 6.1

Four farms are supplied by a canal irrigation system (Figure 6.1). Farms 1 and 2 are connected to Lateral 1 (L1), with Farm 1 covering 100 ha of sorghum and Farm 2 covering 200 ha of sorghum. Farms 3 and 4 are connected to Lateral 2

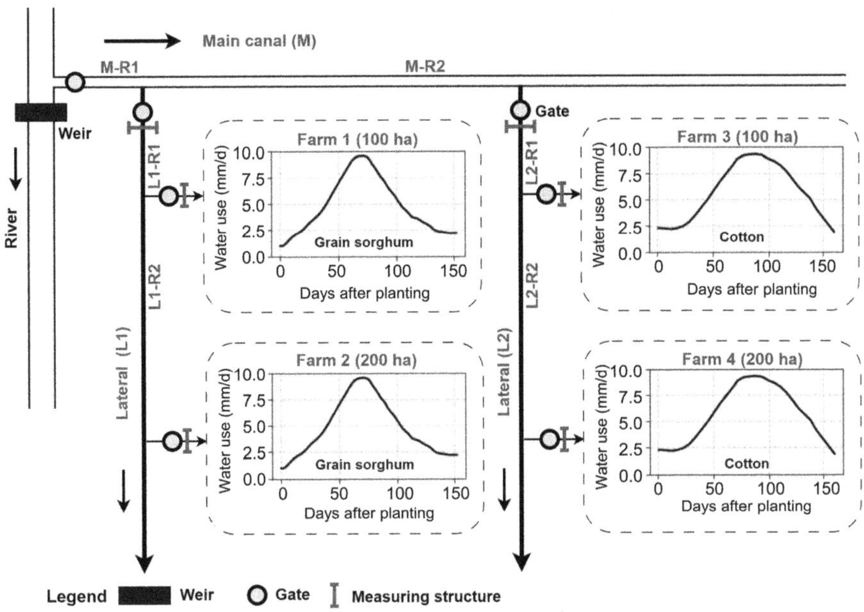

FIGURE 6.1 A schematic diagram of water delivery within an irrigation system.

TABLE 6.1
Average Daily Water Requirements in the Growing Season for Sorghum and Cotton

Month	Sorghum Water Use (mm/day)	Cotton Water Use (mm/day)
April	—	2.0
May	2.3	4.5
June	6.2	8.6
July	8.6	8.6
August	4.6	4.0
September	2.5	0.8

(L2), where Farm 3 grows 100 ha of cotton and Farm 4 grows 200 ha of cotton. Sorghum is planted on May 1, and cotton is planted on April 1. Determine the maximum required capacity of the main delivery canal under the continuous delivery method, assuming a canal conveyance efficiency of 100%. Average daily water requirements for sorghum and cotton during the growing season are shown in Table 6.1.

Solution:

Water is supplied continuously at a rate matching the peak daily crop water requirement.

The daily crop water requirements are provided in mm/day and converted to L/s/ha:

$$1 \text{ mm/day} = \frac{10{,}000 \text{ L/ha/day}}{86,\,400 \text{ s/day}} \approx 0.12 \text{ L/s/ha}$$

The peak demand for sorghum in July is 8.6 mm/day, and the required flow rate is

$$8.6 \text{ mm/day} = \frac{\dfrac{86{,}000\,\dfrac{\text{L}}{\text{ha}}}{\text{day}}}{86,\,400\,\dfrac{\text{s}}{\text{day}}} \approx 1.00 \text{ L/s/ha}$$

The peak demand for cotton in June is 8.6 mm/day, and the required flow rate is 1.00 L/s/ha.

The capacity of Lateral 1 is

$$\left(100 \text{ ha} + 200 \text{ ha}\right) \times 1.00 \text{ L/s/ha} = 300 \text{ L/s}$$

The capacity of Lateral 2 is

$$\left(100 \text{ ha} + 200 \text{ ha}\right) \times 1.00 \text{ L/s/ha} = 300 \text{ L/s}$$

Capacity of the main canal:

$$300 \text{ L/s} + 300 \text{ L/s} = 600 \text{ L/s}$$

The continuous delivery method supplies water to all farms every day at a constant rate, making it easy to manage and requiring minimal scheduling. However, if the system is designed to meet peak seasonal crop demand, a significant volume of water is delivered unnecessarily during off-peak months, resulting in considerable water waste. Alternatively, designing canals at a lower capacity (e.g., matching average demand) would force farmers to store surplus water during low-demand periods—a strategy impractical without reservoirs or ponds, particularly in arid regions with limited storage infrastructure. Thus, while continuous delivery ensures reliability, its inefficiency stems from either excessive infrastructure costs (oversized canals) or reliance on unfeasible on-farm storage solutions.

6.2.2 Rotation Method

The rotation method is probably the most flexible method because it can vary greatly. It involves delivering relatively large irrigation stream sizes for short periods of time. In this method, water is delivered following a pre-arranged schedule to each user in sufficient quantity for a fixed period of time. Rotations may be made (1) between two water users, (2) two or more groups of water users, (3) two or more different laterals, and (4) between definite divisions of the entire project conveyance system. Local conditions and crops determine the type of rotation that is most applicable.

The rotation method has the advantages of reducing the labor cost of applying water to the land, as well as the smaller amounts of water required to cover the land when large irrigation stream sizes are used. However, the advantages are lost if crop and topographic conditions restrict the size of irrigation head required for a large proportion of the growing season to irrigate the average size of the farm, or if the use of irrigation stream sizes larger than those that can be handled results in excess waste of surface water.

To ensure the most economical water distribution, divisions of the conveyance system or the size of groups of water users between which rotation is practiced will probably vary throughout the irrigation season. Good irrigation efficiencies can be obtained under the rotation method with careful management. However, fixed schedules do not permit the irrigator to delay irrigation to a more optimum time, as is permissible with the demand system. When its rotation period is due, the foregoing irrigation might subject crops to severe water stress.

Overall, the rotation method is a highly adaptable irrigation technique that can be tailored to local conditions. With careful management, it can achieve good project irrigation efficiencies and reduce labor and water costs. However, its fixed schedules may limit flexibility in irrigation timing, which could affect crop yields and the amount of water saving.

Example 6.2

Using the same canal irrigation system as in Example 6.1, the total peak water requirement is 600 L/s under a continuous delivery method. The system now adopts a 1-day-in-2 rotational delivery (i.e., each frame receives water every other day). On any given day, only one lateral is operating. The canal conveyance efficiency is assumed to be 100%. Design the rotation schedule for each farm and compare water allocation and flow rates between continuous and rotational methods.

Solution:

The total peak demand for continuous delivery is 600 L/s, with 300 L/s in each lateral. Under 1-day-in-2 rotational delivery each lateral irrigates every other day. To provide the same total water volume over time, the flow rate per lateral must double during its irrigation day. Therefore, the required flow per lateral during irrigation is 600 L/s.

The flow rate to each farm is calculated based on the area share:

The flow rate for Farm 1 (100 ha) is

$$\frac{100\text{ ha}}{300\text{ ha}} \times 600\,\frac{\text{L}}{\text{s}} = 200\text{ L/s}$$

The flow rate for Farm 2 (200 ha) is

$$\frac{200\text{ ha}}{300\text{ ha}} \times 600\,\frac{\text{L}}{\text{s}} = 400\text{ L/s}$$

Similarly, the flow rate for Farm 3 is 200 L/s, and for Farm 4 is 400 L/s. Table 6.2 summarizes and compares the water allocation and flow rates for each delivery

TABLE 6.2
Comparisons of Water Allocation and Flow Rates in July (2-Day Cycle) (Unit: L/s)

Canal/Canal Reach		Main Canal		Lateral 1 (L1)				Lateral 1 (L2)			
		M-R1	M-R2	L1-R1	Farm 1	L1-R2	Farm 2	L2-R1	Farm 3	L2-R2	Farm 4
Continuous	Day 1	600	300	300	100	200	200	300	100	200	200
	Day 2	600	300	300	100	200	200	300	100	200	200
Rotation	Day 1	600	0	600	200	400	400	0	0	0	0
	Day 2	600	600	0	0	0	0	600	200	400	400

Note: The canal reach names (e.g., M-R1, L1-R1) refer to the layout illustrated in Figure 6.1.

method across a 2-day period in July, which represents the peak irrigation demand. The 1-day-in-2 rotational delivery method reduces the number of operational turnouts per day and provides more scheduling flexibility. However, it requires larger lateral canals and farm outlets to handle the increased peak flow rates on irrigation days. While the main canal capacity remains unchanged, the lateral canals must be upsized, increasing the infrastructure cost.

6.2.3 Demand Method

The demand method grants users the right to a certain rate of flow or total seasonal amount of water in acre-feet that can be claimed on demand, subject to restrictions imposed by the operation of the system. While the advantage of this method is its simplicity for users, it requires large canal capacities in proportion to the area served because of fluctuating water supply. Consequently, restrictions must be imposed, such as advance notice of water requests, to ensure an orderly delivery system. Even with restrictions, demands may exceed supply during peak irrigation periods, causing delivery delays that may exceed the period of notice, and requests are fulfilled in the order received. Thus, the demand method often results in an irregular rotation of delivery, using irrigation head sizes comparable to those employed in rotation methods.

The delivery-on-demand method is suited to systems serving orchards with uniformly distributed water demands throughout the year. The method works well where the canal capacity is large in proportion to the area irrigated. This works well where the supply of water is plentiful or is not a problem. However, usually such is not the case. The method is not suitable for long canals and is conducive to significant waste of water. For small canals, if canals receive water from streams of ample flow or from reservoirs, the actual delivery on demand may be fine. However, the cost of increased capacity to deliver water on demand has to be justified, where water is plentiful. Sometimes, the conditions of water supply determine capacities. If water is taken from a stream, then the maximum rate of diversion may be limited by the characteristics of stream flow or water rights. If the water is taken from a storage reservoir, the maximum delivery rate is one of the maximum canal capacities.

From the irrigators' perspective, the demand method is very favorable, as it enables irrigation when needed using the most economical and efficient flow rate. Moreover, it encourages the wise use of water, leading to higher irrigation efficiency. However, it requires capable management to match supply with demand. As it is not economically feasible to design canals and laterals for unlimited capacity, water demands may exceed the available flow capacity during the peak irrigation season. If this occurs, a change to one or both of the other water delivery methods may be necessary.

Example 6.3

Using the same canal irrigation system as in Example 6.1, determine the required capacity of the main delivery canal and each farm outlet every month from April to September, assuming the demand delivery method and 100% canal conveyance efficiency.

Solution:

Under the demand method, water is delivered on request, based on the actual needs of each farm. All farms can potentially receive water at the same time if needed. The peak flow must cover the combined demand. In this case, the capacity of the main canal in July is the same as that of the continuous delivery method. Using the monthly crop water use values from Table 6.1, flow rates for each farm are calculated and summarized in Table 6.3.

For example, crop water use for cotton is 2 mm/day in April, and the required flow is

$$2.0\ \text{mm/day} = \frac{20{,}000\ \dfrac{\text{L}}{\text{ha}}\ \Big/\ \text{day}}{86{,}400\ \dfrac{\text{s}}{\text{day}}} = 0.23\ \text{L/s/ha}$$

TABLE 6.3
Comparisons of Water Allocation and Flow Rates for Each Month (Unit: L/s)

Canal/ Canal	Main Canal		Lateral 1 (L1)				Lateral 1 (L2)			
Reach	M-R1	M-R2	L1-R1	Farm 1	L1-R2	Farm 2	L2-R1	Farm 3	L2-R2	Farm 4
April	69	69	0	0	0	0	69	23	46	46
May	237	156	81	27	54	54	156	52	104	104
June	516	300	216	72	144	144	300	100	200	200
July	600	300	300	100	200	200	300	100	200	200
August	297	138	159	53	106	106	138	46	92	92
September	114	27	87	29	58	58	27	9	18	18

Note: The canal reach names (e.g., M-R1, L1-R1) refer to the layout illustrated in Figure 6.1.

The capacity of Lateral 1 is

$$(100\text{ ha} + 200\text{ ha}) \times 0.23\text{ L/s/ha} = 69\text{ L/s}$$

The demand-based method provides water according to real-time farm needs, offering the highest flexibility to farmers but requiring the canal system to be sized for the maximum simultaneous demand. It can be highly efficient if real-time sensors and metering are available. However, conflicts or shortages may occur during peak periods if the infrastructure is undersized.

6.2.4 Arranged Supply

In the arranged method, water users are required to provide advance notice to the district or command area specifying the flow rate and duration of the water required, typically with an advance notice period of 1–3 days. The district then arranges a supply schedule with the water users, which requires effective communication between farmers and the district. This schedule can work particularly well for upstream users, but coordination is essential to ensure that the downstream users are not adversely affected. A well-designed system with regulatory reservoirs is necessary to avoid spills, especially when water is shut off unexpectedly.

Example 6.4

Using the same canal irrigation system as in Example 6.1, design an arranged supply schedule from July 3 to July 5, based on water requests submitted by farmers on July 2 (Table 6.4). The main canal has a maximum capacity of 600 L/s, and the system operates on a 24-hour delivery schedule. Farmers provide a 2-day advance notice, and the irrigation district is responsible for ensuring equitable allocation, conflict-free delivery, and efficient canal utilization. A regulatory reservoir is available to buffer short-term adjustments.

Solution:

To maximize system use without exceeding the 600 L/s limit, the district schedules:

- Farm 2 and Farm 4 (600 L/s each) on separate days, each receiving full-day delivery.

TABLE 6.4
Water Requests Submitted on July 2

Farm	Area (ha)	Crop	Lateral	Requested Flow (L/s)
1	100	Sorghum	Lateral 1 (L1)	300
2	200	Sorghum	Lateral 1 (L1)	600
3	100	Cotton	Lateral 2 (L2)	300
4	200	Cotton	Lateral 2 (L2)	600

- Farm 1 and Farm 3 (300 L/s each) on the same day, sharing the canal capacity.

The arranged supply schedule is shown in Table 6.5. The district can repeat this 3-day delivery cycle for the remainder of the month or adjust it weekly based on new advance notices. If delivery adjustments or early shut-offs occur, the regulatory reservoir buffers short-term mismatches.

Table 6.6 compares the four water delivery methods based on key operational characteristics, including canal capacity requirements, efficiency, system complexity, and risk of over- or underwatering. Continuous delivery is simple but requires the canal system to sustain a constant rate at all times, making it inefficient and prone to over- and underwatering. Rotational delivery provides water on a fixed schedule, reducing peak demand and infrastructure load, but it offers less flexibility and carries a moderate risk of over- or underwatering if crop needs vary from the schedule. Demand-based delivery is the most flexible method, allowing for real-time scheduling based on actual crop water needs. However, it requires advanced monitoring and management and can lead to capacity overruns if not well coordinated. Arranged supply provides a balance between structure and flexibility by using pre-planned time blocks based on advance notice. It maintains high

TABLE 6.5
Water Allocation and Flow Rate Using Arranged Supply Method (Unit: L/s)

Canal/ Canal	Main Canal		Lateral 1 (L1)				Lateral 1 (L2)			
Reach	M-R1	M-R2	L1-R1	Farm 1	L1-R2	Farm 2	L2-R1	Farm 3	L2-R2	Farm 4
July 3	600	0	600	0	600	600	0	0	0	0
July 4	600	600	0	0	0	0	600	0	600	600
July 5	600	300	300	300	0	0	300	300	0	0

Note: The canal reach names (e.g., M-R1, L1-R1) refer to the layout illustrated in Figure 6.1.

TABLE 6.6
Comparison of the Four Water Delivery Methods

Method	Delivery Pattern	Capacity Needed	Efficiency	Complexity	Over/Under Watering
Continuous	Always on	High (peak month)	Low	Low	High risk
Rotation	Sequential turns	Medium	Medium	Medium	Moderate risk
Demand	Real-time needs	High (variable)	High	High	Low risk
Arranged Supply	Pre-planned blocks	High	Medium-high	Medium	Medium risk

system efficiency, keeps canal use within design limits, and ensures fair and predictable access, although it still requires effective communication and planning.

6.3 DISTRIBUTION CONTROL

Effective control of water deliveries is essential for improving crop productivity by enhancing water use efficiency, minimizing waste, and ensuring reliable service to water users. The control of water deliveries is affected by infrastructure and human-related factors. Good infrastructure is crucial and should consist of appropriate measurements, diversion structures, and control mechanisms, such as gates, checks, and weirs, to regulate water levels and flow rates in canals (see Figure 1.2). Other infrastructure-related factors include regulated reservoirs, pipe systems, and the degree of automation. Human-related factors, such as personnel training, communication equipment, and personnel motivation, play a significant role in operating the system under varying water demands and hydrological conditions. Any limitations in infrastructure or human management may result in water spillage or water shortage, which can reduce crop yield.

The traditional method of controlling and regulating water is conducted manually with sluice gates and weirs. However, this method has limitations, particularly in maintaining a constant flow at the farm turnouts when the upstream water level changes. This usually occurs due to unexpected changes in demand, such as when electric pumps stop irrigation at certain hours due to higher electrical rates, when management changes, or when rain affects the time needed to complete irrigation. Unfortunately, sluice gates cannot be adjusted rapidly, resulting in the delayed stabilization of a constant water supply. Hence, in some situations, automation technology may be necessary.

6.3.1 Automation Control

Advancements in automation technology have revolutionized how we measure and control water levels in canals. With the use of automated systems, the measurement and control of water levels can now be accomplished manually or automatically, locally or remotely. More advanced systems are equipped with a telemetry system that allows remote control of water levels in canals, enabling users to monitor and adjust water delivery in real time. These systems are capable of presenting information on a web page, including water orders, river levels, main canal levels, and canal flows at different metering bridges, and pumping and re-lifting stations (see Figure 6.2). Furthermore, ditch riders can now monitor specific areas of the irrigation delivery system via mobile phones or a traditional web browser, making it easier to track water orders and deliveries. Some systems even utilize graphic user interfaces that allow canal riders to visualize water movement on large district maps, resulting in more efficient management of water orders and deliveries. Overall, automation technology has greatly improved the accuracy, efficiency, convenience, and management of water deliveries in canals.

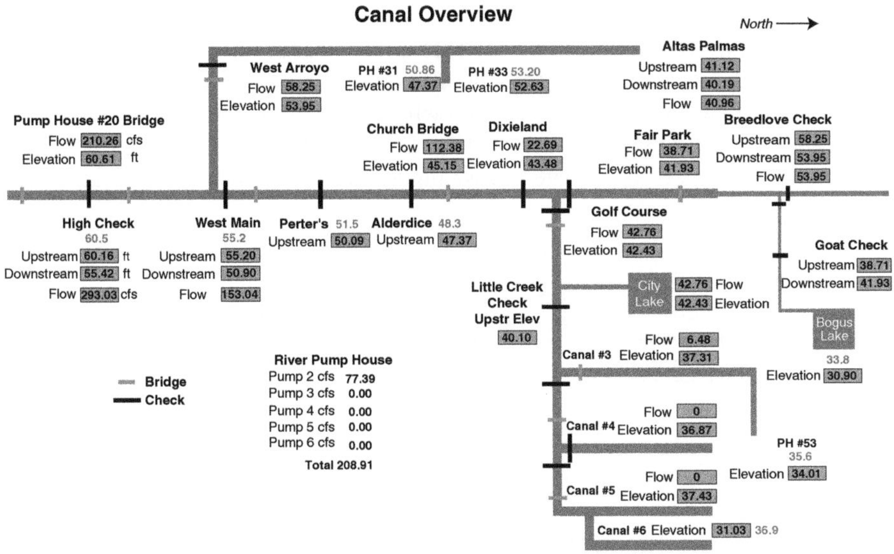

FIGURE 6.2 Computer information of flows and levels in different parts of the irrigation district. (An example of the Harlingen Irrigation District, Texas, United States).

TABLE 6.7
Characteristics Used to Select a Water Control Structure

	Structure to Control Water Level		
Characteristic	**Long-Crested Weir**	**Hydraulically Balanced Gate**	**Overshot Gate**
Ease of use	High	High	High
Energy requirement	No	No	Yes
Installation time	High	Low	Medium
Susceptibility to vandalism	Low	Medium	Medium
Accumulation of sediment	High	Medium	Medium

6.3.2 Water Control Structures

Water control structures are used to control water levels on canals and flow rates to deliver water from secondary to tertiary canals or farm turnouts. They can also be used to maintain constant water levels in the canal and avoid fluctuations that can affect flow deliveries from turnouts.

Some of the structures commonly used to control upstream water levels in irrigated areas are presented in Table 6.7. The selection of a structure is affected by several factors, including those mentioned in Table 6.7, as well as the initial and operating costs of the structures. A short description of these structures is provided.

6.3.2.1 Long-Crested Weir (Duckbill)

The weir is an overflow type dam installed on the canal to raise the water level and provide enough head on a farm turnout. The crest of the weir is elongated to have less variation in upstream water level, allowing the water to overflow the crest. The crest can have several shapes, such as those shown in Figure 6.3. Weirs have sluice gates at the lower part to regulate in case higher flow rates are needed (Figure 6.4).

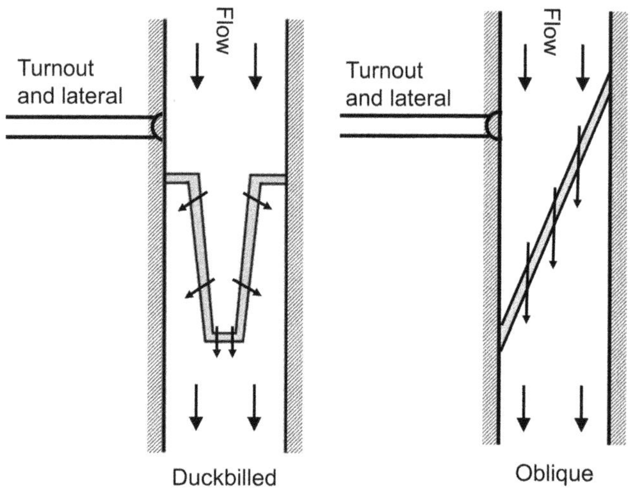

FIGURE 6.3 Forms of long-crested weirs.

FIGURE 6.4 A long-crested weir (inverted duckbill).

The sluice gate is also used to flush silt and avoid accumulation. Weirs are economical, robust, and easy to operate. The long-crested weirs cannot be installed in certain situations, such as when the slope of the canal is low. The weirs should overflow, and this condition happens when the difference between the upstream and downstream levels is higher than one-third of the water head above the crest.

6.3.2.2 Hydraulically Balanced Gates

AMIL gates (upstream control) are hydraulically balanced gates that allow for constant upstream water levels on the upstream side of the gate (Figure 6.5). These gates work by maintaining a balance between the buoyancy float contained on the leaf of the gate and the weight of the gate. Whenever the flow varies and the upstream water level tends to rise or fall, the gate adjusts its opening immediately, automatically passing the exact discharge required to keep the upstream water level constant. During installation, these gates require precise adjustments to work properly. They are simple to operate and reliable. However, supervision is needed to avoid debris from blocking the gates.

6.3.2.3 AVIL Gates (Downstream Control)

These gates maintain constant control of the downstream water level, regardless of upstream level conditions or downstream demands (Figure 6.6). They are used in

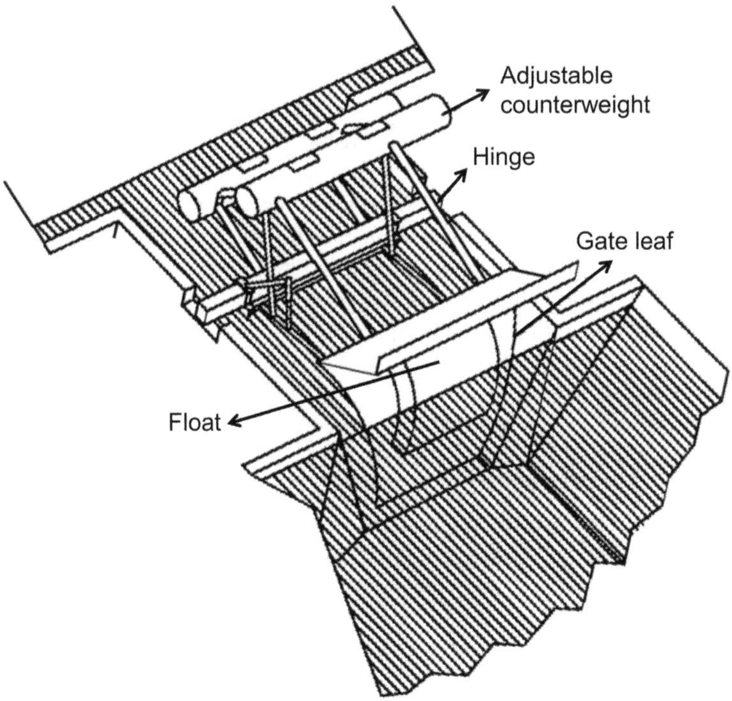

FIGURE 6.5 Sketch of an AMIL gate.

FIGURE 6.6 AVIL gate used to maintain a constant downstream level.

reservoirs and lakes for flood control, as well as in wastewater treatment facilities to regulate flow and water levels efficiently. The gate operates, based on a balancing mechanism, where the moment exerted by the center of gravity about the hinge is counteracted by the moment generated by the upward thrust of the float. Any change in the water level causes the gate to rotate, thereby increasing or decreasing the discharge to restore the water level to the pivot centerline. If the water level falls, the reduced flotation coupling allows the gate to open, increasing the discharge into the downstream channel. On the other hand, if the water level rises, the increased flotation coupling shuts the gate, decreasing the discharge and maintaining the water level at the pivot centerline. This self-regulating system ensures precise flow control and stability, making AVIL gates a reliable solution for managing downstream water levels under varying hydraulic conditions.

6.3.2.4 Overshot Gates

Overshot gates, commonly referred to as leaf gates, are adjustable hydraulic structures used to regulate water flow in canals, rivers, and reservoirs (Figure 6.7). These gates enable precise control of discharge, while minimizing downstream disturbances, making them ideal for applications requiring stable water management.

A typical overshot gate consists of a flat or curved gate leaf anchored to the canal bed by a hinge mechanism, allowing a drawbridge-like pivoting motion. The gate is operated by a hoist system, such as a cable or actuator, which raises or lowers the leaf to modulate flow. The vertical side plates flank the gate to confine the water and prevent lateral spillage. Many designs also incorporate upstream flow measurement devices, such as ultrasonic sensors or weirs, to optimize gate positioning and ensure accurate discharge control.

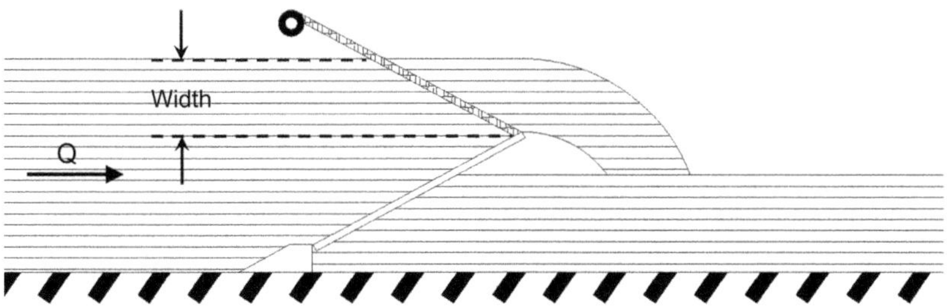

FIGURE 6.7 Dropleaf overshot gate.

When the gate is horizontal (fully open), water flows through the canal uninterruptedly. When the cable raises the gate, water flows over the gate, while the side plates restrict it. Water backs up behind the lift as it is raised. The gate can be lifted at different angles to control water flow. This adaptability allows the gates to handle seasonal flow variations and sudden surges with minimal downstream depth fluctuations.

Modern overshot gates often integrate advanced instrumentation, such as sensors for real-time measurement of flow rates, gate angles, and water levels. These sensors transmit data wirelessly to a Supervisory Control and Data Acquisition (SCADA) system, a centralized platform used to monitor and automate irrigation infrastructure remotely. Motorized actuators further enhance efficiency by enabling preprogrammed or remote adjustments, reducing reliance on manual operation.

6.3.2.5 Electromechanical Controller

Electromechanical controllers, known as "Little man controllers", are automation devices used to regulate canal gates efficiently. These controllers integrate electromechanical components and switches to operate an electrical gate hoist, allowing for automated water level and flow control. These controllers emerged in the mid-20th century as pioneering tools for canal automation, particularly in irrigation networks requiring precise water distribution.

Structurally, these controllers consist of a motor-driven hoist mechanism (e.g., a gear-driven winch), limit switches, relays, and basic sensors, such as float switches or pressure transducers. When sensors detect deviations from target upstream or downstream water levels, the control circuit activates the motor to raise or lower the gate incrementally. For instance, rising water levels trigger gate closure to restrict flow, while falling levels prompt gate opening to increase discharge. This closed-loop feedback mechanism ensures continuous alignment between gate position and hydraulic demand, stabilizing flow with minimal human intervention.

Historically, electromechanical controllers have revolutionized water management by reducing reliance on manual labor and improving operational accuracy. However, their fixed logic (non-programmable) and susceptibility to mechanical wear limited adaptability to complex control scenarios. Now, these controllers have been replaced by programmable logic controllers and SCADA systems. However, they persist as redundant backups in hybrid control setups or as cost-effective solutions for small-scale or legacy infrastructures.

6.3.2.6 Distributors (Constant Adjustable Flow)

The distributor devices are commonly called calibrated gates or Neyrpic modules. They consist of a shaped sill and a fixed baffle plate (Figure 6.8) and deliver constant flows even with small variations in upstream water levels. When the water level is low, the sill operates at free surface flow conditions, and the distributor operates as a weir (Figure 6.9). As the water level rises and the water reaches the lower edge of

FIGURE 6.8 A typical distributor structure used for canal water delivery.

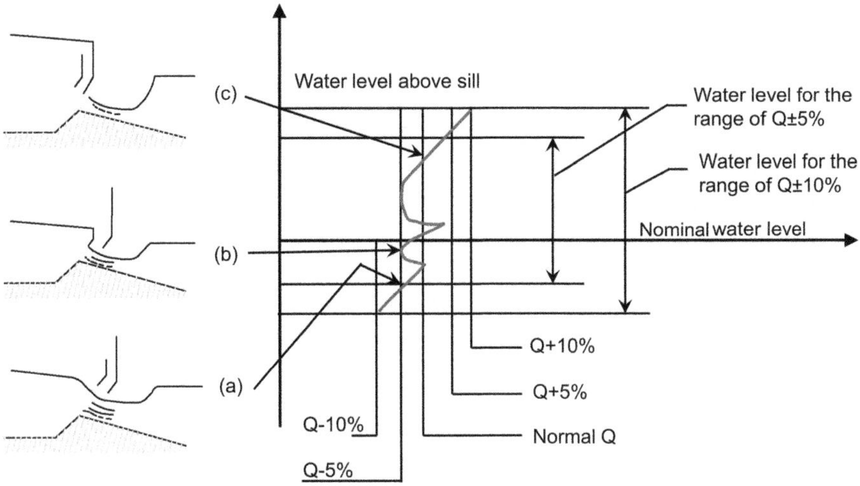

FIGURE 6.9 Distributor to deliver almost constant flows with variable upstream water levels.

FIGURE 6.10 Set of distributors used to deliver constant flows in a tertiary canal.

the baffle plate, the distributor operates as an orifice, thereby reducing the flow. The flow is reduced further as the water level rises over the baffle plate. The distributors come with one or two baffle plates, and they have different numbers of shutters, depending on the flow that they want to deliver. Some canals control the delivery of water using more than one distributor (Figure 6.10). As shown in Figure 6.9, the flow through the distributor can be maintained near its nominal value over a wide range of upstream level conditions. The permissible level ranges are also shown for the following flow conditions: (1) for flow varying by ±5% and (2) for flow varying by ±10% with respect to its nominal value.

6.3.3 Orifices versus Weirs for Controlling Water Levels

The selection of orifices or weirs for water level control depends on their distinct hydraulic behaviors, which are governed by the mathematical relationship between the flow rate and upstream head. Weirs are generally preferred for stabilizing upstream water levels due to their inherent hydraulic sensitivity. The discharge over a broad-crested weir is proportional to the upstream head raised to a power of 1.5. This relationship means that doubling the flow rate results in only a 1.6-fold increase in upstream water level (Figure 6.11), minimizing fluctuations in the main canal. For example, if a weir is designed to maintain a nominal flow, even significant variations in downstream demand will cause relatively small changes in upstream depth, making weirs ideal for applications, such as flood control or navigation channels, where stable upstream levels are critical.

In contrast, orifices are more sensitive to head changes, as their discharge is proportional to the square root of the upstream head. Doubling the flow rate through an orifice requires a fourfold increase in the head, amplifying upstream water level variations (Figure 6.11). Despite this drawback, orifices excel in scenarios requiring constant discharge under variable upstream conditions. At canal turnouts—where water is diverted to secondary channels—orifices are preferred because they maintain near-constant flow rates even as canal levels fluctuate. For instance, during periods of high demand or seasonal water scarcity, an orifice will deliver a consistent volume to irrigation branches, whereas a weir would allow flow to vary with upstream depth.

6.4 WATER DELIVERY SCHEDULING

Irrigation scheduling consists of matching supplies with demands and adjusting the timing, so that better qualities and crop yields are obtained. It also consists of crop needs, soil types, and weather conditions to determine when and how much to

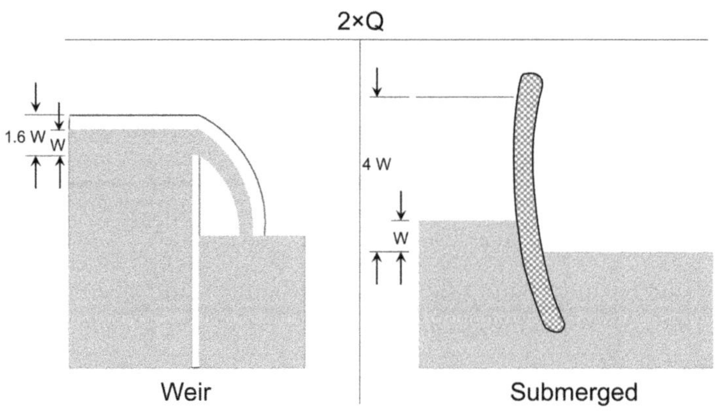

FIGURE 6.11 Fluctuations in the water levels with weirs and submerged structures.

irrigate. The farmers generally make irrigation scheduling decisions. However, they are greatly influenced by the advice of the irrigation district. The irrigation district may have an influence because irrigation system managers have to adapt management decisions to operational and distribution constraints.

6.4.1 Methods for On-Farm Irrigation Scheduling

Various methods can be used for on-farm irrigation scheduling, depending on factors such as the irrigation system in use, the type of crops being cultivated, and the prevailing weather conditions. There are three common approaches: crop appearance and soil feel, estimated evapotranspiration (ET) calculations, and soil water monitoring based on allowable soil water depletion.

Traditionally, farmers have relied on their observations of crop appearance, the feel of the soil, and general irrigation guidelines to schedule irrigation. This method involves visually assessing crop conditions and physically checking the moisture content of the soil.

The ET-based irrigation scheduling method requires ET estimations based on historical weather data. These estimates help in determining the frequency and volume of irrigation required for specific crops and soil types. The districts also use these guidelines to predict water demands per farm outlet. Most advanced irrigation districts have a network of weather stations spaced across the district to measure rainfall and, in some cases, to estimate ET with real-time weather data. For example, the WISE web-based App developed at Colorado State University (Andales et al., 2014) (http://wise.colostate.edu/) can be used to schedule irrigations on individual fields across Colorado, with automated access to daily weather data (the nearest weather stations selectable from a network of 90+).

The soil water depletion methods consist of determining the water holding capacity between field capacity (FC) and permanent wilting point (PWP), commonly known as available water capacity (AWC) (Figure 6.12). Field capacity is when the soil water reservoir is full or when the soil is wet after being saturated and naturally drained; and permanent wilting point is when there is no water in the soil or when the plant wilts, even if more water is added to the soil. Generally, irrigation occurs at one point between field capacity and permanent wilting point at a point known as the

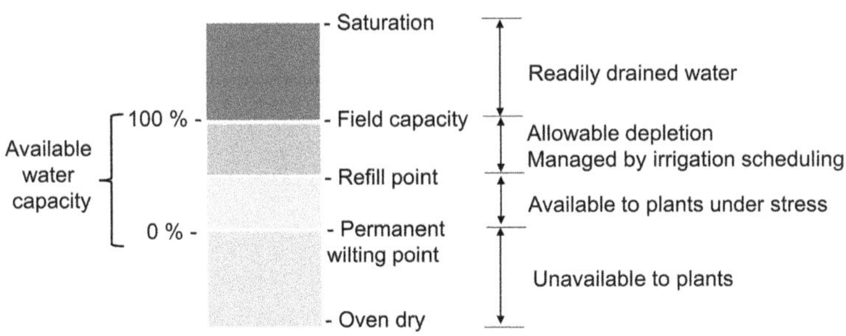

FIGURE 6.12 Illustration of soil water reservoir.

permitted deficit or refill point. This point between field capacity and the permitted point is known as allowable depletion, which can be managed by irrigation scheduling. Some districts may use some sensors connected to a telemetry system to indicate whether irrigation is needed.

Example 6.5

A farmer is growing corn on a loam soil with a root zone depth of 0.9 m (90 cm). The soil's field capacity (FC) is 30%, and the permanent wilting point (PWP) is 15% (volumetric water content). The allowable depletion (f_d) for corn is 50%. Calculate the available water capacity (AWC), allowable depletion (AD), and amount of water needed to be applied during each irrigation event to refill the root zone to field capacity if the field area is 100 ha. The irrigation efficiency is 80%.

Solution:

$$\text{AWC} = (\text{FC} - \text{PWP}) = 0.30 - 0.15 = 0.15 \text{ m/m}$$

$$\text{AD} = \text{AWC} \times \text{Root zone depth} \times f_d = 0.15 \times 90 \times 0.5 = 67.5 \text{ mm}$$

Therefore, irrigation should be scheduled when 67.5 mm of water has been depleted.

Assuming that there is no rainfall, the net irrigation amount required is

$$\text{Net irrigation} = \text{AD} = 67.5 \text{ mm}$$

If the irrigation efficiency is 80%, then

$$\text{Gross irrigation} = \frac{67.5}{0.80} \approx 84.4 \text{ mm}$$

Converting 1 mm to volume per hectare, one gets $1 \text{ mm} = 10 \text{ m}^3/\text{ha}$

$$84.4 \text{ mm} = 844 \text{ m}^3/\text{ha}$$

The volume to apply for the field (100 ha):

$$844 \text{ m}^3/\text{ha} \times 100 \text{ ha} = 84{,}400 \text{ m}^3$$

6.4.2 Adaptation to Different Irrigation Systems

The choice of an irrigation scheduling method depends on various factors, including the irrigation system in use and crop type. ET-based methods and soil water methods can be effectively adapted to sprinkler and drip systems. ET-based methods allow water demands to be precisely matched to application rates, which are the depths of irrigation applied per hour. This flexibility allows for various irrigation intervals, such as daily, twice per week, weekly, or any other desired irrigation interval according to operational capabilities of the district. When employing the ET replacement method with sprinkler and drip irrigation, it is often feasible for canals to continuously supply water to farms. Alternatively, farmers or farms have a small reservoir to store water for a week of irrigation.

Soil water sensors are also used with pressurized irrigation systems. However, they need to be calibrated to determine the relationship between water depletion and the readings of the soil water sensors. The farmer can communicate exactly the amount of water that will be needed to irrigate their crop. Soil moisture sensors are typically stationary, leading to frequent over- or under-estimation of soil moisture due to calibration issues. Moreover, these methods do not account for the spatial variability in soil moisture across a field and can be challenging to install, remove, or manage without interfering with production operations. Alternatively, plant thermal measurements offer a more direct and reliable estimation of crop water stress and are considered a valid approach for determining irrigation timing. Similar to soil water methods, accurately representing spatial variability in plant stress across a field depends on the deployment of sensors.

With surface irrigation, it is easy to control when irrigation is needed, but it is difficult to control the amount of irrigation that needs to be applied according to estimated crop demand. More challenging management is required with surface irrigation systems. Generally, with surface irrigation, the irrigation amount can be determined based on soil water depletion. Irrigation is then managed to approximate this soil water depletion to the application depth. Careful planning and management are required to ensure that water resources are used efficiently and effectively within the limitations of surface irrigation.

According to the National Irrigation and Water Management Survey, more than 70% of irrigation decisions in the United States still rely on traditional methods such as "visual crop condition" or "feel of soil" (USDA-NASS, 2019). In contrast, the adoption of more advanced technologies, such as soil moisture sensors, ET-based water balance methods, and plant status indicators, remains relatively limited. To illustrate the potential of modern approaches, the following examples demonstrate how soil moisture sensors and ET-based irrigation scheduling can be applied to improve irrigation efficiency and decision-making on farms.

Example 6.6

Tensiometers are installed in a sandy loam field to monitor soil water tension in the root zone of soybeans. Daily readings are recorded from July 1 to July 6, and irrigation was applied on July 2. The graphical method was used to predict when the next irrigation should be applied.

Date	Tension (cb)	Note
July 1	48	
July 2	5	Irrigation applied
July 3	7	
July 4	10	
July 5	15	
July 6	20	

Solution:

According to Table 6.8, the recommended soil water tension threshold for sandy loam is 50 cb. After irrigation on July 2, the tension dropped to 5 cb (nearly saturated). It then began increasing as the soil dried. Between July 4 and July 6, the tension increased from 10 to 20 cb over 2 days. The rate of increase is 5 cb/day.

Assuming a rate of 5 cb/day, the next irrigation will be

$$\frac{50\text{ cb} - 20\text{ cb}}{5\text{ cb/day}} = 6\text{ days}$$

Irrigation will be needed in 6 days (July 12) (Figure 6.13).

TABLE 6.8
Soil Water Tension Thresholds for Different Soil Types

Soil Types	Threshold Tension (cb)
Fine sand	30
Loamy sand	40
Sandy loam	50
Fine sandy loam	55
Silt loam	60
Clay loam	70

Source: After USDA-NRCS (1997).

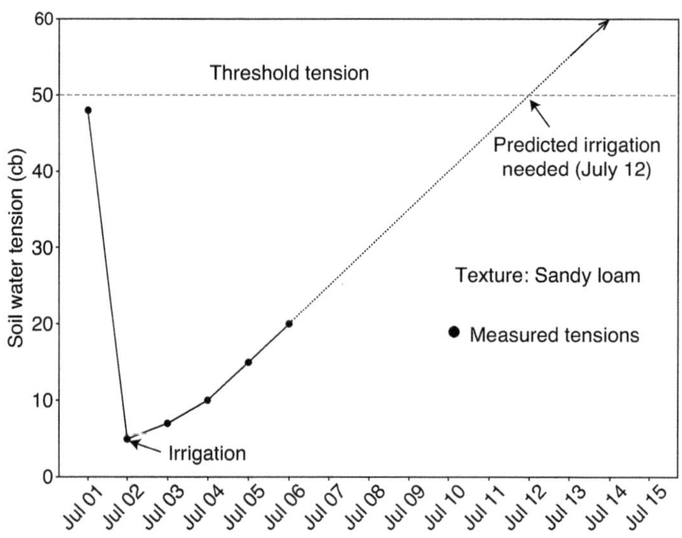

FIGURE 6.13 Graphical method of irrigation scheduling using soil water sensors.

Soil sensor methods are easy to use and provide real-time information on soil moisture status. However, they typically do not indicate how much irrigation is needed. To estimate irrigation requirements more accurately, ET-based methods are commonly used. There are two primary ET-based approaches: one is basic ET-based scheduling (Example 6.7), which estimates crop water use based on weather data and crop coefficients. The other is the ET-based water balance approach (also known as the checkbook accounting method, Example 6.8), which tracks daily water inputs and losses to simulate changes in soil water contents.

Example 6.7

A farmer is growing wheat in the midseason stage. The following data are available:

Daily reference evapotranspiration (ET_o)= 5.5 mm/day
Crop coefficient (K_c) = 1.10
Irrigation interval = 7 days
Effective rainfall = 8.00 mm
Irrigation system efficiency = 85%
How much effective irrigation is needed to meet crop water needs?

Solution:

Step 1: Calculate daily crop evapotranspiration (ET_c)

$$\mathrm{ET}_c = K_c \times \mathrm{ET}_o = 1.10 \times 5.5 = 6.05 \text{ mm/day}$$

Step 2: Estimate the total crop water use over 7 days

$$\mathrm{ET}_c\left(7 \text{ days}\right) = 6.05 \times 7 = 42.35 \text{ mm/day}$$

Step 3: Subtract effective rainfall

$$\text{Net irrigation requirement} = 42.35 - 8.00 = 34.35 \text{ mm}$$

Step 4: Adjust system efficiency

$$\text{Gross irrigation} = \frac{34.35}{0.85} \approx 40.4 \text{ mm}$$

The farmer should apply 40.4 mm of irrigation every 7 days to meet crop water needs.

Example 6.8

Cotton is grown on a sandy loam soil. The field capacity (FC) of the soil is 16.6% (or 0.166 m/m), and the permanent wilting point (PWP) is 6.9% (or 0.069 m/m). The planting date is May 1. Crop coefficients for different growth stages are shown in Figure 6.14. The root zone depths by stage are provided in Table 6.9, and the maximum allowable fraction depletion is 0.65 for cotton. The groundwater table

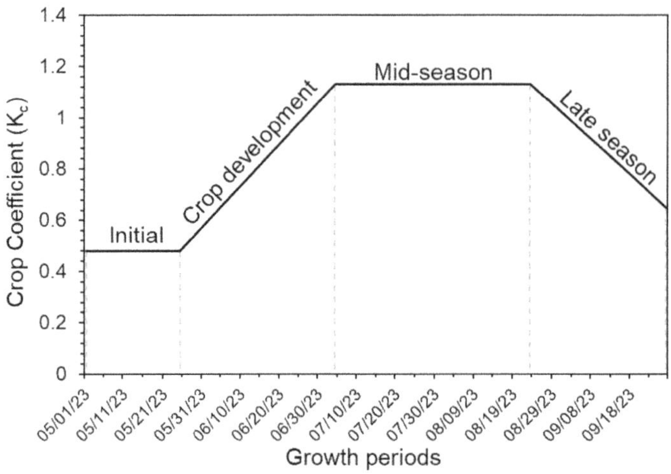

FIGURE 6.14 Crop coefficients of cotton at different growth stages (planted on May 1, 2023).

TABLE 6.9
Example Root Zone Information for Various Annual Crops

	Corn	Grain Sorghum	Soybean	Winter Wheat	Cotton
Growing Stage			**Assumed Root Depth (cm)**		
Initial	30	30	30	30	30
Crop development	>>	>>	>>	>>	>>
Midseason	>>	>>	>>	>>	>>
Late season	100	100	100	100	100

>> indicates that the root depth increases linearly during this stage.

depth is 3 m. Daily reference ET (ET_o) and precipitation data are given in Table 6.10 (the first month is listed as an example). Develop an irrigation schedule throughout the growing season using the ET-based water balance method.

Solution:

The procedure for determining the irrigation schedule is shown below.

Step 1: Calculate daily crop evapotranspiration (ET_c)

Calculate daily ETc by multiplying ET_o by crop coefficients (Figure 6.14):

$$ET_c\left(\text{May } 1\right) = K_c \times ET_o = 0.48 \times 1.95 \approx 0.94 \text{ mm/day}$$

TABLE 6.10
Key Data Required for Calculation

Date	Growth Period	Effective Rainfall (mm)	ET_o (mm/day)	K_c	ET_c (mm/day)	Root Zone (cm)
5/1	Initial period	0	1.95	0.48	0.94	30.00
5/2	Initial period	0	2.11	0.48	1.01	30.00
5/3	Initial period	0	2.25	0.48	1.08	30.00
5/4	Initial period	8.12	2.30	0.48	1.10	30.00
5/5	Initial period	0	2.09	0.48	1.00	30.00
5/6	Initial period	0	1.99	0.48	0.96	30.00
5/7	Initial period	10.05	2.51	0.48	1.20	30.00
5/8	Initial period	0	2.62	0.48	1.26	30.00
5/9	Initial period	0	2.83	0.48	1.36	30.00
5/10	Initial period	0	2.31	0.48	1.11	30.00
5/11	Initial period	0	2.30	0.48	1.10	30.00
5/12	Initial period	2.45	2.12	0.48	1.02	30.00
5/13	Initial period	0	2.39	0.48	1.15	30.00
5/14	Initial period	4.28	2.71	0.48	1.30	30.00
5/15	Initial period	0	2.15	0.48	1.03	30.00
5/16	Initial period	2.73	2.53	0.48	1.21	30.00
5/17	Initial period	1.24	2.28	0.48	1.09	30.00
5/18	Initial period	8.55	1.29	0.48	0.62	30.00
5/19	Initial period	0	2.16	0.48	1.04	30.00
5/20	Initial period	5.87	2.08	0.48	1.00	30.00
5/21	Initial period	0	2.21	0.48	1.06	30.00
5/22	Initial period	0	2.10	0.48	1.01	30.00
5/23	Initial period	0	1.76	0.48	0.84	30.00
5/24	Initial period	0	2.36	0.48	1.13	30.00
5/25	Initial period	0	2.63	0.48	1.26	30.00
5/26	Crop development	10.05	2.17	0.50	1.09	30.78
5/27	Crop development	6.91	1.88	0.51	0.96	31.56
5/28	Crop development	2.48	1.92	0.53	1.02	32.33
5/29	Crop development	0	2.95	0.55	1.62	33.11
5/30	Crop development	0	2.94	0.56	1.65	33.89
5/31	Crop development	0	2.21	0.58	1.28	34.67

Step 2: Determine the total available water (TAW) in the root zone depth (R_d)

$$\text{AWC} = \text{FC} - \text{PWP} = 0.166 - 0.069 = 0.097 \text{ m/m}$$

$$\text{TAW}(\text{May 1}) = R_d \times \text{AWC} = 30 \text{ cm} \times 0.097 = 2.91 \text{ cm} = 29.1 \text{ mm}$$

Step 3: Determine the maximum allowable fraction depletion (f_{dmax})
The maximum allowable fraction depletion is 0.65 for cotton.

Step 4: Calculate allowable depletion (AD)

$$\text{AD}(\text{May } 1) = f_{\text{dmax}} \times \text{TAW} = 0.65 \times 29.1 \text{ mm} \approx 18.92 \text{ mm}$$

Step 5: Determine the upward flow of water from the groundwater table (U_f)

The groundwater table depth is 3 m, and the root zone depth is 30 cm (0.3 m) on May 1. The depth between the groundwater and the root zone is $3.0 - 0.3 = 2.7$ m. For sandy loam soil, U_f=0, as shown in Figure 6.15. There is no upward capillary flow from the water table.

Step 6: Calculate the soil water deficit (SWD) for each day

$$\text{SWD}_i = \text{SWD}_{i-1} + \text{ET}_{\text{ci}} - d_{\text{ei}} - P_{\text{ei}} - U_{\text{fi}}$$

where $\text{SWD}_{(i-1)}$ is the soil water deficit from the previous day; ET_{ci} is the crop evapotranspiration for day *i*; d_{ei} represents the effective irrigation depth on day *i*; P_{ei} is the effective precipitation on day *i*; and U_{fi} is the upward flow from the groundwater on day *i*.

On May 1, assume that the soil is at field capacity (i.e., $\text{SWD}_1 = 0$), then the SWD on May 2 is

$$\text{SWD}_2 = \text{SWD}_1 + \text{ET}_{\text{ci}} - d_{\text{ei}} - P_{\text{ei}} - U_{\text{fi}} = 0 + 0.94 + 0 - 0 - 0 = 0.94 \text{ mm}$$

If SWD > AD, irrigation is triggered, and $d_{\text{ei}} = \text{SWD}_{i-1}$. The daily variations in ET_c, AD, SWD, and required irrigation are shown in Figure 6.16. The number of irrigations required during the growing season is 8, and the total amount is 322.92 mm.

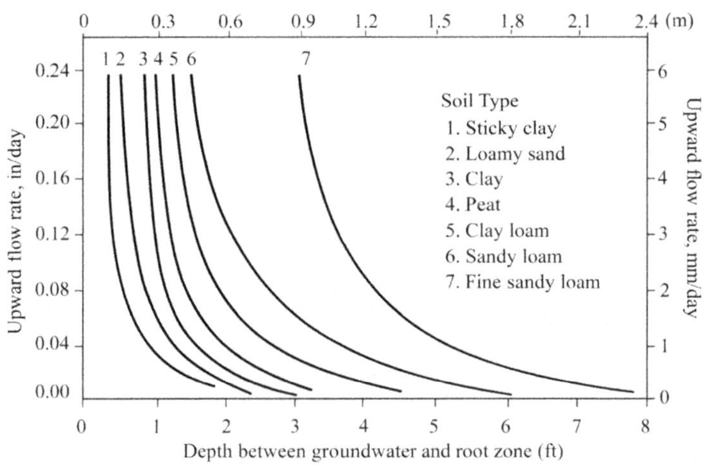

FIGURE 6.15 Upward flow rate versus depth between groundwater and the root zone for various soil types. (Adapted from Doorenbos and Pruitt, 1977.)

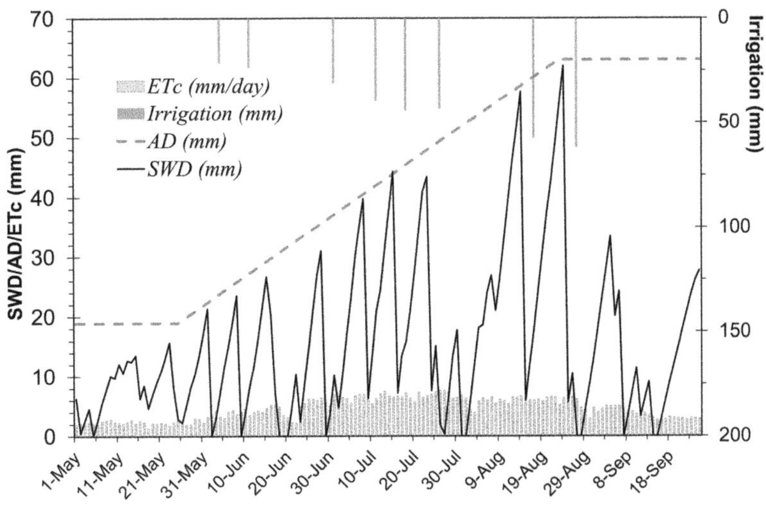

FIGURE 6.16 Variation in soil water depletion (SWD), allowable depletion (AD) crop water use (ET_c), and irrigation during the growing season.

6.4.3 Irrigation District Management

Irrigation districts (or any other organization) often face the challenge of managing water distribution efficiently and equitably. To address this challenge, districts may divide their territory into operation subunits, according to different operational and distribution constraints to schedule irrigation. These subunits can take on the form of tertiary canals or designated areas where particular crops are cultivated, each with distinct water sensitivities. The operational constraints within these may include application efficiencies, crop types, and maintenance requirements. Simultaneously, delivery constraints may be related to the availability of infrastructure or canal capacities.

In response to a changing climate, irrigation districts may adopt strategies based on the crop being grown. For example, after one big irrigation event, most water users would like to delay subsequent irrigation applications as much as possible. However, this may not always be feasible, especially when all producers operate under a free-demand delivery system. In such cases, farmers often find themselves making trade-offs between leaching and allowing a slight water deficit in their crops. Prioritization becomes crucial, and irrigation may favor crops that exhibit lower water sensitivity, as shown in Table 6.11. For instance, consider a scenario where a farmer has sorghum and corn. In this case, the farmer's yield response and productivity per unit of water applied may be higher for corn than for sorghum when it experiences less water stress compared to sorghum. Similarly, crops with shallow root systems, such as vegetables, can witness a significant decrease in productivity when subjected to water stress. In contrast, crops like sugar beets and alfalfa, which have deeper root systems, are better equipped to adapt to water deficit, as they can access water from deeper soil layers. Following a big irrigation event, planning and

TABLE 6.11
Crop Sensitivities to Different Water Shortages

Low	Low-Medium	Medium-High	High
Cotton	Citrus	Beans	Banana
Millet	Grape	Cabbage	Green vegetables
Sorghum	Groundnuts	Corn	Potato
	Soybean	Onions	Rice
	Sugar beet	Tomato	Vegetables
	Sunflower	Melon	Sugarcane
	Wheat		

Source: Adapted from Brouwer et al. (1989).

active engagement with farmers become essential to ensure the efficient delivery of water. This collaborative effort is crucial for balancing the water needs of different crops and ensuring that water is distributed in an equitable and sustainable manner.

6.5 ROTATION SCHEDULES

Rotation schedules in irrigation systems play a crucial role in determining when and how water is allocated to different areas, taking into account various factors such as crop types, design characteristics, and water availability within the canal network. The specific rotation period varies from system to system and is influenced by a range of factors.

6.5.1 Farm Stream Size

At the design stage, the stream size that can be delivered to each farm outlet needs to be selected. It is important to have a good size farm outlet and a high density of outlets per farm so that the system has good flexibility to adapt to changes in demand. In South Texas, United States, some irrigation districts are designed to deliver a maximum stream size of 85 L/s per 16 ha plot, with variations up to 120 L/s. In some irrigation districts in Mexico, the systems are designed to deliver between 60 and 120 L/s per farm outlet. In Pakistan, some districts were designed to deliver fixed amounts of 113 L/s to farms ranging from 60 to 250 ha of land, usually cultivated by 10–150 farmers (Merrey, 1990). In irrigation districts in the United States that serve urban communities, stream sizes can vary from 10 to 200 L/s. In some cases, controlling the farm stream size requires the use of pressure control stands, as shown in Figure 6.17.

FIGURE 6.17 Water stand pipe used to distribute water on the farm with a flexible plastic pipe and a propeller flow meter.

Example 6.9

An irrigation engineer is designing a delivery system for a newly planned irrigation district in South Texas, where farms are organized into 16-ha blocks. Based on regional standards and flexibility needs, the system must be capable of delivering stream sizes between 85 and 120 L/s per outlet. The system must be capable of delivering sufficient water within a 12-hour irrigation shift. The net irrigation requirement is now 50 mm (i.e., 50 L/m^2 or 500 m^3/ha). The application efficiency is 80%. Evaluate whether a stream size of 100 L/s per outlet is sufficient to irrigate a 16-ha block within a 24-hour irrigation shift.

Solution:

The gross irrigation depth is

$$\text{Gross depth} = \frac{50\text{ mm}}{0.8} = 62.5\text{ mm} = 625\text{ m}^3/\text{ha}$$

$$\text{Total volume for 16 ha} = 625 \times 16 = 10,000\text{ m}^3$$

The irrigation time for different stream sizes:

Since 1 L/s = 3.6 m^3/h,

$$\text{At 100 L/s}: \frac{10,000\text{ m}^3}{100\text{ L/s}} = \frac{10,000\text{ m}^3}{3.6 \times 100\text{ m}^3/\text{h}} \approx 27.8\text{ hours}$$

A stream size of 100 L/s is inadequate for irrigating a 16-ha field in a 24-hour shift.

The required stream size of a 24-hour shift is

$$\text{Required flow rate} = \frac{10{,}000\ \text{m}^3}{24\ \text{hour}} = 416.67\ \frac{\text{m}^3}{\text{h}} = 115.\ 7\ \text{L/s}$$

6.5.2 Peak Water Requirements and Seasonal Water Requirements

Peak water use, typically occurring in June and July, represents the period of highest irrigation need. Managing the delivery rate during the peak period is critical to avoid crop water stress and yield loss. Seasonal water demand is the total water use during the growing season. Meeting this demand depends on a combination of rainfall and irrigation to realize the full potential yields of the crops.

Understanding seasonal water demand is essential for planning and determining which areas can be irrigated, based on available water supplies. An example of typical peak and seasonal water demands for different climates is shown in Figure 6.18.

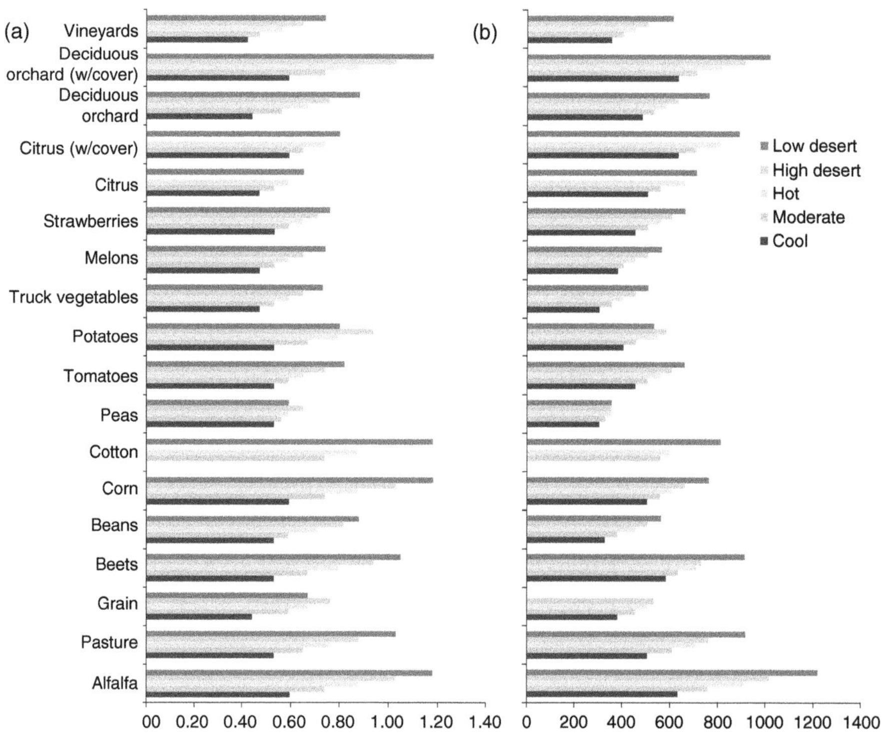

FIGURE 6.18 Crop water demand in different climates. (a) typical peak daily flow rate (L/s/ha) and (b) total seasonal crop water requirement (mm).

The peak water demand multiplied by the number of acres can be used to estimate if the canals have the capacity to irrigate a given area, considering on-farm application and delivery distribution efficiencies.

Example 6.10

An irrigation district in a semi-arid climate is planning water delivery for corn. The total irrigated area is 1,000 ha. Based on historical data, the peak water use in July is 9.2 mm/day. The seasonal irrigation requirement is 700 mm. The canal delivers water at 85% efficiency, and the on-farm application efficiency is 80%. (1) Estimate the required canal flow capacity (in L/s) to meet peak daily demand and (2) calculate the total seasonal gross water demand (m^3).

Solution:

The peak crop water use is provided in mm/day and converted to L/s/ha:

$$9.2 \text{ mm/day} = \frac{\dfrac{9.2 \times 10{,}000 \dfrac{\text{L}}{\text{ha}}}{\text{day}}}{86,400 \dfrac{\text{s}}{\text{day}}} \approx 1.065 \text{ L/s/ha}$$

For 1,000 ha, the peak flow rate is

$$1.065 \text{ L / s / ha} \times 1{,}000 \text{ ha} = 1{,}065 \text{ L / s}$$

Adjusted for canal and on-farm efficiencies:

$$\text{Requried peak flow rate} = \frac{1065 \text{ L / s}}{0.85 \times 0.80} \approx 1{,}566 \text{ L / s}$$

Seasonal irrigation requirement:

$$700 \text{ mm} \times 1{,}000 \text{ ha} = 7 \times 10^6 \text{ m}^3$$

Adjusted for canal and on-farm efficiencies:

$$\text{Seasonal irrigation requirement} = \frac{7 \times 10^6 \text{ m}^3}{0.85 \times 0.80} = 10.3 \times 10^6 \text{ m}^3$$

6.5.3 Irrigation Times

The ability to control irrigation times depends on the irrigation system in use. Sprinkler and drip irrigation systems offer more precise control because pipes distribute the water while sprinklers and emitters regulate the application rate. However, with surface irrigation systems, irrigators have less control over the depth of infiltration, which can vary considerably in both space and time. In surface irrigation, the irrigator typically attempts to apply the required irrigation depth and gains a better understanding of the amount of water applied only after the irrigation process is completed. To manage this, a comprehensive operation plan is needed to anticipate which farm requires water delivery after irrigation on the previous farm is finished.

Example 6.11

A farmer plans to irrigate a 5-ha field of onions and considers two irrigation options: a sprinkler system and a surface furrow system. The field requires a net irrigation depth of 20 mm per application. The sprinkler system has a discharge rate of 15 L/s and operates with an application efficiency of 85%. In contrast, the surface irrigation system delivers water at a relatively high discharge rate of 25 L/s but with a lower application efficiency of 65%. The farmer would like to estimate the gross irrigation depth required for each system, calculate the total volume of water needed to irrigate the entire field, and determine the time required to complete irrigation using each method.

Solution:

For the sprinkler system, the gross depth to be applied is calculated as

$$\text{Gross depth} = \frac{20\text{ mm}}{0.85} \approx 23.53\text{ mm}$$

For a 5-ha field, the total volume of water required is

$$23.53\text{ mm} \times 5\text{ ha} = 1{,}176.5\text{ m}^3$$

With a discharge rate of 15 L/s, the irrigation time can be estimated as

$$\frac{1{,}176.5\text{ m}^3}{15\text{ L/s}} = \frac{1{,}176.5\text{ m}^3}{15 \times 3.6\text{ m}^3/\text{h}} \approx 21.8\text{ hours}$$

For the surface irrigation system, the gross depth to be applied is calculated as

$$\text{Gross depth} = \frac{20\text{ mm}}{0.65} \approx 30.77\text{ mm}$$

For a 5-ha field, the total volume of water required is

$$30.77\text{ mm} \times 5\text{ ha} = 1{,}538.5\text{ m}^3$$

With a canal discharge rate of 25 L/s, the irrigation time can be estimated as

$$\frac{1{,}538.5\text{ m}^3}{25\text{ L/s}} = \frac{1{,}538.5\text{ m}^3}{25 \times 3.6\text{ m}^3/\text{h}} \approx 17.1\text{ hours}$$

The sprinkler system applies water at a lower flow rate but offers uniform and controlled application, which means that irrigation can be paused or adjusted precisely, and all parts of the field receive similar amounts of water. The surface system applies more water faster, but control is limited. The actual infiltration depends on the soil texture, slope, and time, and some areas may be over- or under-irrigated.

6.5.4 Rotation between Laterals

The rotation method is sufficiently flexible in that it permits rotating the delivery among farm outlets of the same lateral or rotating the flow between different laterals. The capacity of the lateral generally determines the number of farm outlets that can be served at the same time in each lateral. Several farm outlets are grouped in the lateral, and the stream size is rotated among these farm outlets. When irrigation is completed in that lateral, the stream size can then be turned into another lateral for irrigation of the land under its command. Each lateral will be a dry part of the time. In this manner, rotation in flow occurs between laterals, as well as between farms, and is sometimes referred to as the periodic method of delivery.

The advantages of rotating the stream size between laterals include reduced canal seepage, reduced aquatic growth, and reduced patrolling of laterals not in use. However, there are disadvantages, such as the requirement of larger capacity of laterals, potentially reducing the lifespan of wooden structures due to alternate wetting and drying, an increased risk of lateral breaks due to burrowing animals, and the possibility of erosion and bank slipping due to rapid filling and emptying of larger laterals. An example of a rotational irrigation schedule between laterals is illustrated in Example 6.2.

6.5.5 Delivery Up and Down Laterals

The delivery of water under rotation is made in two ways: (1) starting delivery at the upper end and proceeding downward along the lateral, and (2) running the head through to the lower end of the lateral and then delivering water in ascending order up the ditch. Delivering water down the lateral has several advantages, such as reducing the temptation for farms further up the lateral to take water out of turn and facilitating the servicing of any missed turnouts. Turnouts that have supplied water can be locked, and the remaining turnouts can be kept unlocked. This method is preferable, if the ditch rider is charged with the responsibility of making changes, and deliveries are made in no less than 24 hours. This is common for large farms and those using small irrigation heads.

The method of delivery up the lateral is preferable when the delivery time to each lateral is short and where the breaks in the lateral are likely to occur. Under this method, the ditch rider can supervise the filling of the lateral and the proper setting of all checks. The closing of turnouts and taking of water by those next in turn can be largely managed by the users themselves, if only one irrigation head is run in the lateral. This is of particular advantage where large heads are used and delivery to each farm is only for a few hours. This also avoids the need to patrol laterals at night.

If a break occurs on a lateral, then a rotating up delivery is made to a farm above the break, avoiding the need to reintroduce the water into the main lateral. In contrast, during rotating down, all farms above the break will have been irrigated and the water due to break will be wasted, even if it is turned to upper farms. Rotating the ditch upwards allows for pre-filling, eliminating time gaps between turns, whereas downward rotation may cause some time delay.

6.5.6 Fixed and Flexible Rotation Schedules

Rotation schedules can be prepared in two ways: (1) they are fixed in advance of the irrigation season and strictly adhered to; and (2) the deliveries are made based on rotation on demand or in the order of notice given by the user. Fixed schedules permit more efficient planning of irrigation operations and better arrangements of farm work. However, they should be strictly adhered to and permit some flexibility. These methods are better suited to crops, such as orchards, where irrigation is not required within any short period of growth. For grain crops and vegetables that require water within short periods of time at certain stages of growth, pre-arranged schedules may not be suitable as the required irrigation timing varies each year, based on climatic conditions and fertigation needs.

6.5.7 Delivery during Shortage

When the water supply cannot meet the demand, the rotation schedule can be modified in one of three ways: (1) reducing the irrigation head, (2) shortening the delivery time, and (3) decreasing the frequency of delivery. Reducing the time of delivery is a typical practice, as reducing the irrigation head may limit the efficiency of its use, and shortening the time that water is delivered per acre may reduce the area irrigated at each turn.

6.5.8 Other Methods

There are other methods of water delivery, including pressure pipelines resembling water supply lines, and sub-irrigation due to elevated groundwater tables caused by seepage in canals with porous soils. These methods offer flexibility and options for efficient water distribution within irrigation systems. For example, pressure pipelines are analogous to water supply lines used in urban areas. These pipelines deliver water under pressure to specific points within the irrigation network. By doing so, they can provide a consistent and controlled supply of water to various parts of the agricultural land. Pressure pipelines are particularly useful when precise water delivery is essential for specific crops or when topographical variations require varying water pressures. This method is known for its reliability and uniform water distribution, making it a valuable addition to modern irrigation systems.

6.6 CHOOSING A METHOD

Selecting the most suitable method for water delivery in irrigation systems is a complex decision influenced by various factors. The choice depends on the specific needs and conditions of each agricultural operation. Here are the key considerations: (1) Types of crops grown: Different crops have distinct water requirements in terms of both the amount and the frequency of irrigation. For example, grain crops may require relatively short but timely irrigation, while forage crops typically need substantial water distributed over an extended growing season. Crops like alfalfa and orchard fruits also have unique irrigation needs; (2) Climate: Climate plays a crucial role in water delivery method selection. In regions where rainfall predominantly occurs during a specific season, irrigation may not be required for most of the year. In such cases, on-demand irrigation is a preferable approach. Additionally, if sufficient rainfall occurs during specific seasons, the delivery of irrigation water immediately after rainfall events might not be necessary; (3) Topography: The topography of the land impacts the size of irrigating heads and the rate at which irrigation can be carried out. Steeper or uneven terrains may require different water delivery strategies compared to flat and uniform landscapes; (4) Agricultural practices: Factors such as the application of fertilizers and inter-cultural operations can affect the choice of water delivery method. Some methods may be better suited to specific agricultural practices, ensuring optimal crop growth and resource utilization; (5) Water supply: The source of water supply influences the timing of irrigation, which should align with crop irrigation needs. The reliability and availability of water sources are essential considerations when selecting a water delivery method; (6) Farm size: Farm size affects the command area of each turnout, the duration of irrigation, and the frequency of head gate adjustments. Larger farms may benefit from more efficient water delivery methods to accommodate their needs; (7) Length of irrigation season: The duration of irrigation season also affects the choice of method. Longer growing seasons may require more efficient and sustainable water delivery techniques to ensure consistent crop health and yield; (8) Social attributes of farmers: The preferences and practices of local farmers play an important role in the acceptance or rejection of a specific water delivery method. Community and sociological factors should be considered; and (9) Other sources of irrigation: The presence of additional sources of irrigation, such as rainfall or groundwater, impacts the necessity and timing of irrigation water delivery.

In general, no one method of water delivery is suitable for all irrigation systems, since the conditions for the use of water differ. Indeed, in a larger irrigation project, different parts of the project may require different methods of water delivery to address local conditions effectively.

6.7 LEGAL PERSPECTIVE

In countries with well-established water laws, such as the United States, there is a legal aspect of water delivery. There are appropriate rights defined in terms of both the maximum rate of use and the total quantity per acre that can be used. Then there are laws for rotation between rights where the extent of each right is small, as well as for prorating the available flow at times of extreme scarcity instead of a strict observance of priorities. The rotation system is recognized as the most efficient and desirable method, producing the highest duty of water of any method of use.

6.8 CASE STUDIES

6.8.1 Texas

The Harlingen Irrigation District Cameron County No. 1 (HIDCC1) was established in 1914 by the Cameron County Commissioner's Court to address the growing need for organized irrigation in South Texas. Today, the district manages water delivery for approximately 38,000 irrigated acres (15,378 ha) and 900 non-agricultural accounts, including municipal, golf courses, and parks. Its operations depend entirely on Rio Grande water, diverted via a main pumping plant near Los Indios, which sources release from Falcon Reservoir. These releases, coordinated by the International Boundary and Water Commission, require a 4-day travel time to reach the district. The pumping facility, originally constructed in the 1920s, operates at a capacity of 470 cfs (13,309 L/s), with a typical peak-pumping rate of 410 cfs (11,610 L/s). To balance supply variability and demand fluctuations, HIDCC1 maintains a robust infrastructure network comprising 57 miles of canals, 200 miles of pipelines, 44 relift pumping stations, and three reservoirs with a combined storage capacity of 1,380 acre-feet (AF) ($1.7 \times 10^6 m^3$), which buffer daily water usage changes.

HIDCC1 employs advanced technologies to enhance water delivery efficiency. Solar-powered Auto-Gates, developed through state grants, use 12-volt actuators to maintain precise canal water levels, preventing overflows and reducing manual labor (HID, 2011). These gates are integrated with a centralized control system via radio modems and antenna systems, enabling real-time adjustments, based on data transmitted from Remote Telemetry Units (RTUs). Operators monitor flows using SCADA-like software, which provides live, 24-hour, and monthly analytics. During peak summer months, a fixed 7-day rotation schedule ensures equitable distribution, while farmers can request 24-hour irrigation slots through mobile platforms, dynamically integrated into canal operations. Automation and telemetry have reduced water waste in critical canal sections by 30–60 AF (3.7×10^5 to $7.4 \times 10^5 m^3$) annually, contributing to the district's 85% delivery efficiency rate. This efficiency is further supported by laser-leveled fields (implemented across 80% of irrigated land) and a tiered pricing structure that penalizes waste while incentivizing conservation through discounts for self-pumping farmers ($3.00/acre).

Irrigation scheduling is tailored to crop requirements and soil conditions. Over 80% of fields are laser-leveled to slopes of 0.05–0.10 ft/100 ft, enhancing uniformity in water application. The district prioritizes furrow irrigation (using polypipe systems) for staple crops such as cotton, corn, sorghum, and small grains, while drip and sprinkler systems support high-value crops (citrus, vegetables, and turf grass). Soil moisture sensors are deployed across fields and transmit real-time data to RTUs, enabling farmers to adopt deficit irrigation strategies during droughts. This precision scheduling reduces water use by 20%–70% for targeted crops without sacrificing yields, aligning with the district's conservation goals.

Governed by Texas's Prior Appropriation Doctrine, HIDCC1 holds senior "Class A" water rights, prioritizing its annual allocation of 98,232 AF ($1.2 \times 10^8 m^3$) from the Rio Grande River during shortages. This legal framework ensures that agricultural and municipal users receive water even in drought conditions. On average, the average annual water diversion is 52,000 AF ($6.4 \times 10^7 m^3$) for irrigation and 15,000 AF ($1.8 \times 10^7 m^3$) for municipal and domestic use.

6.8.2 California

California's Imperial Irrigation District (IID) is responsible for managing one of the largest and most complex water delivery systems in the western United States. As a major agricultural water provider, IID plays a crucial role in distributing Colorado River water to sustain over 445,000 acres (180,086 ha) of farmland in the Imperial Valley.

IID operates an extensive network of canals, reservoirs, and drains to efficiently distribute water to agricultural users. The system includes 1,668 miles of canals and 1,456 miles of drains, allowing the IID to effectively transport and manage water throughout its service area. Farmers in the district receive water through 5,578 delivery gates, ensuring precise and timely irrigation. The district's water delivery is dependent on the All-American Canal, which brings Colorado River water into the Imperial Valley. This water is then routed through IID's lateral canals before being delivered to individual farms. To improve operational efficiency, IID has implemented automated water control systems and mid-lateral regulating reservoirs, such as the newly constructed Lloyd Allen Water Conservation Operational Reservoir, which has a 40-AF (49,339 m^3) capacity. This allows the district to better regulate water supply fluctuations and provide flexibility for on-farm conservation programs. Additionally, IID's SCADA system helps monitor and control water flow in real time, ensuring efficient allocation and delivery.

Imperial Valley farmers rely on scheduled water deliveries to irrigate crops in an arid climate with less than three inches (76 mm) of annual rainfall. Farmers place water orders in advance, and deliveries operate on 12-hour irrigation cycles, optimizing water usage and reducing losses. The district promotes efficient irrigation methods through the On-Farm Efficiency Conservation Program (OFECP), which incentivizes farmers to adopt water-saving techniques, such as drip irrigation, sprinkler systems, land leveling, and tailwater recovery systems. The OFECP has significantly contributed to water conservation in the region. In 2023 alone, participating farmers saved 215,382 AF ($2.7 \times 10^8 m^3$) of water through improved irrigation efficiency. Since the program's inception in 2013, Imperial Valley growers have collectively saved over 1.5 million AF ($1.9 \times 10^9 m^3$) (Imperial Irrigation District, 2024). These conservation measures have helped the district meet its obligations under the Quantification Settlement Agreement (QSA), which requires IID to transfer nearly 500,000 AF ($6.2 \times 10^8 m^3$) of conserved water annually to urban water agencies, including San Diego County Water Authority, Coachella Valley Water District, and Metropolitan Water District of Southern California.

IID's water rights are among the most senior on the Colorado River, granting it an annual entitlement of approximately 3.1 million AF ($3.8 \times 10^9 m^3$). However, ongoing drought conditions and increasing demands on the Colorado River system have required additional conservation measures. In response, IID has actively participated in regional water-saving agreements, including the 2023 System Conservation Implementation Agreement (SCIA) with the U.S. Bureau of Reclamation. Under this agreement, IID voluntarily contributed 106,111 AF of conserved water to Lake Mead to support critical reservoir elevations and ensure long-term system stability.

Additionally, IID has committed to increasing its conservation targets to 750,000 AF ($9.3 \times 10^8 m^3$) annually by 2026, representing 24% of its total entitlement. This ambitious goal includes implementing a Deficit Irrigation Program, which will encourage farmers to temporarily reduce irrigation during peak summer months for forage crops, such as alfalfa, Bermuda grass, and Kleingrass. These measures are designed to balance agricultural water needs with regional water sustainability, while ensuring that IID continues to meet its contractual obligations.

To further enhance conservation efforts, IID is investing in infrastructure upgrades and additional water storage projects. The district has secured $9.5 million in federal funding under the Bipartisan Infrastructure Law to develop a 2,100-AF ($2.6 \times 10^6 m^3$) reservoir adjacent to the East Highline Canal. This new facility is expected to save up to 15,000 AF ($1.9 \times 10^7 m^3$) of water annually, providing additional storage capacity to support irrigation efficiency programs. Moreover, IID continues to expand its system conservation strategies, such as lateral intertie projects, automated canal gates, and interceptor canals, which help reduce water losses and improve overall system reliability.

Despite a record-breaking wet year in 2023, which temporarily alleviated some drought concerns, the Colorado River remains under long-term shortage conditions. IID remains committed to advocating policies that protect its water rights, while actively collaborating with state and federal agencies to develop long-term solutions for sustainable water use. These efforts ensure that the Imperial Valley's agriculture-based economy, which contributes over $5 billion annually, remains viable, while addressing the broader challenges of regional water security.

REFERENCES

Andales, A.A., Bauder, T.A., and Arabi, M. (2014). A mobile irrigation water management system using a collaborative GIS and weather station networks. In: L. Ahuja, L. Ma, and R.J. Lascano, Eds., *Practical Applications of Agricultural System Models to Optimize the Use of Limited Water.* American Society of Agronomy: Crop Science Society of America: Soil Science Society of America, Madison, WI, pp. 53–84.

Brouwer, C., Prins, C., and Heibolem, M. (1989). *Irrigation Water Management.* Training Manual (FAO), no. 4: Irrigation scheduling. FAO, Rome, pp. 43. https://www.fao.org/4/t7202e/t7202e00.htm#:~:text=*Irrigation%20scheduling*%20is%20a%20manual%20prepared%20by,of%20irrigation%20water%20need%20of%20paddy%20rice

Doorenbos, J. and Pruitt, W.O. (1977). *Guidelines for Predicting Crop Water Requirements.* FAO Irrigation and Drainage. Paper no. 24. FAO, Rome.

HID (20211). Innovative technologies for agricultural water management and flow measurement. Harlingen Irrigation District's agricultural water conservation grant from the Texas Water Development Board. https://hidcc1.org/projects/

Imperial Irrigation District (2024). 2023 Water & QSA Implementation Report. Imperial Irrigation District. https://www.iid.com/water/about-iid-water

Keller, J. and Bliesner, R.D. (1990). *Sprinkler and Trickle Irrigation.* Van Nostrand Reinhod, New York.

Merrey, D.J. (1990). Warabundi in Pakistan. In: G.J. Hoffman, T.A. Howell, and K.H. Solomon, Eds., *Management of Farm Irrigation Systems.* American Society of Agricultural Engineers, St. Joseph, MI, pp. 945–957.

USDA-NASS (2019). 2018 Irrigation and Water Management Survey. https://www.nass.usda.gov/Publications/AgCensus/2017/Online_Resources/Farm_and_Ranch_Irrigation_Survey/index.php

USDA-NRCS (1997). Chapter 9: Irrigation water management. In: E. A. Ross, L. A. Hardy, J. Brenner, J. Walker, C. Gustafson, D. Nelson, J. Dalton, and J. Suit, Eds., *National Engineering Handbook Part 652: Irrigation Guide*. USDA, Washington, DC, pp. 9-1–9-223.

7 Operation of Irrigation Systems

NOTATION

AF acre-feet

7.1 INTRODUCTION

The actual operation of an irrigation system is an important consideration. It must be emphasized that the objectives of an irrigation system that we are trying to achieve should always be kept in mind. An irrigation enterprise is designed to accomplish the following objectives: (1) To deliver irrigation water equitably to all water users at the lowest possible cost commensurate with the type of service that the water users are willing to pay and consistent with sound management practices and conservation of water. (2) To maintain the total irrigation system and the drainage system, if present, to the degree necessary to preserve the capacity and the condition of facilities in a manner that prevents undue depreciation; and to store and deliver irrigation water as required by the water users. (3) To keep system seepage and operational waste to a minimum. (4) To be constantly alert to technical advancements and adopt new practices that will improve system operation efficiency.

In short, one might say that the primary responsibility is to operate and maintain a system to equitably deliver water to all its users when needed in the most efficient manner at the lowest possible cost.

7.2 KNOWLEDGE ABOUT IRRIGATION SYSTEM DESIGN

Before one can plan the operation of an irrigation system, it is necessary to know as much detail about its design as possible. This knowledge ensures that operational decisions align with the system's engineered capabilities, preventing inefficiencies, structural failures, and user conflicts. The planners, designers, and construction organization can and should provide the operating organization with information in the form of (1) design criteria, such as maximum flow rate, (2) designers operating instructions, (3) final (as-built) drawings, (4) logs of drill holes (to show geology and water tables), and (5) other documents (photographs, inspection reports, etc.), which are of a great deal of value.

Members of the operating organization must study these documents and become thoroughly familiar with them. Such a procedure will alleviate many worries that will arise later. For instance, if the system is designed and built so that the capacity of the conveyance is capable of delivering water only on a rotational basis (except in rare instances), it would be meaningless to operate on a demand basis. Similarly, if the system is designed and built assuming that water users will apply water to the

 DOI: 10.1201/9781003747444-9

land using sprinklers, radical changes may be necessary if flood irrigation is followed. Disregarding design limitations can lead to catastrophic failures, and there are numerous such examples.

7.3 DELIVERY OF WATER

A reasonable procedure for describing the steps necessary for delivering water to users is to start at the source. To avoid a lengthy discussion on the manner of operation when multiple uses are made of the water, only irrigation water will be considered at this point. When other users are involved, as is generally the case in present-day systems, modifications are necessary, and the operation must ensure that all users are given their due consideration and, in many instances, their legal rights.

7.3.1 Surface Water

If the supply is drawn from surface water sources, it may be necessary for the project operators to establish some sort of hydrologic network in the watersheds from which the source is derived. This network may involve the installation and/or operation of (1) snow surveys to assess the snowpack and predict the timing and volume of snowmelt that contributes to surface water flow. For example, the Airborne Snow Observatory (ASO) from the National Aeronautics and Space Administration (NASA) employs LiDAR and imaging spectroscopy to generate 3D snowpack maps (Painter et al., 2016), enabling accurate prediction of snowmelt in major water basins in the western United States; (2) precipitation gages to measure rainfall and snowfall; (3) stream gaging stations to monitor water flow in rivers and streams; (4) soil moisture measurement devices to track the moisture content of soils; (5) weather stations; and (6) similar hydrologic observation stations. These may be necessary to obtain forecasts of expected water yields. If the government or other agencies operate such services, then the necessary information can be obtained from them.

7.3.2 Groundwater

Effective groundwater management requires meticulous monitoring and data collection to balance extraction rates with natural recharge, particularly in regions that are dependent on shallow aquifers or facing overexploitation. The Salt River Project in Arizona is a good example of an instance where a certain amount of hydrologic data is collected by the project (Phillips et al., 2009). Government agencies, such as the United States Geological Survey (USGS) and others, provide considerable information, especially long-term records on groundwater levels (USGS, 2023).

Understanding the source of water supply, whether from surface water or groundwater, and having access to relevant hydrological data are essential for efficient water delivery and resource management within an irrigation system. These data help in making informed decisions about water allocation and usage.

7.4 WATER SUPPLY FOR THE PROJECT

When considering the water supply for the project, we ask the question: How much storage and what other sources of supply are available? At the beginning of the irrigation season, or as soon as the total storage can be determined, we should be able to apportion the total supply to each farm unit. With this out of the way, the next consideration must be: What is each farm unit's share of system capacity? In other words, it is incumbent upon the operating organization to let the water user know the maximum quantity that can be delivered when everyone entitled to draw water is requested to his per-acre share of the system's capacity. With this information, the water user is then able to plan his cropping pattern and irrigation for the season on the basis of what he can draw on peak demand and his total quantity for the season.

7.5 WATER RELEASE

When the water is released from storage, it passes through the system and is delivered to the highest point or point on each of the farm units. Here, we outline a typical route water follows and discuss the essential structures involved in its control and measurement.

The path of water after it is released from the storage dam is illustrated in Figure 7.1. When water is released, it typically enters the conveyance system, which

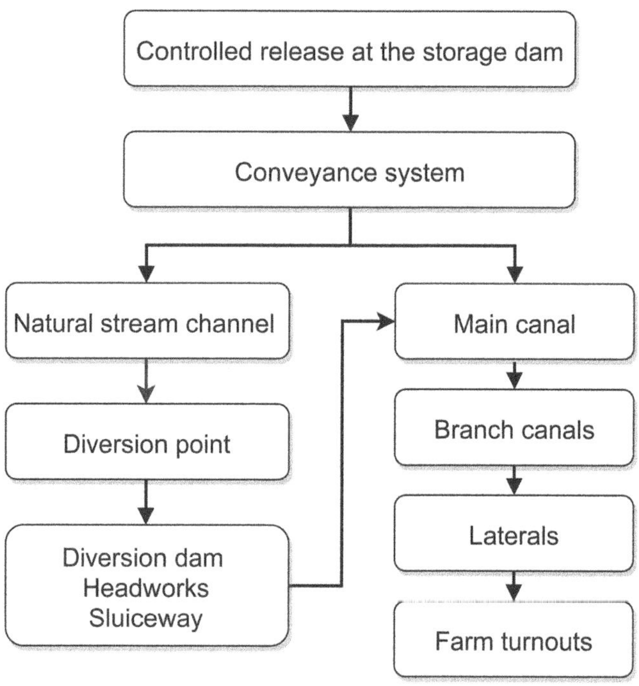

FIGURE 7.1 The path of water released at the storage dam.

can either be the natural stream channel or the main canal leading to the irrigation project. If the water is directed to the natural stream channel, it follows this course to a diversion point, where it is redirected using various structures such as diversion dams, headworks, and sluiceways. Once diverted, the water then enters the main canal.

After water passes through the main canal, it then flows into branch canals, depending on the system's layout. Some water may be diverted away from the main flow to supply farmlands that are adjacent to the canals. This diversion serves the purpose of efficiently distributing the water to where it is needed. Ultimately, the water reaches the laterals, which are smaller channels branching off from the main canal, and from there, it is further distributed to farm turnouts. These laterals play a crucial role in ensuring that the water is delivered to the individual farm units and their respective fields.

The path of water from the storage dam to the farm units involves a network of control structures and measurements designed to optimize water distribution and usage within the irrigation system. These structures can be classified into four categories: (1) Conveyance structures: These include (a) inverted siphons and flumes, (b) tunnels, (c) checks, (d) chutes, and (e) drops, each serving specific purposes in guiding and controlling the flow of water within the system. (2) Water measurement structures: These are essential for accurately quantifying the volume of water being transported and delivered, ensuring that the distribution aligns with the water rights and demands of the users. (3) Cross drainage and protective structures: These structures help manage water flow at points where the system encounters obstacles or requires protection against erosion and other potential issues. (4) Bridges and other crossings: These allow for the safe passage of waterways and roads over the conveyance system.

It is important to note that these structures are designed with the assumption of an open conveyance system. In a closed conduit system, the structures are different. The economy of space does not permit going into various design considerations. Not all of the structures are used for the measurement and control of water. Others must be maintained and will be discussed later under the discussion of that subject. For the operation of the system, we will consider those structures that are used for the measurement and control of water.

7.6 OPERATION

The key objective in the operation of an irrigation system is to consistently deliver irrigation water equitably to all water users at a low cost, with sound management and operation practices, and conservation of water. We have provided a brief overview of (1) the physical system necessary to collect, store, and convey water from the source to the farm; (2) some of the structures along the way and we have noted a classification of structures; and (3) measurement and control structures.

These structures and conveyance systems play a crucial role in the movement of water from its source to the farms while ensuring an equitable distribution over time. However, it's important to note that these are not the only components necessary for operating an irrigation system. Other essential aspects include the development of roads, buildings, fences, communication systems, and a variety of other improvements that are vital for the effective operation of the irrigation enterprise. We will delve into these additional considerations in more detail in the following sections.

7.7 DELIVERY OF WATER

First, recall the methods of delivery. There are three general methods of water distribution to farm units, namely, continuous flow, demand, or rotation. The two most commonly used irrigation projects in the northwestern United States are the demand, which is generally a modified-demand system, and the rotation system. Both of them have certain advantages under certain circumstances.

7.7.1 Demand System

The demand system permits a water user to draw any quantity he may desire. Of course, this must be limited to the system's capacity. As a result, it typically evolves into a modified-demand system to ensure equitable distribution and prevent overloading the system. One of the advantages of the demand system is that it permits the use of larger heads for intermittent periods of time, particularly at the beginning and near the close of the irrigation season. During that period when the system is running at full capacity, most water users would be drawing their share of system capacity and would have to keep it on a continuous basis to meet crop requirements. This system provides flexibility for users but necessitates coordinated scheduling to ensure efficient water distribution.

7.7.2 Rotation System

The rotation method is particularly advantageous in light of sandy soils where it is necessary to use large heads to crowd the water over the ground surface and change it from place to place quite rapidly to avoid the waste of excessive deep percolation.

7.7.3 Combination Method

Next, assume we are to operate under the demand method (variations are necessary if other methods are used), and perhaps we do not have a pure demand method but a combination of methods:

1. **Start at the farm**: The farmer needs to irrigate corn or other crops.
2. **Advance notice for deliver-water orders**: To efficiently dispatch water and avoid excessive operating waste, it is necessary to require advance notice of delivery and changes. This can best be done through the use of water request cards. These cards are an assurance against misunderstandings and provide a record of when and what was ordered. To save time, many users order their requests by telephone or online. If the desired changes are written on a request card and read back to the farmer, errors are avoided. It is the

usual procedure for the ditch rider to have permission from the water user to sign the cards for him. All water request cards should be turned in with the ditch rider's daily water record. Every water change on a daily record sheet should be accompanied and verified by the water request cards signed by the water user or through his special permission by the ditch rider.

3. **Timing, adjustment, and regulation of lateral head:** Timing is a very important factor in avoiding unnecessary radical changes in water surface elevation in the canal system. Accurate timing relies on historical records of previous changes, which enable precise adjustments. Unless there is reasonable knowledge of the time that changes in a flow reach certain division points, it will be difficult to make the required adjustments, and inaccurate regulation can result. In addition to the unnecessary waste of water, it is also a source of difficulty to the water user when variations occur that he didn't know were coming and is not prepared to handle. It is incumbent upon the ditch rider to have reasonable knowledge of farm delivery variations that are caused by inaccurate timing, and that these are not charged to the water user but are charged as losses in regulation.

   The ditch rider should closely regulate and measure the changes in flow required at his lateral head from day to day, so they correspond with the orders he has to fill downstream. With careful handling of all diversions from the main source of the system, water can be dispatched and managed with little operating waste. In addition, there is a need to maintain a constant water surface elevation in the supply canal to avoid variations in distribution to various laterals and farm turnouts. It is also necessary to keep weeds out of head gates to avoid clogging. This is done through various types of devices, such as weed screens, floating diverters to shunt weeds downstream, or mechanical rakes to carry them onto the bank. The type of headgate and the way the headgate is regulated can also help overcome clogging. For example, a pipe opening is more susceptible to clogging when controlled by a circular-type headgate, leaving a crescent opening than with a straight edge on the bottom.

4. **Regulation of canal check structure:** Check structures in the canal system can be and often are misused. Certain water surface elevations are necessary to maintain steady flow through head gates. However, if excessive elevations are maintained, unnecessary losses occur through seepage. This also tends to encourage waterweed growth and subjects the canal bank to greater loads than are necessary. An observant ditch rider will remove boards as necessary and replace them when required to maintain proper water surface elevations.

7.8 CASE STUDIES

This section provides detailed real-world case studies from two U.S. irrigation districts, including the Lower Rio Grande Valley, Texas, and the South San Joaquin Irrigation District, California. These case studies demonstrate the complexities of operational planning, real-time adjustments, and the integration of infrastructure and information systems into irrigation management.

7.8.1 Irrigation Districts in the Lower Rio Grande Valley

The Lower Rio Grande Valley of Texas offers a good case study of how irrigation districts can adapt to increasing water scarcity, aging infrastructure, and operational inefficiencies through integrated planning and innovation in semi-arid regions. The region includes 28 irrigation districts, with 98% of their water supply sourced from the Rio Grande River. However, the river is over-appropriated, which means that existing water rights exceed the sustainable long-term yield of the system. Agricultural users hold approximately 90% of these water rights and account for 80% of annual withdrawals.

Among these districts, the Harlingen Irrigation District Cameron County No. 1 (HIDCC1) is one of the largest irrigation districts and most progressive. Serving over 56,000 acres (22,662 ha), HIDCC1 has been recognized both nationally and internationally for its innovative agricultural water management practices (TWDB, 2025). The district maintains 57 miles (92 km) of main canal and more than 200 miles (322 km) of irrigation pipeline, 44 pump stations, and three storage reservoirs with a total capacity of 1,380 acre-feet (AF) ($1.7 \times 10^6 m^3$). The district holds a Class A irrigation water right of 98,232 AF ($1.2 \times 10^8 m^3$) per year.

Water deliveries are scheduled on a demand basis. Growers request water using water tickets, specifying their preferred delivery date. If this date does not conflict with other requests, the canal rider schedules delivery accordingly. If a conflict arises, the rider coordinates with the grower to determine an alternative. Importantly, the water ticket is not a contractual guarantee, but a scheduling tool used to allocate water based on system capacity.

To improve efficiency and equity, HIDCC1 has made significant investments in infrastructure and automation. Conveyance efficiency has increased from a regional average of 70.8% to approximately 85% (Fipps, 2000), largely due to canal lining, pipe replacement with corrosion-resistant PVC, and automation of delivery structures. A key innovation has been the deployment of solar-powered, aluminum auto-gates that cost less than $3,500 per unit. These gates, operated remotely via smartphones, tablets, or computers, have replaced manually operated wooden gates, reducing labor and increasing response speed. Each irrigation event saves between 40 and 70 AF ($4.9–8.6 \times 10^4 m^3$) of water.

The district has also implemented a Supervisory Control and Data Acquisition (SCADA) system, which networks 69 flow sensors across the infrastructure. SCADA enables real-time monitoring and control, allowing staff to detect and respond to overflows or system failures within 15 minutes, thereby improving reliability and reducing losses.

On-farm efficiency is also a focus. Approximately 80% of irrigated land in the district is laser-leveled to support uniform furrow irrigation using poly-pipe systems. Some fields use drip or micro-jet systems. HIDCC1 encourages the use of soil moisture telemetry systems (e.g., Decagon and AquaSpy) to inform irrigation timing and volume. These systems support deficit irrigation practices that can reduce water use by 20%–70%, depending on crop and field conditions. Weather data from local stations managed by Texas A&M AgriLife Research are used to calculate crop evapotranspiration, providing growers with site-specific irrigation guidance through an online portal.

In 2024, HIDCC1 implemented stricter enforcement against water wastage. Any user found wasting water due to carelessness, inadequate field preparation, or poor maintenance is subject to a $200 fine per incident and must pay three times the normal delivery rate for the volume wasted. Water service is suspended until the issue is resolved, with the amount of wasted water determined by the canal rider.

These operational practices are supported by strong partnerships with local municipalities and the Texas Water Development Board, ensuring both agricultural and municipal stakeholders benefit from long-term water sustainability. Collectively, the operational model of HIDCC1 exemplifies how irrigation districts can modernize aging systems, improve delivery efficiency, enhance grower decision-making, and maintain resilience in water-limited environments.

Example 7.1

A cotton grower in the Harlingen Irrigation District is evaluating whether to continue using conventional furrow irrigation or switch to surge irrigation. The grower trials both systems on similar fields during the same growing season. Based on the following data (Table 7.1), determine which system demonstrates higher water use efficiency and explain the operational implications.

Solution:

The total water applied (acre-feet, or AF):

$$\text{Furrow: } 17.5 \text{ acres} \times 19.53 \text{ in.} = 17.5 \text{ acres} \times 19.53 \text{ in.} \times \frac{1 \text{ ft}}{12 \text{ in.}} \approx 28.48 \text{ AF}$$

$$\text{Surge: } 20.4 \text{ acres} \times 13.48 \text{ in.} = 20.4 \text{ acres} \times 13.48 \text{ in.} \times \frac{1 \text{ ft}}{12 \text{ in.}} \approx 22.92 \text{ AF}$$

Net water use (after tailwater recover, acre-feet, or AF):

TABLE 7.1
Average Daily Water Requirements in the Growing Season for Sorghum and Cotton

Parameter	Furrow Irrigation	Surge Irrigation
Field size (acres)	17.5	20.4
Poly-pipe configuration	18 in.	18 in. + surge valve
Average inflow rate (gpm)	1,500	750
Seasonal water applied (in.)	19.53	13.48
Tailwater recovered (AF)	1.31	0.85
Irrigation duration per set	19 hours	78 hours (3-hour cycles)
Cotton yield (pound/acre)	950	950

$$\text{Furrow: } 28.48 \text{ AF} - 1.31 \text{ AF} = 27.17 \text{ AF}$$

$$\text{Surge: } 22.92 \text{ AF} - 0.85 \text{ AF} = 22.07 \text{ AF}$$

Total yield (pounds):

$$\text{Furrow: } 17.5 \text{ acres} \times 950 \text{ lbs/acre} = 16{,}625 \text{ lbs}$$

$$\text{Surge: } 20.4 \text{ acres} \times 950 \text{ lbs/acre} = 19{,}380 \text{ lbs}$$

Water use efficiency (pound/AF):

$$\text{Furrow: } \frac{16{,}625 \text{ lbs}}{27.17 \text{ AF}} \approx 612 \text{ lbs/AF}$$

$$\text{Surge: } \frac{19{,}380 \text{ lbs}}{22.07 \text{ AF}} \approx 878 \text{ lbs/AF}$$

Surge irrigation improves water use efficiency by approximately 43% compared to furrow irrigation (878 vs. 612 lbs/AF). Although surge irrigation requires longer application durations and more management, it significantly reduces water consumption while maintaining yield. These results support broader adoption of surge systems in regions facing water scarcity, especially when paired with remote automation and SCADA technologies.

7.8.2 South San Joaquin Irrigation District (SSJID)

The South San Joaquin Irrigation District (SSJID) serves over 72,000 acres of farmland in California's Central Valley, providing irrigation water to the agricultural areas near Escalon, Manteca, and Ripon. Established in 1909, SSJID was established to ensure a reliable and economical supply of irrigation water, transforming the region's agricultural productivity. Water is sourced from the Stanislaus River and stored in key reservoirs, including New Melones (2.4×10^6 AF, or $3.0 \times 10^9 \text{m}^3$), Tri-Dam ($2.0 \times 10^5$ AF, or $2.5 \times 10^8 \text{m}^3$), and Woodward reservoir (3.3×10^4 AF, or $4.1 \times 10^7 \text{m}^3$). The district conveys water through a network of over 3.5 miles of main supply canals, 362 miles of laterals, and 315 miles of pipelines, and operates 32 groundwater wells as supplementary sources.

A key milestone of SSJID's development was its collaboration with the Oakdale Irrigation District to develop the Tri-Dam Project—a self-financed system of reservoirs, dams, and hydropower plants. This initiative significantly increased water storage reliability and added 135 MW of renewable hydropower generation. Annual revenue from hydropower sales ($4–17 million) provides financial resilience and supports infrastructure modernization.

In recent decades, SSJID has prioritized system efficiency and service improvement. These efforts include installing solar arrays, expanding into domestic water service, and transitioning to pressurized pipeline delivery systems through its irrigation enhancement project. The project aims to improve precision, reduce tailwater losses, and support water conservation under drought and land-use shifts (SSJID, 2022).

To support equitable and efficient service, SSJID implemented a capacity analysis tool that divides the delivery network into Distribution Service Areas (DSAs)—zones served by common laterals or pipelines (SSJID, 2022). For each DSA, the district calculates a Service Ratio, defined as

$$\text{Service ratio} = \frac{\text{On-farm demand}}{\text{Distribution system supply capacity}} \tag{7.1}$$

A lower service ratio indicates that the system has excess capacity relative to demand, allowing more operational flexibility. For example, a service ratio of 0.7 implies that the system must run at capacity for 7 days of a 10-day irrigation cycle to meet demand, leaving 3 days of downtime. Conversely, service ratios above 1.0 suggest that the system is undersized relative to peak demand. SSJID identified areas with high service ratios as targets for infrastructure enhancement, helping prioritize capital improvement projects.

Technology upgrades, including SCADA systems, automated gates, and in-field flow sensors, support real-time delivery management. These investments enhance the district's responsiveness to water shortages, land conversion, and adoption of high-value, water-intensive perennial crops such as walnuts and almonds.

Example 7.2

A grower irrigates 150 acres of pasture that relies on a 10-day rotational schedule. Each acre of pasture requires 8 mm/day during peak evapotranspiration periods. The lateral serving this DSA has a maximum flow capacity of 9 cfs. Calculate the service ratio and determine whether the current system allows sufficient delivery flexibility.

Solution:

On-farm demand (during 10-day cycle):

$$8\,\frac{\text{mm}}{\text{day}} \times 10\text{ day} \times 150\text{ acres} = 8\,\frac{\text{mm}}{\text{day}} \times \frac{1\text{ ft}}{304.8\text{ mm}} \times 10\text{ day} \times 150\text{ acres} \approx 39.4\text{ AF}$$

Since $1\text{ cfs} = 1.983\text{ AF/day}$, the system capacity for 10 days is

$$9 \times 1.983 \times 10 \approx 178.5\text{ AF}$$

The service ratio is

$$\frac{39.4}{178.5} \approx 0.22$$

With a service ratio of 0.22, this lateral has substantial excess capacity. It only needs to operate at full flow for approximately 2.2 days out of a 10-day cycle to meet demand. This enables significant scheduling flexibility, allowing the district to accommodate additional users, extend run times, or conserve energy through strategic system idling. This example demonstrates how low service ratios promote operational flexibility and help irrigation districts prioritize infrastructure investment. SSJID can use such data to identify undersupplied areas (e.g., with ratios >1.0) that would benefit from targeted upgrades to pipelines, gates, or automation systems.

SSJID employs a hybrid irrigation delivery strategy. In older canal-lateral systems, deliveries follow a 10-day rotational schedule, which is ideal for traditional flood irrigation. However, modernized IEP zones support demand-based delivery, better suited for pressurized systems like drip and sprinkler irrigation. While the 10-day standard remains the district policy, studies by the Irrigation Training and Research Center (ITRC) show that more flexible delivery would better serve users with pressurized systems, enabling smaller, more frequent irrigations aligned with crop evapotranspiration needs.

Despite policy constraints, irrigation enhancement project zones have shown marked improvements in delivery efficiency, reducing losses and enhancing water use flexibility. These results underscore the value of infrastructure reinvestment and real-time system monitoring in advancing sustainable irrigation.

Example 7.3

A walnut grower in the SSJID district irrigates 100 acres of orchards. Walnut trees require about 3.0 AF of water per acre per season, with a total seasonal demand of 300 AF. Due to drought restrictions, the district provides the grower with a reduced allocation of 2.5 AF per acre, for a total of 250 AF.

The grower is evaluating two irrigation methods: (1) Furrow irrigation (60% application efficiency) delivered via open canal delivery (75% conveyance efficiency) and (2) sprinkler irrigation (85% application efficiency) delivered via pressurized pipeline (95% conveyance efficiency).

If the grower schedules 20 AF of water for an irrigation event, determine

1. How much usable water is delivered under each method?
2. How much total water can be saved during the 250 AF allocation?
3. The implication for deficit irrigation planning.

Solution:

1. The net on-farm water delivered per 20 AF event:

$$\text{Furrow (Canal): } 20 \times 60\% \times 75\% = 9.00 \text{ AF}$$

$$\text{Sprinkler (Pipeline): } 20 \times 85\% \times 95\% = 16.15 \text{ AF}$$

2. Usable water over full season (250 AF allocation):

$$\text{Furrow (Canal): } 250 \times 60\% \times 75\% = 112.5 \text{ AF}$$

$$\text{Sprinkler (Pipeline): } 250 \times 85\% \times 95\% \approx 201.9 \text{ AF}$$

$$\text{Water saved using sprinkler: } 201.9 - 112.5 = 89.4 \text{ AF}$$

3. Implication for deficit irrigation:

$$\text{Furrow shortfall: } 300 - 112.5 = 187.5 \text{ AF}$$

$$\text{Sprinkler shortfall: } 300 - 201.9 = 98.1 \text{ AF}$$

Sprinkler irrigation, delivered through SSJID's modern pipeline system, enables significantly higher water use efficiency, delivering 80% more usable water over the season compared to traditional furrow irrigation. This improved efficiency enables the grower to better manage water under drought conditions, apply water more strategically, and reduce the severity of deficit irrigation stress.

REFERENCES

Fipps, G. (2000). Potential Water Savings in Irrigated Agriculture for the Lower Rio Grande Planning Region (Region M). Final Report. TWRI TR-183. The Texas Water Resources Institute (TWRI).

Painter, T.H., Berisford, D.F., Boardman, J.W., Bormann, K.J., Deems, J.S., Gehrke, F., Hedrick, A., Joyce, M., Laidlaw, R., Marks, D., Mattmann, C., McGurk, B., Ramirez, P., Richardson, M., Skiles, S. M., Seidel, F.C., and Winstral, A. (2016). The airborne snow observatory: Fusion of scanning lidar, imaging spectrometer, and physically-based modeling for mapping snow water equivalent and snow albedo. *Remote Sensing of Environment*, 184, 139–152. https://doi.org/10.1016/j.rse.2016.06.018

Phillips, D.H., Reinik, Y., Skarupa, T.E., Ester III, C.E., and Skindlov, J.A. (2009). Water resources planning and management at the Salt River Project, Arizona, USA. *Irrigation and Drainage Systems*, 23, 109–124.

SSJID (2022). 2022 Water Master Plan, the South San Joaquin Irrigation District Board of Directors on December 13, 2022. https://www.ssjid.gov/water-master-plan/

TWDB (2025). Texas Project for Ag Water Efficiency. Texas Water Development Board. https://www.twdb.texas.gov/conservation/agriculture/demonstration/lower-rio-grande-valley.asp

USGS (2023). USGS Groundwater Data for the Nation. U.S. Department of the Interior, U.S. Geological Survey. https://waterdata.usgs.gov/nwis/gw

8 Water Measurement

NOTATION

Q	discharge rate or flow rate (m^3/s)
C_d	discharge coefficient
A (Equation 8.1)	area of the orifice opening (m^2)
g	gravitational constant (= 9.81 m/s^2)
h_1	head on the upstream side of the orifice (m)
h_2	head on the downstream side of the orifice (m)
d	diameter of the orifice (m)
t	transit time (s)
c	sound speed in the water (m/s)
D	diameter of the pipe (m)
θ	central angle of the notch
$\frac{D}{\tan\theta}$	distance between the ultrasound transducers (m)
v	flow velocity (m/s)
H (Equation 8.2)	total energy head on the weir (flow depth plus the approach velocity energy) (m)
L	length of the crest weir (m)
C_v	velocity discharge coefficient
h_u	head above the orifice on the upstream side (m)
h_d	head above the bottom of the orifice on the downstream side (m)
Q_T	total flow rate (m^3/s)
q	flow rate in each strip (m^3/s)
A (Equation 8.8)	transversal area of the section (m^2)
w	width of the water flow (m)
H (Equation 8.8)	maximum flow depth (m)
H (Example 8.10)	measured water depth (m)
k	flume-specific coefficient
n	empirically derived exponent based on the flume geometry

8.1 INTRODUCTION

Measuring water is necessary to manage water more efficiently. Irrigation districts need to measure water to distribute it among several users of an irrigation district. Water should be delivered fairly according to each water user's allocation. Irrigation districts need to charge water to their water users to receive service and pay for the operation and management of the irrigation system. At the farm level, water measurement is necessary to match crop water requirements with water supply and promote its efficient use. Water at rest, or stored water, is measured in units of volume (m^3). Water in motion is measured as a flow rate, expressed as volume per unit of

DOI: 10.1201/9781003747444-10

time (m^3/s) or (L/s). This chapter introduces various methods and technologies used for water measurement.

8.2 MEASURING WATER FROM WELLS

There are several methods to measure water from wells. The most common method is using propeller flow meters. There are some other methods that use certain properties of water, such as ultrasonic and magnetic.

8.2.1 Propeller Flow Meters

Propeller flow meters are the most common devices used for measuring water flow rate. Water meters help irrigators better manage and schedule irrigation. They are also a tool for estimating irrigation water use. A propeller flow meter measures the velocity inside a pipe and shows the flow rate reading on a dial. There are three main types of propeller flow meters. The saddle type can be welded or clamped (Figure 8.1a and b), open flow (Figure 8.1c), or flanged (Figure 8.1d). The weld in line flow meter shown in Figure 8.1b may also be fitted with straightening vanes.

Installing a meter: The meter should be installed and placed correctly to ensure that readings will be accurate. It is also important to prevent debris from collecting on the propeller. Water should be clean, but if it contains sediment, the meter should

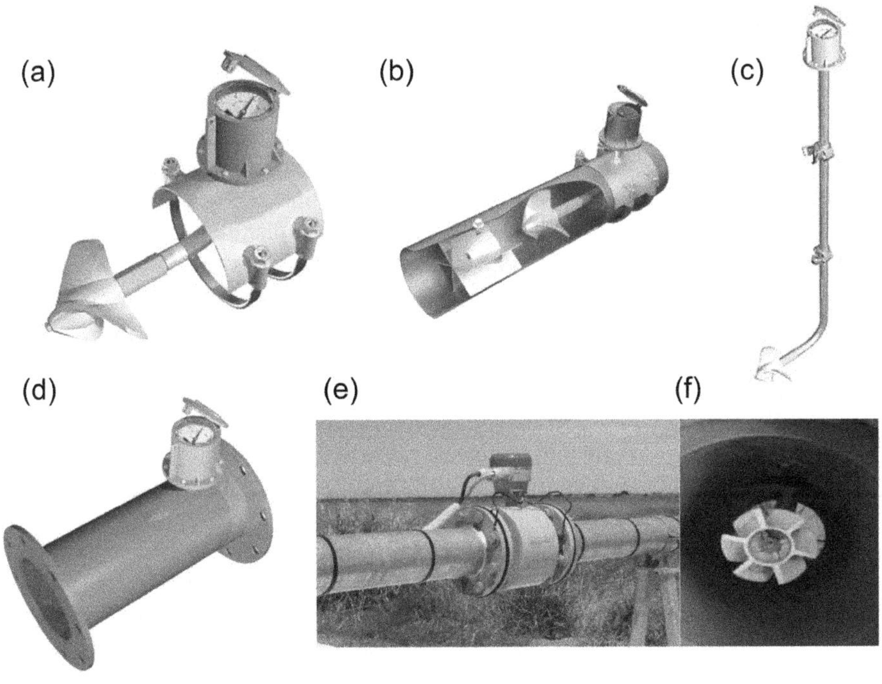

FIGURE 8.1 Propeller flow meter types. (a) clamped saddle meter, (b) welded-in saddle meter, (c) open flow meter, (d) flanged meter, (e) propeller meter coupled to aluminum or PVC pipe for furrow irrigation, and (f) small propeller meter for water with excessive trash.

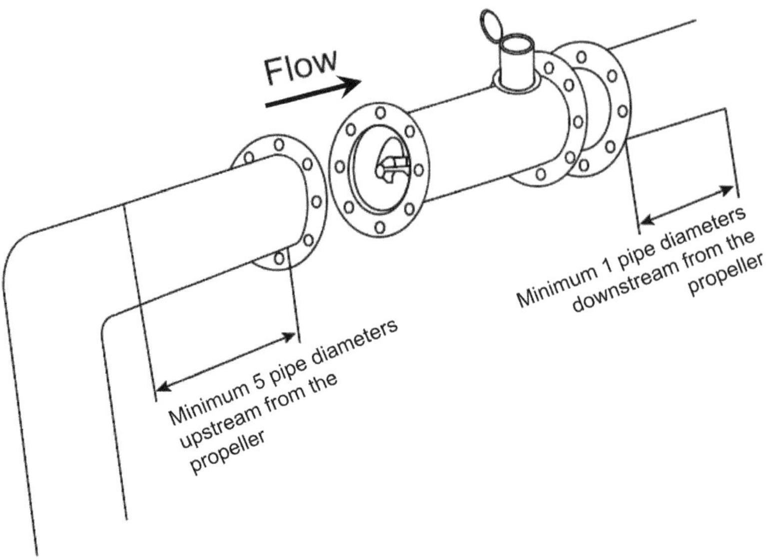

FIGURE 8.2 Distance requirements for installing propeller flow meters.

be appropriately located so that settling sediment will not obstruct the flow. Some of these meters are coupled to aluminum or PVC pipe, usually when they will be used in furrow irrigation (Figure 8.1e). When there is excessive trash in the water, a small propeller can be installed (Figure 8.1f). Some obstructions before the meter, including elbows, valves, pumps, or changes in diameter, can cause disturbances in flow measurements. To avoid this, the meter should be at minimum distances upstream and downstream of any obstructions, as shown in Figure 8.2. A minimum of five pipe diameters upstream from the propeller and one diameter downstream from the flange is usually sufficient, although the manufacturers' requirements may vary with different meter models and sizes. If five diameters are required upstream and one diameter downstream, and the pipe diameter is 10 in. (25.4 cm), then the length of the pipe upstream before any obstruction should be at least 50 in. (127 cm), and the length downstream should be 10 in. (25.4 cm). If there is not an enough length either upstream or downstream, meters should have straightening vanes, as shown in Figure 8.1b. Adding vanes will reduce the undisturbed length requirement to about 1½ pipe diameters upstream and ½ diameter downstream.

8.2.2 Orifice Plate

Orifices consist of an opening in a plate (Figure 8.3). The orifice can be free-discharging or submerged. The orifice is freely discharged when it discharges into the air and submerged when it discharges into water. The head-discharge relation for circular, sharp-edged, submerged orifices is

$$Q = C_d A \sqrt{2g(h_1 - h_2)} \tag{8.1}$$

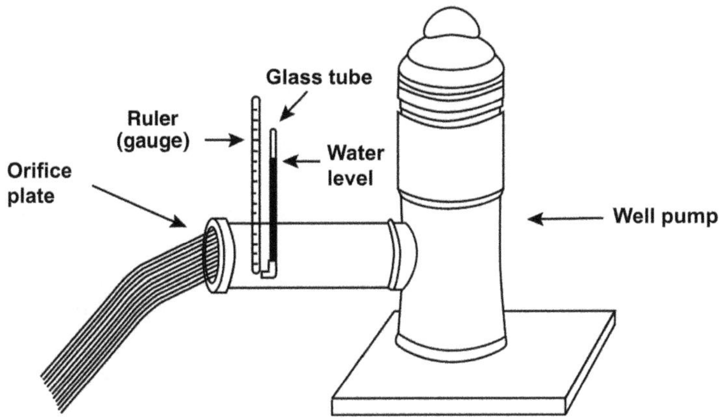

FIGURE 8.3 Orifice plate water meter.

where Q is the discharge rate, m^3/s; C_d is the discharge coefficient ranging between 0.57 and 0.61; A is the area of the orifice opening, m^2; g is the gravitational constant, $9.81\ m/s^2$; and $h_1 - h_2$ is the head differential across the orifice, m. For free discharging orifices h_2 = zero, and h_1 is measured from the center of the orifice, which is further illustrated in Section 8.3.6, measuring water in open channels.

Example 8.1

A circular orifice plate is used to measure water flowing through a submerged opening in an irrigation system. The orifice has a diameter of 10 cm, and the head on the upstream side of the orifice is 1.20 m, while the head on the downstream side is 0.70 m. The discharge coefficient for the sharp-edged, submerged orifice is 0.60. Using an acceleration due to gravity of $9.81\ m/s^2$, calculate the discharge rate of water through the orifice in both cubic meters per second and liters per second.

Solution:

Given the diameter of the orifice $d = 10\ cm = 0.10\ m$,

$$\text{Calculate the orifice area: } A = \frac{\pi d^2}{4} = \frac{\pi(0.10)^2}{4} = 0.00785\ m^2$$

The head differential across the orifice: $h_1 - h_2 = 1.20 - 0.70 = 0.50\ m$

$$Q = C_d A\sqrt{2g(h_1 - h_2)} = 0.60 \times 0.00785\sqrt{2 \times 9.81 \times 0.50} \approx 0.01475\ m^3/s$$

Convert to liters per second: $0.01475\ m^3/s = 14.75\ L/s$

8.2.3 Ultrasonic Flow Meters

Ultrasonic flow meters measure the velocity of liquid within a pipe using high-frequency acoustic sensors. Unlike mechanical meters, such as propeller types, ultrasonic meters have no moving parts. This design minimizes mechanical wear, reduces maintenance costs, and extends the lifespan of the device. However, their accuracy can be slightly affected by factors, such as fluid temperature, density, and electrical conductivity.

Ultrasonic flow meters are generally classified into two main types: (1) Transit-time (Transmission) flow meters and (2) Doppler (Reflection) flow meters. The most widely used ultrasonic flow meter is the transit time type. A typical transit-time flow measurement system utilizes two ultrasonic transducers mounted on opposite sides of a pipe, each functioning as both transmitter and receiver (e.g., transducers A and B in Figure 8.4). These transducers alternately send and receive bursts of sound waves through the flowing liquid—one pulse in the direction of flow and the other against it. The pulse moving with the flow is accelerated, while the one moving against it is delayed. The difference in transit time between the two directions ($t_2 - t_1$) is proportional to the average flow velocity along the sound path.

By using the absolute transit time, the distance between the ultrasound transducers ($\frac{D}{\tan\theta}$), and the sound speed in the water (c), the flow rate is easily found (Figure 8.4). Transit-time meters are highly accurate under clean flow conditions, but their performance can be affected by the presence of air bubbles or suspended solids.

Transit-time flowmeters can be of the wetted or non-intrusive (clamp-on) type. Clamp-on types are attached externally to the pipe and do not come into contact with the fluid, which eliminates the risk of corrosion and is ideal for retrofitting. The wetted type has components that are in direct contact with the fluid and are further categorized as insertion-type or spool-piece (flow cell) meters.

Doppler flow meters are named after the Austrian physicist and mathematician Christian Johann Doppler in 1842. A Doppler ultrasonic flow meter operates on the Doppler shift principle. One transducer transmits an ultrasonic beam of ~0.5 MHz into the flow stream (see Figure 8.5). The flow must contain sonically reflective

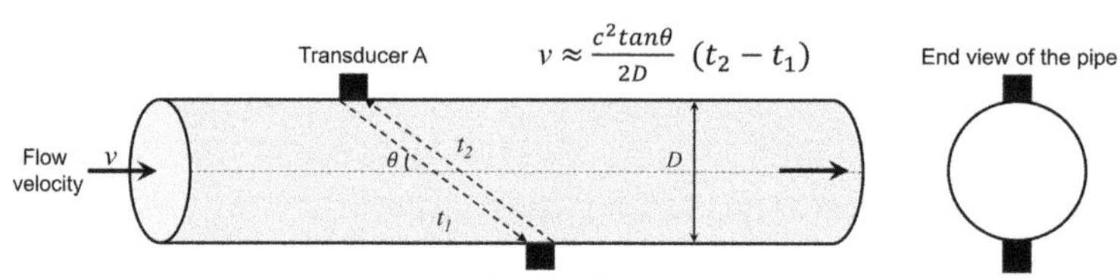

FIGURE 8.4 Transit-time flow meters.

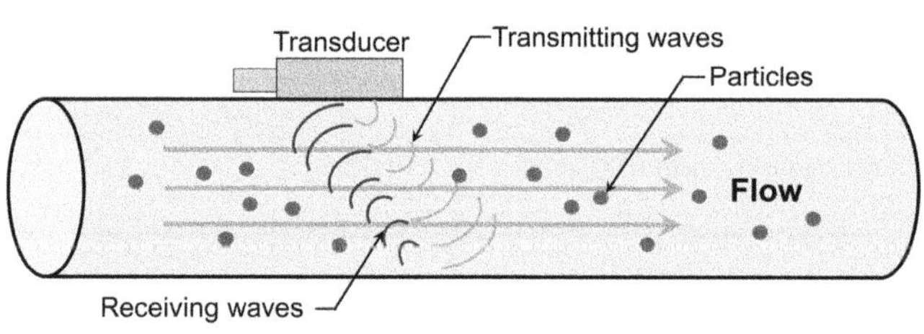

FIGURE 8.5 Illustration of a Doppler ultrasonic flow meter equipped with a transducer on the surface of a pipe.

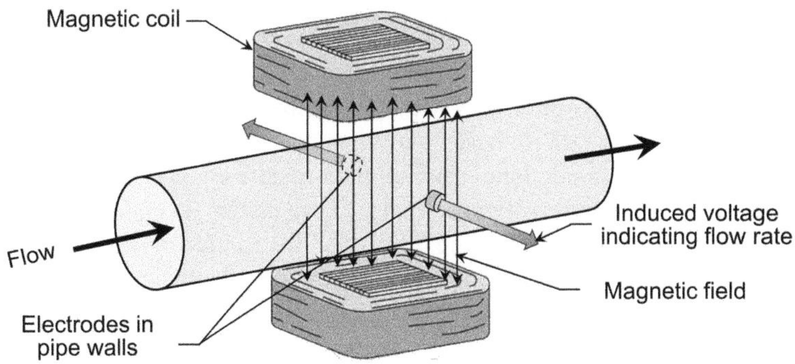

FIGURE 8.6 Magnetic flow meter.

materials, such as suspended particles or entrained air bubbles. These reflectors cause a frequency shift in the returned signal, which is detected by the transducer. The magnitude of this frequency shift is directly linked to the velocity of the moving particles and therefore the velocity of the fluid. When the internal diameter of the pipe is known, the Doppler meter can calculate the volumetric flow rate. Doppler meters are well-suited for fluids with high levels of solids or bubbles but are not effective in clean or single-phase flows.

8.2.4 Magnetic Flow Meters

Magnetic flow meters (commonly referred to as mag meters) operate based on Faraday's law of electromagnetic induction. In a magnetic flow meter, two magnetic coils generate a magnetic field perpendicular to the fluid movement (Figure 8.6). As the conductive liquid passes through the magnetic field, it generates a voltage that is proportional to the flow velocity. This voltage is picked up by electrodes embedded in the pipe wall. The electronic transmitter processes this voltage signal to compute the flow rate. Magnetic flow meters remain unaffected by changes in fluid temperature, pressure, viscosity, or density. However, they require the liquid to have a minimum electrical conductivity, typically around 3 μS/cm, to generate a measurable signal. Like ultrasonic meters, mag meters have no moving parts, making them low-maintenance and highly durable.

8.3 MEASURING WATER IN OPEN CHANNELS

Some water flow meters used for measuring water in wells can be used for canals. Special attention should be placed to sedimentation and trash. For example, propeller flow meters can be used on a farm outlet, but algae, trash, and other debris can clog the propeller. For this reason, the most common devices used to measure flow rate in canals are weirs, flumes, and orifices. These flow measurement structures are designed to allow flow rate estimation from the upstream water depth (head) in front of flumes or weirs, using an established correlation between head and discharge.

8.3.1 Ultrasonic Flow Meters for Open Channels

Unlike the confined conditions of a pipe, open channels present a variable environment where the dynamics of flow are affected by many factors. Ultrasonic flow meters can measure flow rate in these conditions with a high degree of precision and minimal intrusion. Ultrasonic open channel flow meters are widely used to measure flows in streams, rivers, and irrigation canals. These meters do not physically intrude into the water column, thus eliminating the risk of obstruction or damage from debris and minimizing maintenance requirements. They are adaptable to a wide range of channel shapes and sizes and can be installed with relative ease. When installed and calibrated correctly, ultrasonic flow meters can provide highly accurate flow measurements. For optimal performance, certain installation guidelines must be followed: (1) a sufficiently long stretch of straight channel upstream of the sensor is necessary to ensure a fully developed flow profile; and (2) the transit time of ultrasonic signals needs to be corrected for the speed of sound changes that are affected by temperature fluctuations. Ultrasonic flow meters are often less expensive than other types of flow meters, especially when considering installation and long-term maintenance costs. In addition, they are not affected by silting, suspended solids, or suspended grease, which is particularly useful in open channels.

8.3.2 Weirs

Weirs are structures in which water overflows on the top of the structure. Considering that all the water flows over the weir and the geometry of the weir is known, the depth of water behind the weir can be converted to a rate of flow. The calculation of the flow relies on the fact that fluid will pass through the critical depth of the flow regime in the vicinity of the crest, and the flow should not be submerged. Weirs can be categorized into several types, including sharp-crested, broad-crested, suppressed, and contracted weirs.

Sharp-crested weirs are widely used for small to medium discharge measurements. They are made from thin plates with edges of less than 2 mm (Figure 8.7a and b). The accuracy depends on avoiding submersion and forming a nappe that must be ventilated on the downstream side of the weir plate, so that pressure there is atmospheric. The height of weir, measured from the channel bottom to the crest, should be at least twice the maximum head (H) of flow, which is crucial to reduce the velocity of the approach. The head measurement point should be at a distance of no less than three times the maximum head to make sure that the flow is stable. Additionally, the downstream water level should be at least 5 cm below the crest of the weir. This necessitates a high head loss as the water passes over the weir, which can be a substantial drawback in scenarios such as irrigated agriculture. Sharp-crested weirs can be designed with various shared openings to suit different measurement needs and channel conditions, including rectangular, trapezoidal, and triangular (V-notch) (Figure 8.7c).

8.3.2.1 Sharp-Crested Rectangular Weirs

The discharge to a rectangular sharp-crested weir when the width of the canal equals the width of the crest can be approximated with a power function of the form:

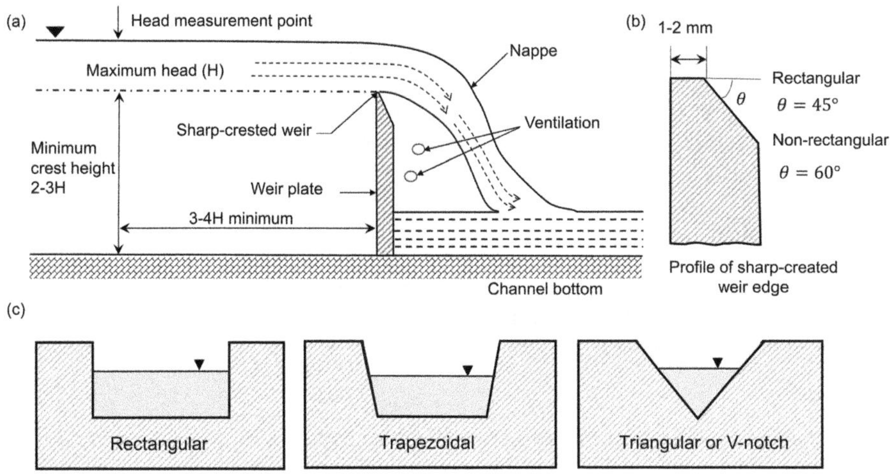

FIGURE 8.7 (a) Illustration of a sharp-crested weir, (b) profile of sharp-crested weir edge, and (c) three types of sharp-crested weirs.

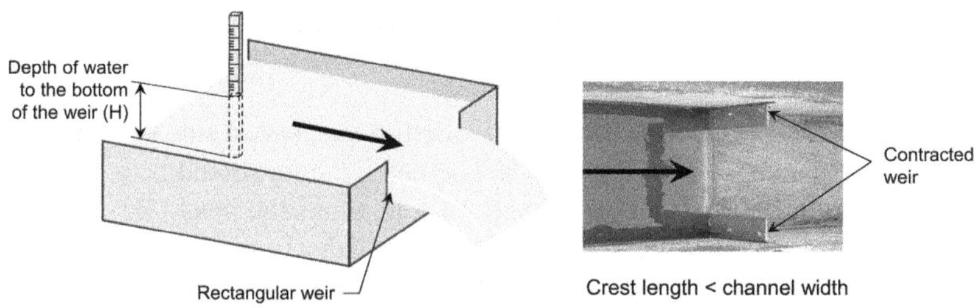

FIGURE 8.8 Field installation of a rectangular weir.

$$Q = C_d L H^{3/2} \tag{8.2}$$

where Q is the discharge rate, m^3/s; C_d is the discharge coefficient and is approximately equal to 1.84; L is the length of the crest weir, m; and H is the total energy head on the weir (flow depth plus the approach velocity energy), m.

When the weir has a contraction, as shown in Figure 8.8, then Equation (8.2) becomes

$$Q = C_d (L - 0.2H) H^{3/2} \tag{8.3}$$

Example 8.2

Find the flow over a rectangular weir in a canal with an 80 cm width, where an upstream water depth from channel bottom to the crest of the weir is 30 cm and the weir has a crest length of 80 cm.

Solution:

Since the weir is suppressed (i.e., crest length = channel width), Equation (8.2) is used.

Given $H = 30$ cm $= 0.3$ m and $L = 80$ cm $= 0.8$ m,

$$Q = 1.84LH^{3/2} = 1.84 \times 0.8 \times (0.3)^{3/2} \approx 0.2419 \text{ m}^3/\text{s or } 241.9 \text{ L/s}$$

Example 8.3

Find the flow over a rectangular weir in a canal with an 80 cm width, where an upstream water depth from channel bottom to the crest of the weir is 30 cm and the weir has a crest length of 35 cm.

Solution:

Since the weir is contracted (i.e., crest length < channel width), Equation (8.3) is used.

Given $H = 30$ cm $= 0.3$ m and $L = 35$ cm $= 0.35$ m,

$$Q = 1.84(L - 0.2H)(H)^{3/2} = 1.84 \times (0.35 - 0.2 \times 0.3) \times (0.3)^{3/2}$$

$$\approx 0.0877 \text{ m}^3/\text{s or } 87.7 \text{ L/s}$$

8.3.2.2 Sharp-Crested Trapezoidal Weirs

Trapezoidal weirs, commonly known as Cipolletti weirs, have a side slope of 1:4 horizontal to vertical, as shown in Figure 8.9. The notch length should be a minimum of three times the head (3H), though a length of four times the head (4H) or greater is preferable. These weirs, being fully contracted, have a discharge formula similar to the rectangular weirs, with a discharge coefficient approximately equal to 1.86, as follows:

$$Q = 1.86LH^{3/2} \tag{8.3}$$

Example 8.4

Find the flow over a trapezoidal weir in a canal with an 80 cm width, where an upstream water depth from channel bottom to the crest of the weir is 30 cm and the weir has a crest length of 80 cm.

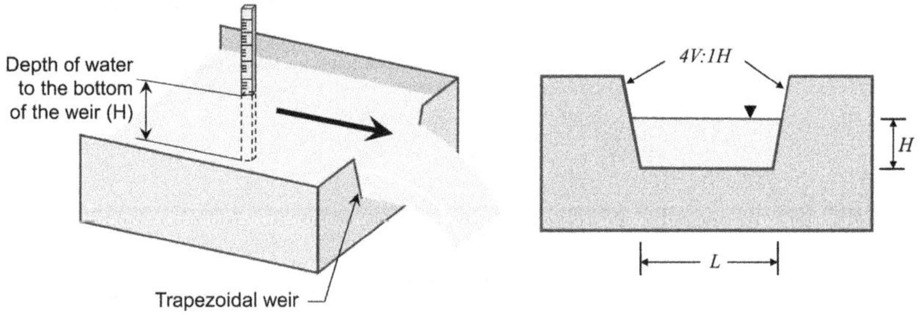

FIGURE 8.9 Field installation of a trapezoidal weir.

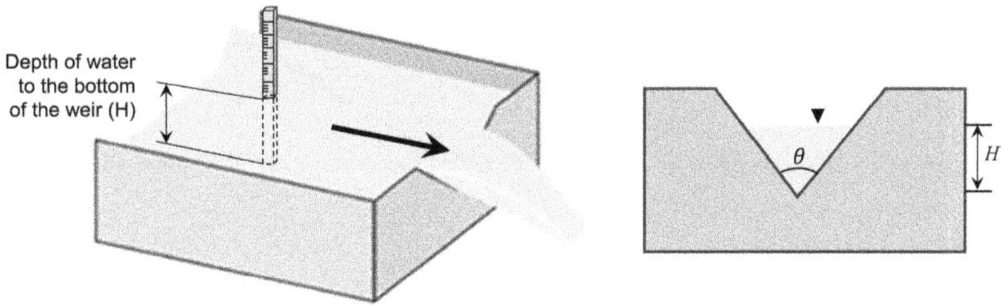

FIGURE 8.10 Installation of a V notch weir.

Solution:

Given $H = 30$ cm $= 0.3$ m and $L = 80$ cm $= 0.8$ m,

$Q = 1.86LH^{3/2} = 1.86 \times 0.8 \times (0.3)^{3/2} = 0.2445$ m^3/s or 244.5 L/s

8.3.2.3 Sharp-Crested V Notch Weirs

It is the most accurate open-channel constriction for measuring discharge. The equation can be as follows:

$$Q = C_d \tan\frac{\theta}{2} H^{5/2}$$

where θ is the central angle of the notch (Figure 8.10), and C_d is the discharge coefficient, which can be approximated to 1.4.

Example 8.5

Find the flow for triangular sharp-crested weir with an angle of 90° installed in a trapezoidal channel if the head of water over the crest of the weir is 17 cm.

Solution:

Given $H = 17$ cm $= 0.17$ m and $\theta = 90°$,

$$Q = 1.4\tan\frac{\theta}{2} H^{5/2} = 1.4 \times \tan\left(\frac{90°}{2}\right) \times (0.17)^{\frac{5}{2}} \approx 0.0167 \text{ m}^3/\text{s or } 16.7 \text{ L/s}$$

8.3.3 Flumes

These structures are well suited to measure irrigation water in open channels because they are economical to construct, flexible to adapt to different canal configurations besides producing a small head loss. The most common are the long-throated flumes and broad-crested weirs. Their operation is based on developing critical depth flow in a throat section created by a raised sill, a contracted width, or a combination of floor and sidewall contractions (Figure 8.11). Under critical flow, the discharge through the critical section can be computed as a function of the measured upstream head.

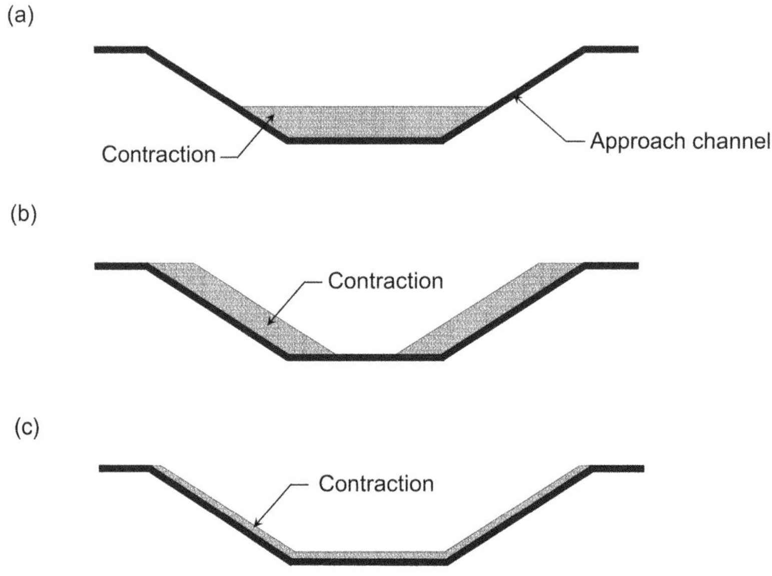

FIGURE 8.11 (a) broad-crested weir with raised sill, (b) long-throated flume with contracted width, and (c) usual flume with both floor and sidewall contractions.

Flumes that are solely width-contracted are particularly suitable for sites where sediment must be transported through the measurement structure.

Flumes can have several configurations, depending on the type of approach channel, the shape of the throat section, the location of the gaging station, and the use or lack of a diverging transition section. These configurations are crucial, as they impact the measurement accuracy and adaptability of the flume to different environmental conditions. Modern flume designs often use specialized computer software for precision. An example of such software is WinFlume from the USDA Agricultural Research Service, which is available for free download at https://www.usbr.gov/tsc/techreferences/computer%20software/software/winflume/index.html (Wahl et al., 2000). This software assists in various aspects of flume design, including consideration of construction methods, selection processes, and detailed design procedures. Additionally, it encompasses the underlying hydraulic theory and provides instructions for using the WinFlume software, as detailed by Clemmens et al. (2001).

There are some flumes that are laboratory-calibrated, such as the Parshall (Figure 8.12) and trapezoidal flumes (Figure 8.13), and in order to measure the discharge with good degree of accuracy, the construction plans should be carefully followed. The flow of the Parshall flume through its control sections is curvilinear. In long-throated flumes, the streamlines are essentially parallel in the throat section, making it possible to establish an analytical head-discharge relation. The long-throated flumes also have better accuracy over laboratory-calibrated type structures and have the ability to make field modifications and perform computer calibrations using as-built dimensions.

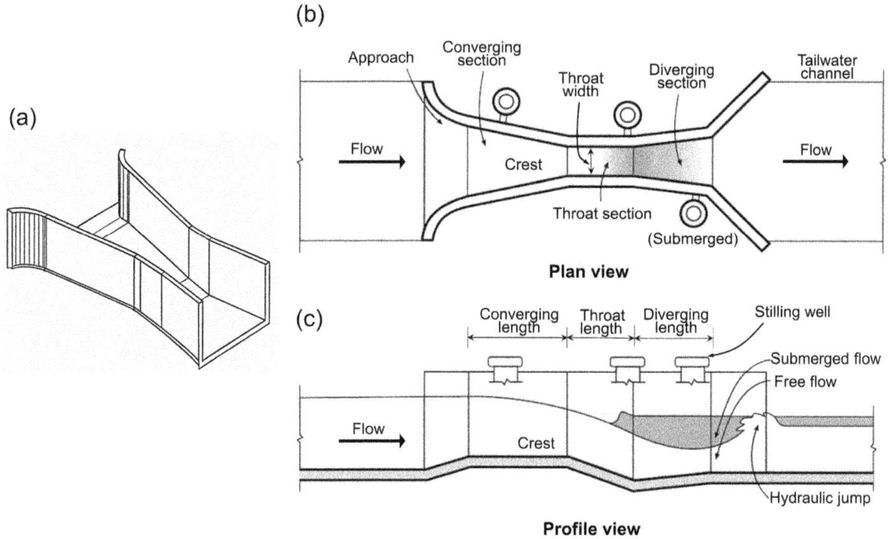

FIGURE 8.12 Parshall flume: (a) Example of a constructed Parshall flume, (b) plan view, and (c) profile (side) view illustrating flow through the flume.

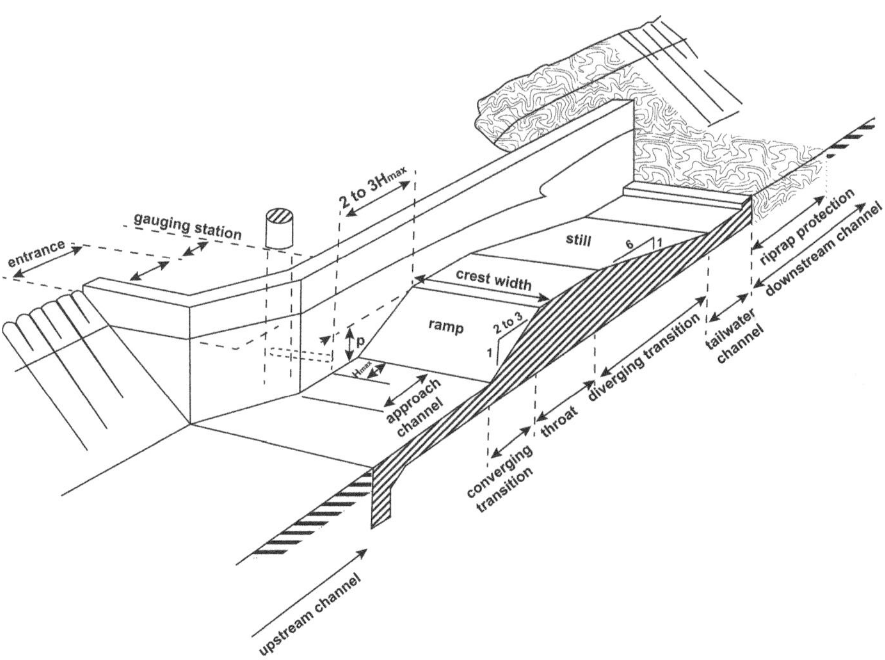

FIGURE 8.13 Standard trapezoidal flume.

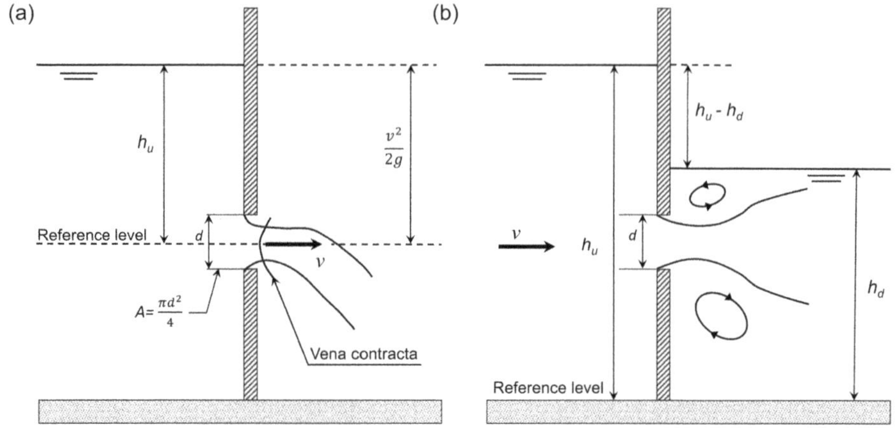

FIGURE 8.14 (a) Free flow orifice and (b) submerged orifice.

8.3.4 Orifices

Any gate, structure or opening in which the water level is higher than the opening, is referred to as an orifice (Figure 8.14). The orifices can have free flow (Figure 8.14a) or submerged conditions (Figure 8.14b). The equations are as follows:

For freely flowing orifice, the flow through the orifice can be expressed as follows:

$$Q = C_d C_v A\left(2gh_u\right)^{\frac{1}{2}} \tag{8.4}$$

For fully submerged orifice:

$$Q = C_d C_v A\left[\left(2g(h_u - h_d\right)\right]^{\frac{1}{2}} \tag{8.5}$$

where Q is the discharge rate, m^3/s; h_u is the head above the orifice on the upstream side, m; h_d is the head above the bottom of the orifice on the downstream side, m; C_v is the velocity discharge coefficient and $C_v = 1$ for velocities of less than 1 m/s; C_d is the discharge coefficient that depends on the geometry of the orifice structure, which varies between 0.6 and 0.8; and A is the area of the orifice.

Example 8.6

A small gated pipe outlet in an irrigation channel has a circular orifice with a diameter of 12 cm. The water level upstream is 0.60 m above the center of the orifice, and the downstream side is open to the atmosphere, so the orifice flows freely. Assume a discharge coefficient $C_d = 0.62$, and a velocity coefficient $C_v = 1.0$. Calculate the flow rate.

Solution:

Given the diameter of orifice $d = 12\text{ cm} = 0.12\text{ m}$, the head above orifice center $h_u = 0.60\text{ m}$,

Calculate the orifice area: $A = \frac{\pi d^2}{4} = \frac{\pi(0.12)^2}{4} \approx 0.0113\ \text{m}^2$

$$Q = C_d C_v A\left(2gh_u\right)^{\frac{1}{2}} = 0.62 \times (1.0) \times (0.0113) \times \sqrt{2 \times (9.81) \times (0.60)}$$

$$\approx 0.0240\ \text{m}^3/\text{s or } 24.0\ \text{L/s}$$

Example 8.7

A circular orifice in a vertical sluice gate controls the flow between two sections of an irrigation canal. The orifice has a diameter of 30 cm and is fully submerged on both sides. The water level upstream is 1.20 m above the center of the orifice, while the water level downstream is 0.90 m above the center. Assume a discharge coefficient $C_d = 0.63$, and a velocity coefficient $C_v = 1.0$. Calculate the flow rate.

Solution:

Given the diameter of the orifice $d = 30\ \text{cm} = 0.30\ \text{m}$, upstream head $h_u = 1.20\ \text{m}$, and downstream head $h_d = 0.90\ \text{m}$,

Calculate the orifice area: $A = \frac{\pi d^2}{4} = \frac{\pi(0.30)^2}{4} \approx 0.0707\ \text{m}^2$

$$Q = C_d C_v A\left[2g\left(h_u - h_d\right)\right]^{\frac{1}{2}} = 0.63 \times (1.0) \times (0.0707)\sqrt{2 \times (9.81) \times (1.20 - 0.90)}$$

$$\approx 0.1081\ \text{m}^3/\text{s or } 108.1\ \text{L/s}$$

8.4 MEASURING WATER WITH CURRENT METERS

There are several portable propeller meters that can be placed directly into the flow to measure water velocity. The meters measure the average and maximum speed of the current. The flow meters use different expandable rods to place the meter at the point where the velocity needs to be measured. The cross-sectional area of the flow stream should be measured and calculated by measuring the water depth at several points across the channel. The average velocity measured with the current meter times the cross-sectional area equals the flow rate.

The maximum velocity (v_{max}) tends to be located at the surface (about 0.05–0.25 times the depth) and the minimum velocity (v_{min}) at the edges and the bottom of the channel, due to the friction of the water particles with the walls of channel (Figure 8.15). In open channel flow, the velocity distribution typically follows a logarithmic distribution. The velocity at 0.6 times the depth is equal to the average velocity (v_{med}) (Figure 8.15a–c). The velocity is more variable as the shape of channel is more irregular.

Considering that the velocity varies across the transversal section and along the depth, the section should be divided into vertical strips and the velocity measured in each strip, repeating the measurements at least three times. There are three methods to measure the average speed in each vertical:

i. By averaging speeds obtained at 0.2 times (0.2d) and 0.8 times (0.8d) the depth, a very accurate estimation can be obtained. This method is recommended when the velocity is very variable or when the depth of water is greater than 1 m.

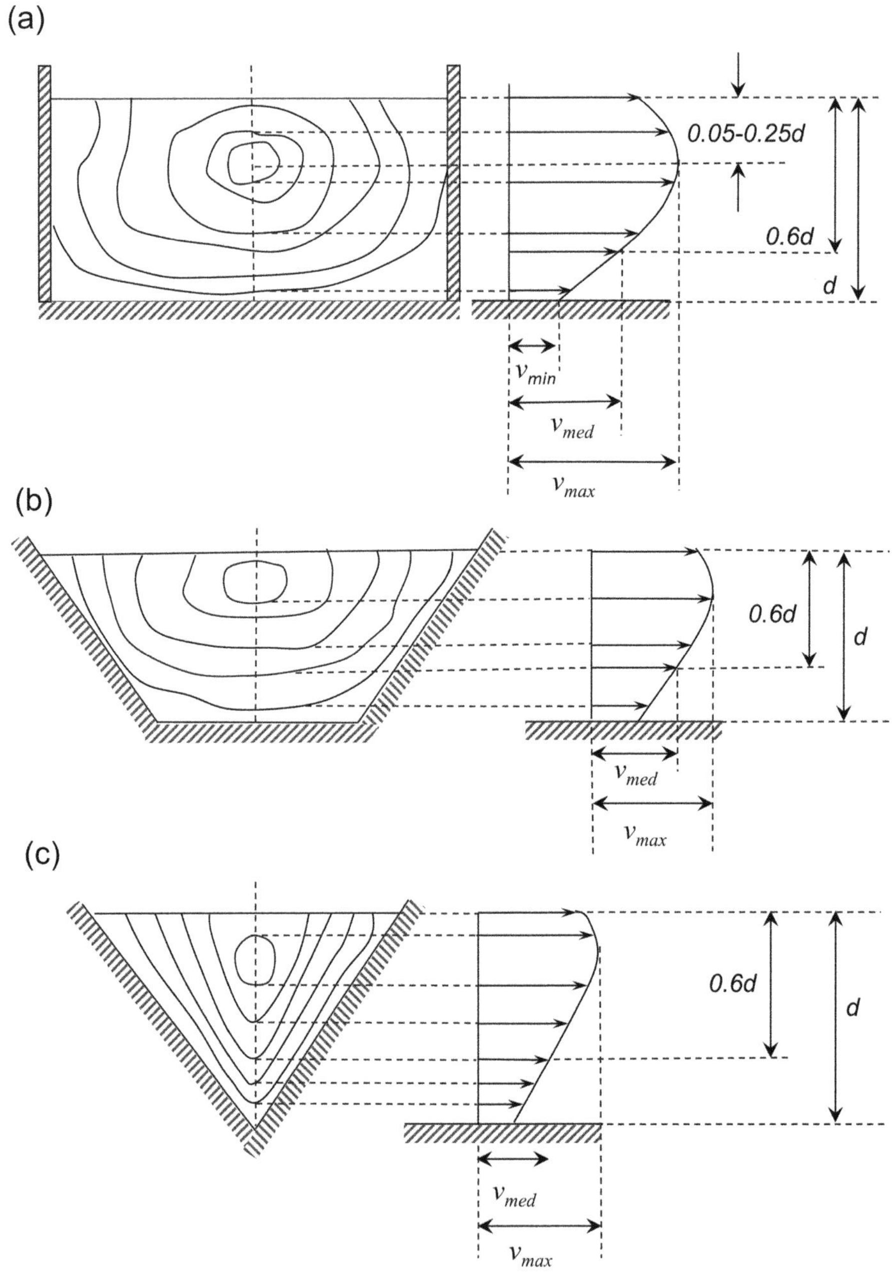

FIGURE 8.15 Water velocity on the horizontal and vertical cross-section of the stream.

ii. When the water depth is less than 1 m and there is little variation in the vertical velocity, it is recommended to measure the velocity at 0.6 times the water depth (0.6d).

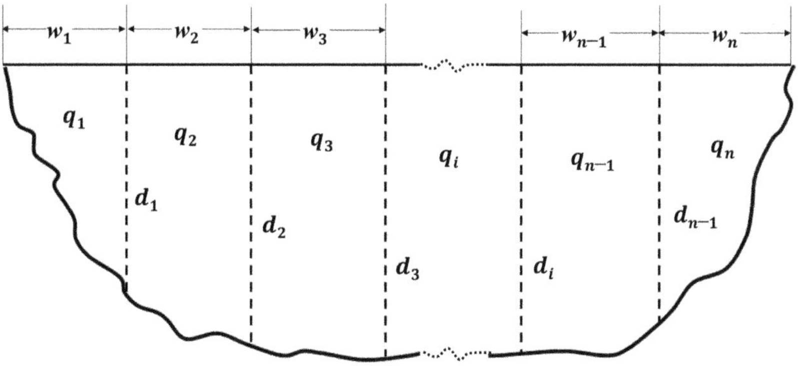

FIGURE 8.16 Division of transverse section into strips for flow measurements.

iii. For small ditches with transversal areas of less than 0.8 m^2, and when the velocity doesn't vary a lot, the surface velocity can be multiplied by a coefficient that varies from 0.85 to 0.95.

Then, the transversal section is divided into strips to have more accuracy in area and velocity measurement (Figure 8.16).

The flow rate in each strip (q_i) can be calculated by multiplying by the velocity of each strip:

$$q = A * v \tag{8.6}$$

The total flow rate (Q_T) is simply the sum of all the partial flow rates:

$$Q_T = q_1 + q_2 + q_3 + \cdots + q_n \tag{8.7}$$

Example 8.8

An irrigation canal is 2.4 m wide, with a relatively regular cross-section and a maximum depth of 0.9 m. To estimate the flow, the cross-section is divided into three vertical strips of equal width (0.8 m each). A portable current meter is used to measure velocity at 0.6 times the depth in each strip (since the depth is less than 1 m and velocity variation is low).

The following measurements are obtained:

Strip	Width (m)	Depth (m)	Velocity at 0.6d (m/s)
1	0.80	0.70	0.25
2	0.80	0.90	0.33
3	0.80	0.80	0.30

Calculate: (1) The cross-sectional area of each strip; (2) the flow rate (q) for each strip using; and (3) the total flow rate (Q_T).

Solution:

The area and partial flow rate for each strip are calculated using Equation (8.6):

Strip	Area, A_i (m^2)	Velocity, v_i (m/s)	Partial Flow Rates (m^3/s)
1	$0.80 \times 0.70 = 0.56$	0.25	$q_1 = A_1 \cdot v_1 = 0.56 \times 0.25 = 0.1400$
2	$0.80 \times 0.90 = 0.72$	0.33	$q_2 = A_2 \cdot v_2 = 0.72 \times 0.33 = 0.2376$
3	$0.80 \times 0.80 = 0.64$	0.30	$q_3 = A_3 \cdot v_3 = 0.64 \times 0.30 = 0.1920$

The total flow rate:

$$Q_T = q_1 + q_2 + q_3 = 0.1400 + 0.2376 + 0.1920 = 0.5696 \text{ m}^3/\text{s or } 569.6 \text{ L/s}$$

A simple floating device can be used in small water ditches to estimate flow when flow rate is less than 100 L/s. The float is an object that is light, and it can be a sponge ball or a piece of wood. The velocity is estimated by measuring the time of advance of the float in a given distance, making sure that wind doesn't affect the movement of the float.

If the area of small streams has the shape of a parabola, the area can be estimated as follows:

$$A = \frac{2}{3} wH \tag{8.8}$$

where A is the transversal area of the section, m^2; w is the width of the water flow, m; and H is the maximum flow depth, m.

Example 8.9

A small irrigation ditch has a parabolic cross-section, with a top width of 1.2 m and a maximum depth of 0.30 m. To estimate the flow, a sponge ball is floated over a 10-m distance, and an average time recorded for the float to travel this length is 25 seconds. Based on field experience, a correction factor of 0.85 is used to convert surface velocity to average velocity for this type of ditch. Calculate the flow rate in this ditch.

Solution:

Calculate the surface velocity:

$$v_{\text{surface}} = \frac{\text{Distance}}{\text{Time}} = \frac{10 \text{ m}}{25 \text{ seconds}} = 0.40 \text{ m/s}$$

The average velocity: $v_{\text{average}} = 0.85 v_{\text{surface}} = 0.85 \times 0.40 = 0.34$ m/s

The cross-section area: $A = \frac{2}{3} wH = \frac{2}{3} \times (1.2) \times (0.30) = 0.24 \text{ m}^2$

Flow rate: $Q = Av = 0.24 \times 0.34 = 0.0816 \text{ m}^3/\text{s}$ or 81.6 L/s

8.5 MEASURING AGRICULTURAL RUNOFF

Agricultural runoff measurement is an important aspect of water management in farming. Accurate measurement helps assess water use efficiency, nutrient losses, and environmental impacts. Runoff can be monitored with portable, temporary or permanent installations, depending on the purpose and site conditions. For temporary or short-term measurements, pre-calibrated structures made of lightweight materials such as aluminum, PVC, or plastic are commonly used. These structures are easy to install, reposition, and remove, making them ideal for research plots or seasonal monitoring efforts. Among portable measurement devices, flumes like the Palmer-Bowlus, H-flumes, trapezoidal, or circular designs are prevalent. These are advantageous especially due to their adaptability to the shape of ditch and their ability to operate efficiently under a high degree of submergence. Additionally, they are designed to allow the passage of trash, sediment, and vegetation, without frequent clogging. Such features reduce maintenance needs and increase data reliability in real-world agricultural environments.

In most field settings, irrigation return flows, the water that drains off the field after irrigation, are collected at the lowest elevation point, typically into a drainage ditch. In these areas, the available water head is often very small, making traditional measurement devices impractical. In such low head situations, a circular flume made of PVC pipe, as shown in Figure 8.17, is a practical solution. These flumes can be

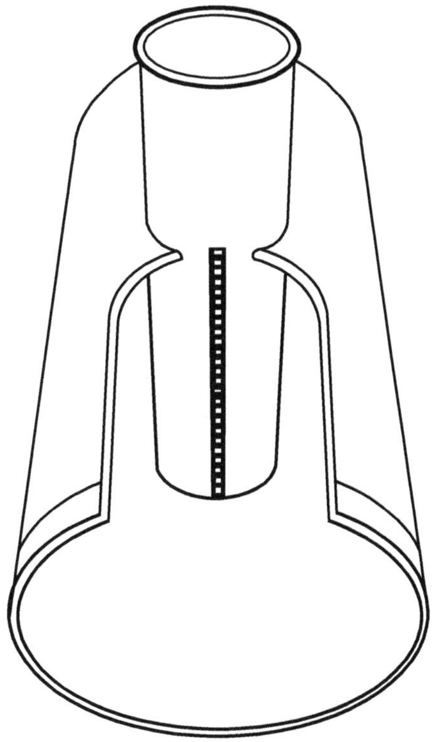

FIGURE 8.17 Circular flume. (Adapted from Samani and Herrera, 1998.)

equipped with a pressure transducer and an internal data logger. This setup allows for the accurate measurement of flow rates even in conditions where traditional methods might not be feasible due to low water pressure.

Example 8.10

A circular flume made from PVC pipe is installed at the outlet of an irrigated field to measure return flow. A pressure transducer is mounted inside the flume and records a water depth of 0.06 m during a runoff event of 30 minutes. The flume was pre-calibrated, and the head-discharge relationship is known to follow the equation: $Q = kH^n$, where Q is the flow rate (m^3/s), H is the measured water depth (m), $k = 0.48$ is a flume-specific coefficient, and $n = 1.6$ is an empirically derived exponent based on the flume geometry. Calculate the runoff flow rate and total runoff volume.

Solution:

The flow rate: $Q = kH^n = 0.48 \times (0.06)^{1.6} \approx 0.0053\ \text{m}^3/\text{s or } 5.3\ \text{L/s}$

The total runoff volume for 30 minutes: $0.0053\ \dfrac{\text{m}^3}{\text{s}} \times 30\ \text{minutes} \times \dfrac{60\ \text{seconds}}{1\ \text{minute}} = 9.54\ \text{m}^3$

REFERENCES

Clemmens, A.J., Bos, M. G., and Replogle, J.A. (2001). *Water Measurement with Flumes and Weirs*. Publication 58. International Institute for Land Reclamation and Improvement/ILRI, Wageningen, The Netherlands, 382 p.

Samani, Z. and Herrera, E. (1998). A Low Cost Water Measuring Device. College of Agriculture and Home Economics New Mexico State University, Guide M-226. https://share.google/5naQ2UBQ4NsqtXZTL

Wahl, T.L., Clemmens, A.J., Replogle, J.A., and Bos, M.G. (2000). WinFlume-Windows-based software for the design of long-throated measuring flumes. *The 4th Decennial National Irrigation Symposium*, Phoenix, AZ.

9 Drainage

NOTATION

K	hydraulic conductivity of the soil (cm/day)
DC	drainage coefficient (in./day, mm/day, or cm/day)
P	long-term average rainfall (mm) during the period from one month before planting to the end of the growing season
T	hydraulic transmissivity (cm^2/day)
D_p	drain depth (m or cm)
Q	design discharge, flow rate (ft^3/s)
C	appropriate drainage curve based on land use and regions
M	drainage area (mi^2)
A	cross-sectional area (L^2)
v	flow velocity (ft/s or m/s)
R	mean hydraulic radius (ft)
s	hydraulic gradient or energy loss per foot of length or channel slope (ft/ft)
n	coefficient of roughness
z	side slope of channel (ft/ft, or m/m)
b	bottom width of channel (ft)
d	depth of flow (ft)
P (Example 9.3)	wetted perimeter (ft)
i	hydraulic gradient
Δh	effective vertical depth (m)
S_d	spacing between parallel drains (L) (m)
L_d	length of the drain (L) (m)

9.1 INTRODUCTION

Drainage problems usually develop as a consequence of improper irrigation. Historical evidence attesting to the occurrence of such problems can be found on every continent of the world. A major contribution to the decline and disappearance of some ancient civilizations can be attributed to their failure to heed to the drainage hazard. Efficient drainage systems must be provided on all irrigation projects when natural conditions are inadequate to remove water in excess of crop needs. Of all the water that is brought into or originates within an irrigation system, some is consumed by crops and vegetation, while nature aids in the disposal of a portion through evaporation and deep percolation. The remainder of the water must be cared for in some other manner.

DOI: 10.1201/9781003747444-11

Surplus water may include waste from irrigated farms, surface runoff from snow and rainfall, seepage and leakage from irrigation canals and distribution systems, artesian water, and percolation from farm irrigation systems. The timely performance of preventive and regular maintenance on drainage systems is absolutely necessary if the systems are to function as designed and built. Drainage systems should be thoroughly examined periodically to determine if they are functioning properly and to identify any necessary maintenance.

Very likely, planners and designers of irrigation systems anticipate and include provisions for addressing many of the drainage problems. When these problems have not been foreseen by these groups, have not been adequately addressed, or when unforeseen problems arise, then operation and management personnel may be called upon to design and construct drainage facilities. In any event, those who manage the irrigation system are responsible for the operation, maintenance, and improvement of all such facilities.

9.2 DRAINAGE FOR CONTROL OF GROUNDWATER LEVELS AND SOIL SALINITY

Prior to the development of irrigation systems, the soil, underlying geologic formations, topography, natural vegetation, and regional precipitation patterns exist in a state of equilibrium. This natural balance supports the existing plant cover and hydrological conditions. However, when land is cleared, leveled, and repeatedly irrigated for agricultural production, this equilibrium is often disrupted. Such alterations to the landscape frequently lead to rising groundwater levels and the accumulation of salts in the root zone—conditions known as waterlogging and soil salinization. These issues typically occur together and, if not addressed, can severely reduce soil productivity. To maintain soil health and ensure long-term agricultural viability, subsurface drainage systems must be implemented to manage excess water and control salinity. However, effective drainage design is inherently complex, influenced by soil characteristics, topographic variations, geological conditions, and climate. Ultimately, drainage should be viewed as an essential component of land management, aimed at preserving the sustainability of irrigated agriculture.

Irrigation districts are primarily concerned with the drainage of irrigated lands, and sometimes with lands that are not irrigated but require drainage due to nearby irrigation. The management's principal objective should be to predict and prevent waterlogging and to keep salts and alkalis away from crops. It should try to avoid the necessity of having to reclaim waterlogged lands by not allowing them to become that way in the first place. Management personnel must be familiar with the design, construction, operation, and maintenance of subsurface drains.

The solution of drainage and salinity problems may require the installation of several types of drain devices, such as (1) open drains, (2) covered drains, (3) wells, and (4) sumps. Drains are an important and necessary part of nearly all irrigation enterprises, not only to remove excess surface and subsurface water but to control soil salinity.

Like canals, drains must be properly maintained and kept at grade; otherwise, they will fail in their intended purpose. Drains must be kept clean and free of silt and vegetation. Where erosion is too rapid, control structures of concrete or rock and rock riprapping may be necessary. Proper maintenance from the beginning helps prevent a drain system from getting out of control. Closed drains are susceptible to clogging from silt deposits, root and vegetation growth, and the presence of water-oriented animals. Regular inspection, accurate record keeping, and prompt maintenance, even at the first indication of distress, are essential for the successful operation of drainage systems.

9.3 BASIC DEFINITIONS

Some basic definitions used in this chapter, as shown in Figure 9.1, include the following:

Ditch: An open waterway excavated in the earth to collect and/or convey drainage water is a ditch, which may carry flow from both surface and subsurface drainage.

Channel: This is that part of a stream or ditch where the flow of water is carried.

Drain: A drain is a conduit, such as tile, pipe, or tubing, installed beneath the ground surface, which collects and/or conveys drainage water.

The outlet: The outlet is the terminal point of the drainage system, ditch, or drain under consideration.

9.4 TYPES OF DRAINAGE

While many areas benefit from natural drainage, where excess water flows from fields to swamps, lakes, or rivers, this is often not sufficient for agricultural needs. Natural drainage systems can be limited in capacity and may not effectively remove

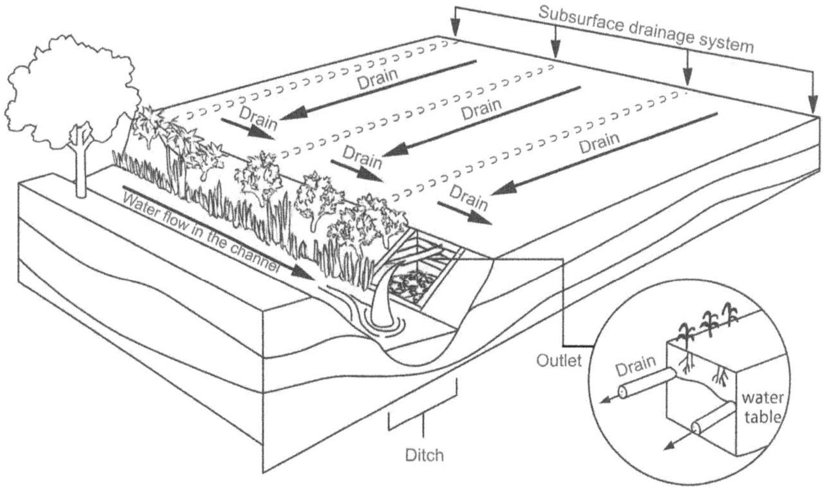

FIGURE 9.1 Schematic diagram of a subsurface drainage system in agricultural fields.

water during periods of heavy rainfall or from certain soil types. This inadequacy necessitates the implementation of artificial or man-made drainage systems. Artificial drainage is designed to supplement or replace natural drainage, ensuring better control of water levels in agricultural fields. The main types of drains are surface and subsurface.

9.4.1 Surface Drainage

Surface drains (also referred to as open drains) are typically in the form of a ditch. They remove excess surface water from agricultural land and operate under low gravity conditions. When deep enough, the ditch can provide relief to adjacent areas. Surface drainage can be achieved by constructing new ditches, improving natural channels, or modifying the land's contours. Surface drainage is also used as an outlet (disposal and collection) of subsurface drainage. Open ditches have a low initial cost and are easy to inspect; however, they reduce the cropping area, require a right-of-way, and have a high maintenance cost.

9.4.2 Subsurface Drainage

Subsurface drains, also known as closed drains, are installed underground to remove excess groundwater below the ground surface. Quite often, subsurface drainage systems are referred to as tile drains. In the past, perforated clay tile and concrete pipe sections were used to facilitate the drainage of agricultural lands. Today, perforated corrugated polyethylene pipe is used in place of tiles (Figure 9.2). To avoid drainage clogging by silt and sand, and to increase water flow through the pipe, laterals are

FIGURE 9.2 Installation of subsurface drains.

surrounded by a nylon envelope or "sock". Subsurface drainage should be planned to complement open drainage surface removal.

9.5 FUNCTION OF DRAINS

From a functional perspective, both surface and subsurface drainage can be categorized into two classes: relief drainage and interception drainage. Relief drains are used when the water table is consistently close to the ground surface and the area is static and flat. Interception drains are designed to intercept and reduce the flow to the problem area. When planning a subsurface drainage system, the designer must evaluate the site conditions and determine whether to use a relief or interception type drainage system.

9.5.1 Relief Drains

Relief drains for subsurface drainage are typically designed to lower a high water table using a network of polyethylene pipe laterals. These laterals are installed below the ground and used to drain the field by gravity. At the lower part of the field, the laterals are connected to a collector drain (Figure 9.3). The collector drain receives the flow from all of the laterals and generally discharges into an open drainage ditch. If the outlet point is at a lower elevation than the water level in the drainage ditch, it will

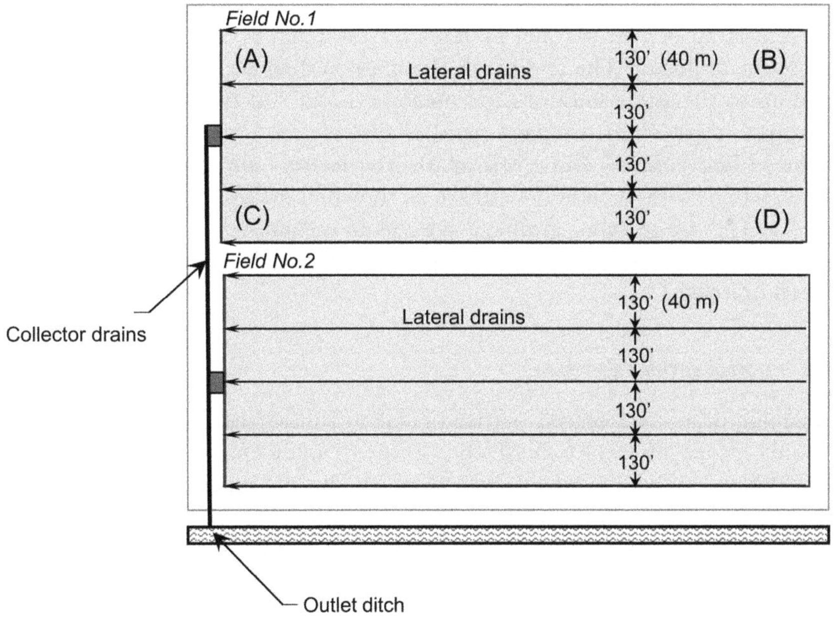

FIGURE 9.3 Typical design layout of a subsurface drain system that shows the spacing, size, and grade of laterals and collectors and the outlet of ground waters.

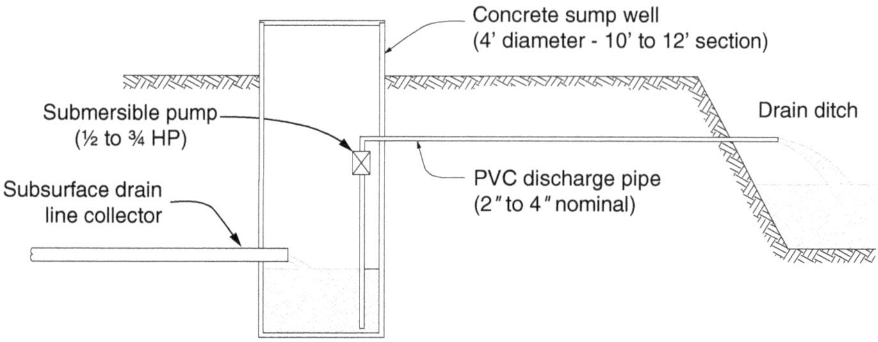

FIGURE 9.4 Typical sump well schematic showing the transfer of groundwater into a drainage ditch.

be necessary to install a sump well to temporarily hold the groundwater and pump the water to the drainage ditch (Figure 9.4). The intention is to maintain the groundwater at a level below the root zone for a given crop. Sometimes, low-lying points are located in isolated areas, and the cost of using electricity to run the pumps is high for permanent pumping of small drainage quantities. Therefore, in some cases, it is recommended to use solar or wind pumps to pump these small quantities.

A minimum drain installation depth of 5 ft is specified by the Natural Resources Conservation Service. The most common relief drain system consists of parallel lateral drains located perpendicular to the main drain (Figure 9.3). The drain's arrangement can vary according to the site location. The arrangement can be random, consisting of two parallel systems, or the laterals can be connected at an angle to the collector. The laterals in the main system are spaced at any interval according to the site conditions, permeability, and soil type. Depending on the soil texture, most relief drainage parallel systems are composed of laterals that are spaced between 100 and 150 ft apart. The laterals are installed at a grade of between 0.025 ft/100 ft and 0.1 ft/100 ft, as shown in the example of Figure 9.3 and Table 9.1. The use of relief drainage systems in conjunction with other best management practices, such as land leveling, can improve the overall effectiveness of artificial drainage.

9.5.2 Interception Drains

Interception drains are placed perpendicular to subsurface flows to intercept and reduce the propagation of excessively wet areas. Interception drains are installed at a right angle to the flow of subsurface water. This placement is strategic for capturing water moving through the soil via subsurface flow, which can be caused by natural underground currents or seepage from irrigation systems. Quite often, interceptor drain lines are installed on agricultural land alongside earthen irrigation canals that have high seepage potential. In this condition, an open drain can be used to intercept excess water from the leaky canal. In the case of closed conduct drains, the depth of the interceptor line varies in relation to the depth of the

TABLE 9.1
Existing Ground Elevations, Proposed Elevations of Subsurface Lines, and Depth of Cover for Field Locations Shown in Figure 9.3 (Field No.1)

Location	Ground Elevation ft (m)	Proposed Elevation ft (m)	Depth ft (m)
A	99.70 (30.39)	93.34 (28.45)	6.36 (1.94)
B	99.05 (30.19)	94.05 (28.67)	5.00 (1.52)
C	98.60 (30.05)	93.08 (28.37)	5.52 (1.68)
D	99.30 (30.27)	93.78 (28.58)	5.52 (1.68)

water table. For closed drains, the depth is an important factor determined by the level of the water table. The interceptor drain lines must be sufficiently deep to effectively lower the water table or intercept subsurface flow, but not so deep that they become ineffective or too costly.

9.6 MONITORING WATER TABLES

It is important to monitor the water tables before installing a subsurface drain system to determine if drainage is needed or to evaluate the performance of an existing drainage system. An observation well can help study the fluctuation of water tables and monitor salinity in the water during the growing season. Observation wells consist of open auger holes drilled at various locations into which a perforated PVC pipe (approximately 1–2 in. in diameter) is pushed. The PVC pipe is commonly referred to as a piezometer. Piezometers help in studying the fluctuations of the water table over time. By monitoring these fluctuations, especially during the growing season, farmers can gain a better understanding of the dynamics of groundwater in relation to their crop water requirements and the efficiency of their drainage systems. Figure 9.5 shows the installation of a piezometer. Topographic maps and soil surveys are also useful while monitoring the water table. Understanding the field topography is essential in determining where water is likely to accumulate. Low areas may require more intensive drainage, and the placement of piezometers in these areas can provide targeted data for managing these challenges.

9.7 DRAINAGE DESIGN CONSIDERATIONS

A drainage system should be designed to remove excess gravitational water and lower the water table sufficiently far from the ground surface, so that it does not interfere with plant growth. The designer should consider the desired depth to which the water table should be lowered (Figure 9.6) and also take into account the rainfall received and the irrigation that will be applied. During the design process, the depth and spacing of the relief and collector lines are selected. Other design considerations are the maximum length of laterals, the material and diameter of the pipe, and the slope grade at which the lines should be installed. The following soil characteristics

FIGURE 9.5 Installing a piezometer (2-in. PVC tube) up to 9 ft deep with a 2-in. auger to monitor fluctuations of the water table level.

are important for the design: (1) permeability and hydraulic conductivity, and (2) drainage coefficient or water relief outflow rate.

9.7.1 Permeability and Hydraulic Conductivity

Permeability is a property that expresses the capacity of soil to transmit water. The ease to transmit water can be low, moderate, or high, as shown in Table 9.2. The hydraulic conductivity (*K*) of the soil is a numerical value of the ability of soil to conduct flow of water, and it represents the velocity of water in the soil and depends on several properties, including pore size, soil structure, and soil chemistry, as well as water properties such as density and viscosity. It is a product of two factors—one is a function of soil called permeability and the other is a function of fluid properties. The sandier soils have higher permeability and higher hydraulic conductivities, as shown in Table 9.2. The conductivity of the soil is important in determining the size of drains and is affected by soil texture.

Example 9.1

A farmer in a humid region wants to install subsurface drainage for a field with silt loam soil. The goal is to achieve good drainage. Using the hydraulic conductivity values and drain spacing guidelines provided in Table 9.2, determine the appropriate lateral drain spacing and drain depth.

Solution:

From Table 9.2, silt loam soil has a relatively low permeability. To achieve good drainage ($K = 0.38$ in./day or 1.0 cm/day), the lateral drain spacing should be maintained at approximately 90 ft (27 m). The drain depth should be set between 3.5 and 4.0 ft (1.1–1.2 m).

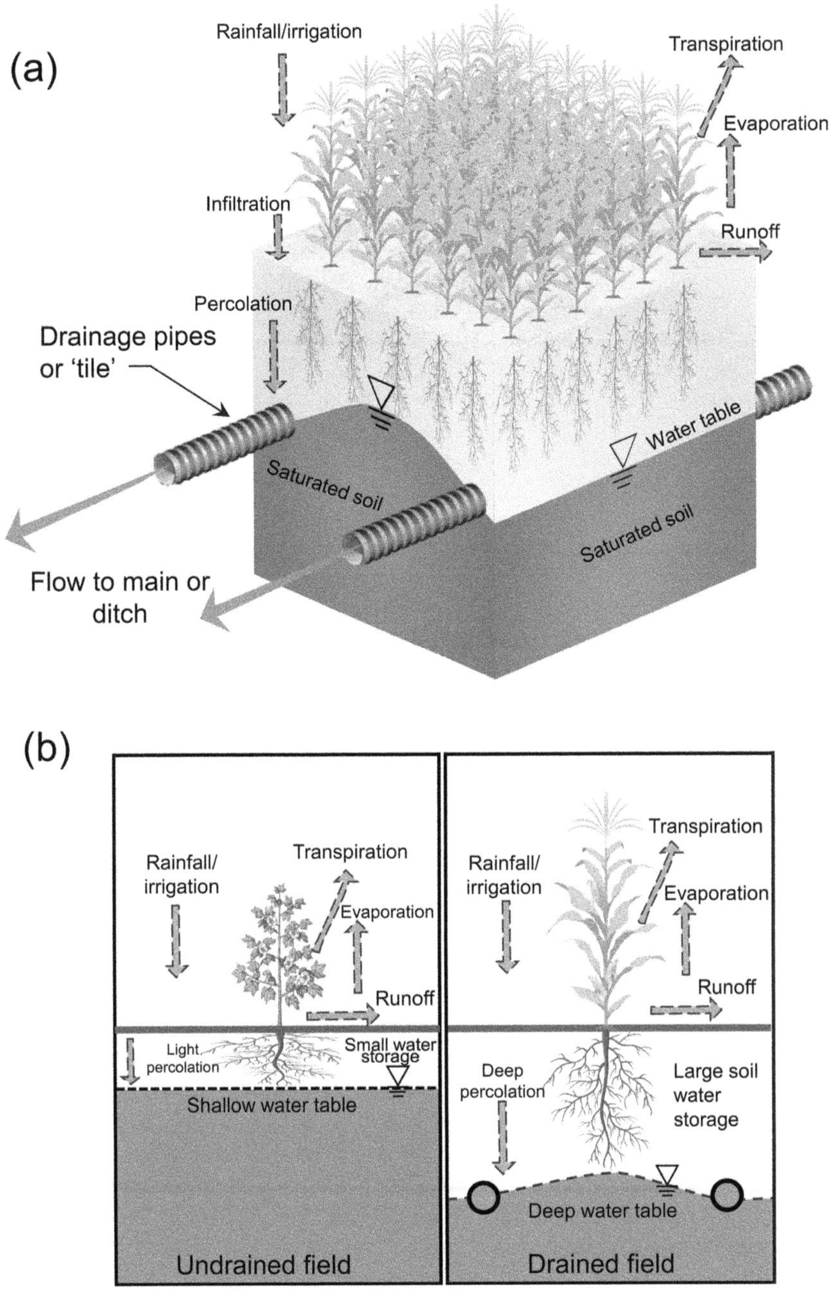

FIGURE 9.6 (a) Illustration of a soil profile with subsurface drainage, and (b) comparison of water table level under drained/undrained conditions with root and plant development and water flux exchanges (water balance).

TABLE 9.2
Examples of Parallel Drain Lateral Spacing and Depths for Different Soil Types in Humid Areas

		Drain Spacing (ft) for			
Soil Type	**Soil Permeability**	**Fair Drainage (Hydraulic Conductivity of 0.25 in./day or 0.6 cm/day)**	**Good Drainage (Hydraulic Conductivity of 0.38 in./day or 1.0 cm/day)**	**Excellent Drainage (Hydraulic Conductivity of 0.50 in./day or 1.3 cm/day)**	**Drain Depth (ft)**
Clay loam	Very low	70 (21 m)	50 (15 m)	35 (11 m)	3.0–3.5 (0.9–1.1 m)
Silt clay loam	Low	95 (29 m)	65 (20 m)	45 (14 m)	3.3-3.5 (1.0–1.1 m)
Silt loam	Moderately low	130 (40 m)	90 (27 m)	60 (18 m)	3.5–4.0 (1.1–1.2 m)
Loam	Moderate	200 (61 m)	150 (46 m)	95 (29 m)	3.8–4.3 (1.2–1.3 m)
Sandy loam	Moderately high	300 (91 m)	210 (64 m)	150 (46 m)	4.0–4.5 (1.2–1.4 m)

Source: Adapted from ASABE EP260.5 Standard (ASABE, 2024)

9.7.2 Drainage Coefficient or Water Relief Outflow Rate

The drainage coefficient (DC) is the rate of water removal to obtain the desired protection of the crop from excess water. It is typically expressed as the vertical depth of water removed per 24 hours, in units of inches per day (in./day) or millimeters per day (mm/day). The selection of an appropriate DC is influenced by several factors, including rainfall frequency, intensity, and duration, soil porosity and hydraulic conductivity, surface drainage efficiency, and crop sensitivity to waterlogging. In humid areas, where excess water primarily comes from precipitation, drainage coefficients commonly range from 3/8 to 3/4 in./day (9.5 to 19.1 mm/day), which depends on soil type, drainage conditions, and crop type. For example, poorly drained areas or sensitive crops may require a higher DC. In arid regions, the source of excess water is typically over-irrigation or seepage from adjacent fields. Therefore, lower DC values are often sufficient, ranging from 0.12 to 0.24 in./day (3.0 to 6.1 mm/day). These values depend heavily on irrigation practices and the design of the irrigation system.

The American Society of Agricultural and Biological Engineers (ASABE) EP260.5 Standard (ASABE, 2024) provides detailed DC values for humid areas, as listed in Table 9.3. A comparative analysis of soil drainage in both humid and arid areas can be found in Table 9.4. If surface drainage is poor, DC should be increased by 15%–20%. In areas where surface water is admitted into subsurface drainage (via inlets), higher DC values are necessary to account for direct runoff entry. The designer selects the drainage coefficient based on the expected deep percolation, the

TABLE 9.3
Drainage Coefficients (DC) for Pipe Drains in Humid Areas (USDA-NRCS, 2001)

Crops and Degrees of Surface Drainage		Drainage Coefficient, in./day (mm/day)	
		Mineral Soil[a]	Organic Soil
Field crops	Surface drainage[b]	$\frac{3}{8}$ to $\frac{1}{2}(10-13)$	$\frac{1}{2}$ to $\frac{3}{4}(13-19)$
	Subsurface drain with open surface inlets	$\frac{1}{2}$ to $1(13-25)$	1 to $1\frac{1}{2}(25-38)$
Truck crops	Surface drainage[b]	$\frac{1}{2}$ to $\frac{3}{4}(13-19)$	$\frac{3}{4}$ to $1\frac{1}{2}(19-38)$
	Subsurface drain with open surface inlets	1 to $1\frac{1}{2}(25-38)$	2 to $4(51-102)$

[a] These values may vary, depending on special soil and crop conditions. Where available, local recommendations should be followed.

[b] No surface water is discharged directly into the drain.

TABLE 9.4
Comparison of Drainage of Soils in Humid and Arid Areas

Description	Humid Areas	Arid Areas
Source of excess water	Precipitation	Excess irrigation and groundwater from surrounding areas
Surface drainage	Surface drainage is required	Surface drainage is usually a part of the irrigation system
Water table control	Water table is controlled so it does not interfere with plant growth	Water table is controlled so it does not interfere with plant growth
Salinity and alkalinity control	—	Water table is controlled to control salinity and alkalinity in the soil and groundwater
Drain depth	0.6–1.5 m	1.5–3.0 m
Drain spacing	10–50 m or even 90 m	50–200 m
Depth to water table from the ground surface	60–75 cm	Deeper than 75 cm. A shallow water table over longer periods of time can cause the accumulation of salts in the root zone

Source: Adapted from USDS-NSCS (2001).

rainfall received, and the irrigation depths applied. The drainage coefficients and the area to be drained (which depends on the length and spacing of drains) are used by the designer to select the diameter of lateral and collector drains. The drainage coefficient can also be estimated from field measurements, including water table drawdown between laterals, drainable porosity of the soil, discharge measurements, and system geometry, such as lateral spacing and length. Where local data is not available, default values provided in Table 9.3 are typically used.

Skaggs (2007) provided an equation for the drainage coefficient (DC, cm/day) based on rainfall (P), drain depth (D_p), and hydraulic transmissivity (T):

$$\mathrm{DC} = 0.004P - 0.0086D_p + 0.0095T - 0.396 \quad (9.1)$$

or, if T is not known:

$$\mathrm{DC} = 0.004P - 0.0086D_p - 0.141 \quad (9.2)$$

where P is the long-term average rainfall (mm) during the period from 1 month before planting to the end of the growing season; D_p is the depth of the drain (cm); and T is the hydraulic transmissivity of the soil profile (cm^2/day).

Example 9.2

A drainage engineer is designing a subsurface drainage system for a 20-ha vegetable (truck crop) farm in a humid region with organic soil and adequate surface drainage. The field receives 600 mm of rainfall from one month before planting through the end of the growing season. The planned drain depth is 100 cm, and the hydraulic transmissivity is estimated at 70 cm^2/day. Estimate the drainage coefficient (DC) using the empirical equation. Compare this value with the design value from Table 9.3. Discuss the implications for drainage system design.

Solution:

Method 1: Given $P = 600$ mm, $D_p = 100$ cm, and $T = 80$ cm^2/day, using Equation (9.1),

$$\begin{aligned}\mathrm{DC} &= 0.004P - 0.0086D_p + 0.0095T - 0.396\\ &= 0.004(600) - 0.0086(100) + 0.0095(70) - 0.396\\ &= 2.40 - 0.86 + 0.665 - 0.396 = 1.809\ \text{cm} = 18.09\ \text{mm/day}\end{aligned}$$

Method 2: From Table 9.3, the DC range for humid areas, truck crops, organic soil, and normal surface drainage is 19–38 mm/day.

The calculated value is lower than the tabular recommendation. In practice, the higher tabular value should be used unless strong field evidence justifies a lower design coefficient. For truck crops (which are often sensitive to waterlogging), a DC of at least 19 mm/day is recommended.

9.8 DESIGN OF OPEN DRAINS

Open drains can consist of lateral and main drains. Lateral drains usually have a "V" shape, and main drains are typically designed with a trapezoidal cross-section for increased capacity and structural stability. Open drains can collect and intercept runoff water from adjacent fields or can lower their water table when designed at an adequate depth. Open drains can also serve as a means of disposing of drainage water, conveying it to a convenient outlet. The design of open drains must consider hydrologic, topographic, and soil characteristics of the site to ensure effective performance under varying climatic conditions. The purpose of this section is to provide a comprehensive overview of the general steps used in the design of drainage systems. More details can be found in USDA-NRCS (2021). The general steps for designing open drains include (1) obtaining site-specific data, including topography, soil borings, existing drainage features, and water table elevations; (2) establishing the hydraulic grade-line, i.e., slope of the water surface in the channel; (3) determining the drainage area or watershed areas contributing to each drain; (4) establishing the layout plan for the main and lateral drains; (5) obtaining the profiles and cross-sections of the drain alignment; and (6) designing the channel capacity using hydraulic principles.

9.8.1 Layout

The layout of an open drain system should begin with the identification of main and lateral ditch locations, using aerial imagery, topographic maps, or digital elevation models (DEMs), supported by on-site visual inspections. Additionally, Global Positioning System (GPS) tools can be used for precise geolocation and mapping. The configuration of the drainage system can be random (adapted to natural drainage patterns) or parallel (common in uniform topography).

Drainage infrastructure is typically owned and managed by irrigation districts or drainage districts, which hold the legal right-of-way for the construction, operation, inspection, and maintenance of the drainage systems. Therefore, the system layout should be designed with accessibility and maintenance in mind. To ensure proper documentation and ease of reference, the NRCS recommends a standardized naming convention for identifying drains (USDA-NRCS, 2021). For example, main drains are numbered (e.g., Main No. I, Main No. II). Laterals are labeled with Roman numerals and capital letters, where the first lateral above the outlet of Main No. I is designated as Lateral No. IA, the second as Lateral No. IB, and so forth. Sub-laterals can be identified using Arabic numerals and lowercase letters (e.g., Lateral No. IA1a). In some cases, local stream names or traditional identifiers may also be used. An example of this naming scheme is illustrated in Figure 9.7.

9.8.2 Profiles and Cross Sections

Accurate profiles and cross sections are essential for both the preliminary design and detailed engineering of open drainage channels. These surveys help define the alignment, slope, and dimensions of the drain, as well as identify topographic

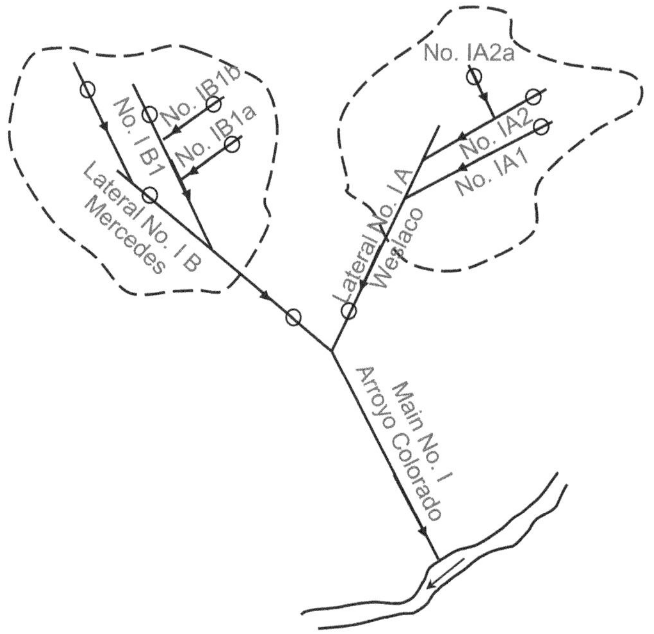

FIGURE 9.7 Example of main and lateral names.

constraints. Profiles are longitudinal elevations taken along the proposed drain path. These should include readings at regular intervals, as well as at critical locations, such as road and railroad crossings, utility lines, topographic breaks, natural depressions, and existing drainage structures. The cross sections can run perpendicular at right angles to the directions of the drainage and at predetermined spacing. The spacing can vary from about 500 ft to 1 mile, and special judgment should be placed to avoid excessive surveying. A complete profile should show the following, including existing ground surface, proposed and existing ditch bottoms, the hydraulic grade line (water surface elevation), right-of-way limits, proposed and existing structures (culverts, flumes, outlets), benchmarks and datum elevations, and seasonal or design water table elevations (if available). Cross-sectional data should be used to calculate flow areas, design slopes, and check for potential obstructions or erosion-prone segments.

9.8.3 Channel Design

The design of an open drainage channel involves determining the required flow capacity and then designing a cross-section (trapezoidal or otherwise) that can convey this flow under stable conditions.

9.8.3.1 Estimating Design Discharge (Q)

The capacity of the drain is influenced by the size of the drainage area, the amount and intensity of precipitation, soil infiltration and natural runoff, and the amount of

FIGURE 9.8 Drainage curves for different regions in the United States. (Adapted from USDA-NRCS, 2021.)

return flows from irrigation or leaching. Due to the variability and complexity of these factors—particularly in large or irregular watersheds—the design of agricultural drainage channels often relies on empirical formulas. One widely used method in the United States is the Cypress Creek Formula, which is recommended by the Natural Resources Conservation Service:

$$Q = CM^{0.83} \tag{9.3}$$

where Q is the design discharge in ft^3/s; C is the appropriate drainage curve based on land use and regions; and M is the drainage area in miles^2.

The constant drainage curve coefficient depends on regional hydrology, land use, and design drainage requirements (Figure 9.8). Representative values from the USDA-NRCS (2021) are shown in Table 9.5.

9.8.3.2 Hydraulic Design Using Manning's Equation

Once the required flow capacity is known, the designer determines the cross-sectional dimensions that can carry the flow under steady, uniform conditions. The required channel dimensions are calculated by solving Manning's equation for a given flow rate (Q), hydraulic gradient (s), and channel roughness. Uniform flow is generally assumed, and the hydraulic gradient is assumed to be equal to the slope of the channel length.

The rate of flow (Q) passing a given cross-sectional area of saturated soil (A) is the product of area and the effective velocity of flow through the section (v):

TABLE 9.5
Recommended Drainage Curve Coefficients (USDA-NRCS, 2021)

Region	Land Use	Drainage Curves (*C*)
Northern humid region	Pasture	25
	Cultivated (grain crops)	37
Southeastern region	Woodland (coastal plain)	10
	Pasture (coastal plain)	30
	Cultivated (coastal plain)	45
	Cultivated (delta)	40
	Riceland	22.5
Red River Valley (MN and ND)	Cultivated	20
Southwestern region	Rangeland	15
	Riceland	22.5
Western region[a]		

[a] The capability of open drains is typically determined by the return flow from irrigation water and leaching requirement. When accounting for surface runoff from rainfall, the Cypress Creek formula can be applied, incorporating a value of *C* as recommended by the relevant state drainage guides.

$$Q = Av \tag{9.4}$$

The Manning equation is usually written as follows:

$$v = \frac{1.486}{n} R^{2/3} s^{1/2} \tag{9.5}$$

where v is the mean velocity of water in ft/s; R is the mean hydraulic radius in ft, which is calculated as the cross-sectional area of the channel divided by its wetted perimeter; and s is the hydraulic gradient or energy loss per foot of length (ft/ft). For open channels with very small slopes, it may also be defined as the slope of the energy gradient, and n is the coefficient of roughness used in the Manning equation.

The roughness coefficient will depend on the material of the channel and the maintenance of the channel. Table 9.6 shows the value of Manning's n for drainage ditch design recommended by the Natural Resources Conservation Service (USDA-NRCS, 2021). These values can be used if no specific site studies are available.

Example 9.3

A drainage engineer is tasked with designing a trapezoidal open ditch to drain surface runoff from a cultivated 640-acre (1.0 miles^2) field located in the Southeastern Coastal Plain region. The slope of the land allows for a channel bed slope of 0.001 ft/ft, while the soil and vegetation conditions suggest using a Manning's roughness coefficient of 0.035. The channel will have a bottom width of 4.0 ft

TABLE 9.6
Values of the Coefficient of Roughness Used in the Manning Equation for Drainage Ditch Design

Hydraulic Radius	*n*
<2.5	0.040–0.045
2.5–4.0	0.035–0.040
4.0–5.0	0.030–0.035
>5.0	0.025–0.030

Source: Adapted from USDA-NRCS (2021).

and side slopes of 2:1 (horizontal:vertical). Estimate the design discharge using the Cypress Creek formula. Determine the required depth of flow (*d*) so that the trapezoidal channel can safely carry the flow using Manning's equation.

Solution:

Step 1: Estimate the design discharge.

Given $M = 1.0$ miles², and $C = 45$ for cultivated land in the Southeastern region (Table 9.5). Using Equation (9.3):

$$Q = CM^{0.83} = (45)\times(1.0)^{0.83} = 45 \text{ ft}^3/\text{s}$$

Step 2: Determine flow depth using Manning's equation.

Given bottom width $b = 4.0$ ft, side slope $z = 2$, channel slope $s = 0.001$, and roughness coefficient $n = 0.035$, try trial-and-error using Manning's formula and solve for depth *d*:

Try trial depth $d = 2.0$ ft:

Flow area: $A = d(b + z\times d) = 2.0\times(4 + 2\times 2.0) = 16.0 \text{ ft}^2$

Wetted perimeter: $P = b + 2d\sqrt{1+z^2} = 4 + 2\times(2.0)\times\sqrt{1+2^2} \approx 12.944 \text{ ft}$

Hydraulic radius: $R = \dfrac{A}{P} = \dfrac{16.0}{12.944} \approx 1.236 \text{ ft}$

Using Manning's equation: $v = \dfrac{1.486}{n} R^{2/3} s^{1/2} = \dfrac{1.486}{0.035}\times(1.236)^{2/3}$

$$\times(0.001)^{1/2} \approx 1.55 \text{ ft/s}$$

Discharge: $Q = Av = 16.0\times 1.55 = 24.8 \text{ ft}^3/\text{s}$ (Too low)

Try trial depth $d = 3.0$ ft:

Flow area: $A = 3.0\times(4 + 2\times 3.0) = 30 \text{ ft}^2$

Wetted perimeter: $P = 4 + 2\times(3.0)\times\sqrt{1+2^2} \approx 17.42 \text{ ft}$

Hydraulic radius: $R = \dfrac{30}{17.42} \approx 1.722 \text{ ft}$

Using Manning's equation: $v = \dfrac{1.486}{0.035}\times(1.722)^{2/3}\times(0.001)^{1/2} \approx 1.93 \text{ ft/s}$

Discharge: $Q = Av = 30\times 1.93 = 57.9 \text{ ft}^3/\text{s}$ (Slightly high)

Try trial depth $d = 2.68$ ft:

Flow area: $A = 2.68 \times (4 + 2 \times 2.68) \approx 25.09 \text{ ft}^2$

Wetted perimeter: $P = 4 + 2 \times (2.68) \times \sqrt{1 + 2^2} \approx 15.98 \text{ ft}$

Hydraulic radius: $R = \frac{25.09}{15.98} \approx 1.570 \text{ ft}$

Using Manning's equation: $v = \frac{1.486}{0.035} \times (1.570)^{2/3} \times (0.001)^{1/2} \approx 1.82 \text{ ft/s}$

Discharge: $Q = Av = 25.09 \times 1.82 \approx 45.7 \text{ ft}^3\text{/s (Acceptable)}$

This depth allows the trapezoidal channel to safely carry 45.7 ft³/s with the given geometry and roughness. To ensure the long-term functionality and resilience of the drainage system, a freeboard of 0.5 ft is recommended for final construction, consistent with guidelines for small farm drainage ditches.

9.9 DESIGN OF CLOSED CONDUCT DRAINS

The purpose of subsurface drainage is to lower the water table to a level that does not interfere with crop root development, aeration, or yield. Proper design ensures that excess water is removed efficiently, particularly during periods of high rainfall or irrigation. The installation depth of lateral drains depends on several factors, including drain spacing, soil texture and permeability, crop rooting depth, the desired drop in the water table, climate conditions, and the required management of salinity and alkalinity. In practice, drains are typically placed at a minimum depth of 2 m in arid areas (with a minimum of 1.5 m at the upper end) and at a depth of 1.5 m in humid areas. Deeper drains are preferred in arid climates to aid in controlling salinity and alkalinity, as these areas typically receive less rainfall, experience higher evaporation rates, and have more saline soils and groundwater.

The spacing between subsurface lateral drains can vary widely, generally ranging from 15 to 50 m, and is affected by soil type, drain depth, hydraulic conductivity, crop type, and desired drainage response time (Table 9.2). In soils with moderate permeability, the drains can be spaced between 30 and 45 m. Soils with low permeability will require closer spacing. A closer spacing reduces the time to drain a certain volume of water but increases the cost of the system. The choice of pipe diameter for interceptor lines may also impact spacing due to limitations in flow capacity.

Subsurface drain design is based on Darcy's Law, which defines the velocity of water flow through soil as

$$v = Ki \tag{9.6}$$

where v is the flow velocity (L/T), K is the saturated hydraulic conductivity of the soil (L/T), and i is the hydraulic gradient (dimensionless).

The hydraulic gradient (i) is the ratio of the effective vertical head (Δh) to the horizontal flow path, which is typically half the drain spacing:

$$i = \frac{\Delta h}{S_d/2} = \frac{2\Delta h}{S_d} \tag{9.7}$$

where Δh is the effective vertical depth, defined as the vertical distance between the bottom of the drain and the elevation of the water table at the midpoint between adjacent drains. This head difference is the driving force for saturated flow through the soil and is used to calculate the hydraulic gradient in Darcy's Law; and S_d is the spacing between parallel drains (L).

Combining Equation (9.4) with Equation (9.6), the capacity of the drain is

$$Q = AKi \tag{9.8}$$

where Q is the flow rate (L^3/T) and A is the cross-sectional area intersected (L^2).

The area of flow depends on the type of subsurface drainage systems. For relief drains (parallel drains), the area is

$$A = S_d L_d \tag{9.9}$$

For interception drains (cutting across a slope):

$$A = \Delta h L_d \tag{9.10}$$

where S_d is the spacing between laterals (L); L_d is the length of the drain (L); and Δh is the effective vertical depth, i.e., from the bottom of the drain to the water table.

Example 9.4

A subsurface drainage system will be designed for a loamy soil field in a humid region. Drains are spaced at 25 m, with a drain depth of 1.4 m. The saturated hydraulic conductivity of the soil is estimated to be 0.50 m/day, and the desired drop in the water table between drains is 0.4 m. The length of each lateral drain is 200 m. (1) Calculate the hydraulic gradient; (2) estimate the flow rate (Q) for one lateral drain using Darcy's Law; and (3) determine the required minimum pipe capacity for the lateral drain.

Solution:

Given $S_d = 25$ m, $K = 0.50$ m/day, $\Delta h = 0.4$ m, and $L_d = 200$ m,

$$\text{Hydraulic gradient:} \quad i = \frac{2\Delta h}{S_d} = \frac{2 \times (0.4)}{25} = 0.032$$

$$\text{Flow rate:} \quad Q = AKi = S_d L_d Ki = 25 \times (200) \times (0.50) \times (0.032) = 80 \text{ m}^3/\text{day}$$

$$80 \frac{\text{m}^3}{\text{day}} = \frac{80 \times (1{,}000)}{24 \times (3{,}600)} \text{L/s} \approx 0.93 \text{ L/s}$$

The required minimum pipe capacity is 0.93 L/s (14.7 gpm)

9.10 MAINTENANCE OF OPEN DRAINS

The greatest advantages of open drains are the low initial cost and their ability to convey large quantities of water. However, this benefit is often offset by higher long-term maintenance requirements. Open drains are widely used to manage various drainage problems, primarily by intercepting surface and subsurface flows and serving as conveyance pathways for water from fields and subsurface (closed) drainage systems. When used in combination with covered drainage systems, it is essential to maintain the open channel to ensure water levels remain below the outlets of the closed drains. These closed systems cost the landowner between \$100 and \$700/acre to install. If not properly maintained, the effectiveness of subsurface systems can be severely compromised.

Open drains require regular maintenance to function as designed. The frequency and degree of cleaning depend on the climate, amount of rainfall, and the depth at which the groundwater table must be kept below the ground surface. In stable materials, shallow surface drains may only require spot cleaning annually and a complete cleaning approximately every 5 years. In unstable soils, yearly cleaning may be required at the bottom of the drains to maintain depth, particularly if the pipe drain outlet is located in an open channel. In the more stable, deep, open drains, chemicals used periodically will prevent the growth of weeds, willows, and tules. These plants should be removed (e.g., by burning) once treated. Post-storm maintenance is critical, as large storms often introduce debris, such as tumbleweeds, which can clog flow and obstruct structures. In some regions, burrowing animals, such as muskrats and beavers, can damage channel integrity, requiring monitoring and control.

To minimize erosion and ensure structural integrity, all drain banks should be seeded with grass—ideally during construction—and reseeded or leveled after major cleanings. Vegetated banks help stabilize the soil, reduce erosion, and provide access for maintenance. Grassed areas, particularly those above the water line, should be fertilized every other year. Maintenance roads adjacent to drains should be inspected and repaired, particularly in the spring and after significant rainfall events.

Surface water entry into open drains should be facilitated through well-constructed pipe inlets or lined channels. These structures generally require inspection only after major storms or during scheduled cleanings. Unlined bank cuts should be avoided to prevent erosion. Pipes smaller than 18 in. in diameter used as surface inlets must be checked frequently in the spring for blockages.

Grade control structures should be routinely inspected to ensure they are functioning properly and free from damage or blockage. Trash racks and baffles must be kept clear of vegetation and debris. Livestock watering access points should be reinforced with riprap or concrete and fenced to prevent damage to the bank. Fences across drain sections must also be regularly inspected and cleared of obstructions.

Wide-bottomed, shallow floodway channels should be grassed on the bottom and sides as soon as practical if this has not been done in the construction stage. The grass should be clipped to a height of 10–13 cm around early June and fertilized with approximately 20 kg of nitrogen per acre every two years. Grazing should be limited, especially in early spring, to avoid damage to newly established vegetation.

Natural waterways used as drains should be left in their natural condition as much as possible. Localized erosion can be addressed with spot filling using rock or gravel. Sections experiencing sustained flows may need to be lined with rock along the channel bottom. Inlets for excess irrigation or stormwater should be designed using pipe inlets with riprap-protected bases.

9.11 MAINTENANCE OF COVERED DRAINS

Covered (or closed) drains, typically constructed using tiles or perforated pipe, are placed below the ground surface at desired depths and spacings. When properly installed, these systems generally require very little maintenance to keep them operating satisfactorily, except for a short period after construction. However, failures or partial failures can occur due to unstable soil conditions, leading to pipe misalignment, pulled joints, outlet blockage, or sediment accumulation.

All drain outlets should be inspected in spring and after heavy storm events. This ensures that the outlet pipes continue to discharge freely into open drains and that side slope erosion has not displaced the outlet structures. Flap gates, when used, must be checked monthly to prevent backflow. Rodent screens should also be inspected regularly to ensure they remain in place and functional, as small animals can nest in or damage outlet pipes. Fencing is recommended to restrict livestock access to drain outlets.

Other elements of a good closed drain system include manholes, junction bores, and silt traps, which serve the same purpose at all junction points and at each change in vertical or horizontal alignment. These are valuable to operation and management personnel because they provide better inspection, easier and cheaper cleaning and repair, and a means of finding blockages, measuring discharges, and obtaining water quality samples of individual segments of the lines. It is very important that these facilities are kept clean at all times. They should also be kept covered to prevent trash from entering the lines and for safety.

Flow monitoring should be initiated immediately after the construction is complete. Record discharge rates at manholes and outlet points to establish baseline conditions. This can be done by measuring the depth of water in the pipe that discharges into the manhole and by actually measuring discharge at the drain outlet. If the discharge at any of the measuring points drops suddenly, further investigation should be conducted because there is a good possibility that a segment of the drain is partially or completely blocked. Field inspections for surface indicators such as sinkholes, wet spots, or vigorous vegetation growth above drain lines can help identify problem areas.

If a small sinkhole is discovered, it should be backfilled and checked later for any additional settlement. If a large sink hole is found, a good-sized hole should be dug down to the tile. Holes like these often develop over broken tile or joints that have become separated. If a tile is broken, it should be replaced immediately. Joints that have been pulled apart can be repaired satisfactorily by replacing tile butts (broken pieces of tiles) over the joint and backfilling around the joint with coarse gravel.

Wet spots that suddenly appear over tile lines are good indicators that the tile line may be plugged or partially plugged. If the drain is only partially plugged, the plug may be removed by placing a ball somewhat smaller than the pipe in the pipe upstream from the wet spot. Balls have been used very successfully to remove sand and silt from tile lines. Sewer rods can also be used to probe and clear the pipe.

If these measures fail, the plug will have to be located by uncovering and opening up a section of the tile line. Broken pipes, pulled joints, or clogged pipes should be repaired as soon as possible so that the drain system functions as intended. Plugs in older pipe drains are usually caused by tree or plant roots. Copper sulfate injected into the system will usually kill the roots. Then, by using a power cleaning rod and cable operated from the downstream side, the dead roots can be broken off and washed out to the nearest manhole and removed. When manholes are not available, a hole should be excavated to the drain downstream from the plug, and one or more pipe joints should be removed so that the power cleaning rod can be inserted into the pipe. A screen should always be placed over the pipe opening on the downstream side so that water can flow down the drain, but no roots or material falling into the hole will get into the drain.

Periodic checks should be made along the lines to ensure that trees and willows have not started to grow over or near the drains. New growth should be sprayed with appropriate chemicals. If there are trees and willows growing near the drains that cannot be removed, the drain should be treated with copper sulfate. The first treatment should be made in late spring. If the tree and shrub roots are a serious problem, a second treatment should be made in late summer. The copper sulfate will not kill or harm the trees and shrubs, so the treatment will need to be applied annually. Established water quality standards must be followed closely when drains are treated with copper sulfate or other chemicals. Regular inspection, early problem detection, and timely repair are key to the long-term functionality of covered drainage systems. With proper care, these systems can provide decades of reliable performance in managing waterlogged soils and supporting agricultural productivity.

9.12 DRAINAGE WELLS AND PUMPS

Drainage wells are employed to lower and maintain the water table by pumping groundwater from deep wells. This method is particularly useful in areas with high water tables where surface and subsurface drains are impractical or ineffective. When water quality is suitable (low salinity and contaminants), the pumped groundwater may also be reused for irrigation. The pumping action creates a cone of depression in the water table, thereby reducing its elevation. Determining the optimal number and placement of wells requires the installation of test wells and an evaluation of drawdown effects and hydraulic conductivity. Due to the high cost of well installation and operation, this method is typically reserved for high-value cropland where drainage significantly improves productivity.

9.12.1 Drainage Observation Wells

Drainage observation wells are usually part of a subsurface drainage system. Properly installed observation wells require minimal maintenance, but significant or sudden changes in water table levels over a stable period (e.g., 3–4 months) may indicate a clogged or damaged well. The amount of work involved in cleaning the well can vary from pumping out silt and sand to pulling the pipe and installing it in a new hole. The most common need for maintenance results from the pipe in the well-being bent over or pulled out by farm or highway equipment. In order to keep a reliable, complete record of the water table, these wells should be reinstalled and protected by a 4 by 4 painted post. All automatic recorders installed on observation wells require constant maintenance to keep the clock and recorder operating properly.

9.12.2 Sumps

Sumps are used to collect water in low-lying, waterlogged areas and typically reach depths of up to 20 ft (6.1 m). Small pumps are employed to evacuate water from sumps, which can increase operational costs and require routine maintenance. A typical sump well schematic showing the transfer of groundwater into a drainage ditch is shown in Figure 9.4. Sumps are often installed at the outlet of tile drainage systems when gravity discharge is not feasible. Proper sump management ensures the continued efficiency of subsurface drains and prevents waterlogging in flat or poorly drained areas.

9.13 ENVIRONMENTAL CONSIDERATIONS

Artificial drainage systems are designed to remove excess water from the crop root zone, thereby supporting healthy plant growth and maximizing yields. Excess surface water produces runoff, also known as tailwater, while excess subsurface water produces deep percolation, also referred to as seepage. It is common to refer to runoff and deep percolation water as return flow or drainage effluents. The quality of drainage water may be lower than that of the supply water due to the transportation of sediment, as well as, in some cases, nutrients and pesticides. The nutrient loadings from agricultural runoff and irrigation return flow sometimes result in low dissolved oxygen levels in natural streams, which fail to meet the requirements for aquatic life designated by the State of Texas and described in the Texas Water Quality Standards (TAC §§307.1–307.10). An indication of low quality could be an increase in fish kills in natural streams.

Since water is a precious resource, it is important to consider reusing or appropriately managing drainage water. This approach can help mitigate environmental impacts, ensuring that agricultural practices do not detrimentally affect the surrounding ecosystems. Implementing strategies for the efficient reuse or treatment of drainage water can significantly reduce the adverse environmental effects, promoting a more sustainable and responsible agricultural system.

Best management practices (BMPs) should be considered to reduce nutrient, residue, and sediment runoff from agricultural lands. Nutrient management is necessary to avoid overuse of fertilizers and control the placement and timing of fertilizer to reduce the risk of pollution. Pest management is necessary and should be managed responsibly by monitoring thresholds, taking into account beneficial and non-beneficial pests. Other practices include using crop rotation and residue management to prevent the transport of sediment, which can attach nutrients and pesticides. Leaching irrigation depths should be applied, but over-irrigation and waste should be avoided by scheduling irrigation. Where necessary, the following additional practices should also be considered to reduce erosion and runoff, such as land leveling for irrigation, the construction of grade stabilization structures, reduced tillage, and the establishment of filter strips between drainage ditches and irrigated fields. Filter strips are areas of herbaceous vegetation situated between cropland, grazing land, or disturbed land (including forestland) and environmentally sensitive areas. Additionally, drainage practices must not encroach on existing or potential wetlands. Adherence to regulatory guidelines and environmental stewardship ensures that agricultural drainage contributes to long-term productivity without compromising surrounding ecosystems.

REFERENCES

ASABE EP260.5 Standard (2024). Design and Construction of Subsurface Drainage Systems on Agricultural Lands in Humid Areas, FEB2015(R2024).

Skaggs, R.W. (2007). Criteria for calculating drain spacing and depth. *Transactions of the ASABE*, 50, 1657–1662.

USDA-NRCS (2021). Chapter 14 Water management (drainage). In: R. L. Marlow, P. H. Willey, R. D. Wenberg, J. Rice, W. K. Twitty, G. Stroup, E. Ross, V. Backlund, M. Mattinson, W. J. Ochs, R. D. Wenberg, B. G. Bishay, J. E. Ayars, and N. R. Fausey, Eds., *Part 650 Engineering Field Handbook, National Engineering Handbook.* United States Department of Agriculture, Washington, DC, pp. 14.1–14.147.

Section 3

Maintenance of Irrigation Systems

10 Maintenance of Irrigation Systems
General Considerations

10.1 INTRODUCTION

The maintenance of irrigation systems plays an important role in ensuring their effective and sustainable operation. Once the system is constructed and placed in service, ongoing maintenance activities are required to preserve its functionality and prevent deterioration. These activities include regular inspections, minor repairs, cleaning, vegetation control, and structural reinforcements, all of which help minimize operational disruptions and prolong the system's service life.

The maintenance of irrigation systems can be classified into four groups: (1) normal maintenance, (2) catch-up maintenance, (3) essential structural maintenance, and (4) preventive maintenance (Table 10.1). Normal maintenance includes routine, day-to-day tasks, such as debris removal and minor repairs that ensure uninterrupted operation. If neglected, it will result in an accumulation of deferred maintenance needs. Catch-up maintenance refers to the priority-deferred maintenance that needs to be met to improve the irrigation system's performance. It corrects deficiencies in deferred maintenance resulting from unmet normal maintenance needs. Essential structural maintenance refers to the maintenance of regularly scheduled activities vital for maintaining system functionality, such as flow control structures required

TABLE 10.1
Comparison of Maintenance Types

Maintenance Type	Focus	Timing	Trigger	Labor Type	Examples
Normal	Daily minor tasks	Routine	Operational need	In-house	Debris removal, minor fixes
Catch-up	Backlog/ deferred repairs	On-time/ intermittent	Deferred tasks	In-house/ contractor	Major crack repair, sediment removal
Essential structural	Scheduled critical tasks	Regular (planned)	Fixed schedule	Mixed	Gate lubrication, inspections
Preventive	Early failure detection	Conditional	Risk/condition	Skilled/technical	Seepage repair, rust protection

DOI: 10.1201/9781003747444-13

for discharge measurement and gate lubrication and inspections, which, in turn, are crucial for managing the water delivered and that of the entire irrigation system. Preventive maintenance, on the other hand, focuses on identifying and mitigating potential failures before they occur, using condition-based triggers like seepage, corrosion, or structural wear. Together, these maintenance types form a comprehensive strategy for sustaining infrastructure performance, reducing emergency repairs, and optimizing long-term costs.

The need for an organized, well-funded maintenance strategy becomes even more crucial in large-scale irrigation districts or where multiple users depend on shared infrastructure. Without regular maintenance, issues such as canal sedimentation, leakage, vegetation overgrowth, structural failures, and inaccurate flow measurements can severely compromise irrigation efficiency and water equity among users.

Successful examples from the United States, including those in Texas and California, demonstrate that well-maintained irrigation systems make a significant contribution to crop yield, resource conservation, and long-term economic sustainability. Internationally, similar lessons have been learned in countries, such as Australia, Israel, and India, where government agencies and user associations collaborate to maintain and modernize irrigation infrastructure. These cases underscore the value of preventive and participatory maintenance frameworks. Irrigation system maintenance is not an optional activity but a core component of water resources management. It safeguards infrastructure investments, supports agricultural productivity, and ensures equitable water distribution across users and seasons.

A well-structured maintenance program aims to keep an irrigation system fully operational, prolong the life and functionality of its facilities, and achieve these goals in the most cost-effective manner. Maintenance of irrigation systems reduces system downtime, prevents costly repairs, and helps maintain irrigation uniformity. Proper upkeep also enhances energy efficiency and minimizes water losses resulting from leaks, clogs, or mechanical failures. Inadequate maintenance can lead to uneven water application, increased labor costs, and diminished yields.

10.2 MAINTENANCE PROGRAM

10.2.1 Inspection

Routine inspection is the key to a successful irrigation system maintenance program. Regular, systematic evaluations of canals, pipelines, pumps, gates, reservoirs, and drainage structures help detect early signs of wear, erosion, blockages, leakage, vegetation encroachment, or mechanical failures. Early identification and correction of these issues allow operators to prevent major repairs and minimize system downtime.

Inspections should follow a standardized schedule—daily, weekly, seasonal, or annual—depending on the component's vulnerability and usage frequency. For example, pump stations and canal gates require more frequent attention during the irrigation season, while sedimentation levels in reservoirs may be surveyed annually.

In addition to scheduled inspections, event-triggered inspections are essential following major storms, floods, or mechanical breakdowns. These reactive inspections help evaluate damage and guide immediate repair actions.

Effective inspection programs also include detailed documentation. Field observations should be recorded systematically, often using digital tools or GIS-enabled mobile applications to track maintenance needs spatially and temporally. Reports should detail the condition of structures, any detected faults, and recommended actions.

Ultimately, thorough inspections enable preventive maintenance, extend the lifespan of infrastructure, reduce emergency repair costs, and improve water delivery reliability. Training field personnel to recognize early warning signs and encouraging user participation in reporting problems further enhances the irrigation system's effectiveness.

10.2.2 Acquisition of Data and Information

Comprehensive data collection is essential for effective irrigation system maintenance. A complete inventory should be developed that encompasses all physical infrastructure, from headworks and diversion structures to farm turnouts. This inventory should include detailed descriptions of canals, pipelines, reservoirs, pump stations, control gates, culverts, drainage outlets, and all associated equipment and facilities. In addition to physical assets, auxiliary infrastructure, including buildings, machinery, and communication systems, should also be cataloged.

Once the inventory is compiled, detailed technical records must be created for each component. These records should include construction specifications, installation dates, maintenance histories, design capacities, and any modifications over time.

Annual performance reports should document the operational status and efficiency of the system. Tracking this data over time enables operators to identify trends, diagnose problems, and prioritize repairs or upgrades. Where feasible, digital asset management systems or GIS-based platforms should be used to integrate data across spatial and temporal scales.

Moreover, it is essential to maintain a chronological history of system operations and maintenance. This documentation not only supports troubleshooting and audits but also facilitates informed decision-making and long-term planning. Ultimately, data-driven maintenance planning enhances the system's sustainability, ensures efficient water delivery, and reduces the risk of unforeseen breakdowns.

10.2.3 Feedback from Farmers

Farmers are the primary beneficiaries and daily users of irrigation systems; therefore, incorporating their feedback into maintenance planning is both logical and essential. Their insights offer real-time, ground-level observations on system performance, including early signs of problems such as blockages, leaks, low-pressure zones, or malfunctioning components. By establishing open communication channels with water users, system managers can identify issues early, prevent serious failures, and ensure timely maintenance interventions.

Engaging farmers also enhances transparency and trust in system management. Holding regular forums, surveys, or community meetings provides structured opportunities to collect feedback and suggestions. Encouraging farmers to report system anomalies, which is supported by clear reporting procedures, strengthens participatory maintenance approaches. Equally important is farmer education and capacity building. Organizing training sessions on basic irrigation system operation, water conservation practices, and minor troubleshooting empowers users to take preventive measures themselves. This sense of ownership reduces reliance on external technical assistance for minor repairs and increases the long-term sustainability of the system. In successful irrigation schemes across Texas and California, for instance, user associations, such as the Lower Rio Grande Valley Water Districts (TWDB, 2016) and the Central Valley Water User Associations (https://cvpwa.org/financial-affairs-committee/), have been instrumental in system upkeep by reporting maintenance needs, supporting small repairs, and advocating for infrastructure upgrades. Integrating farmer feedback and training into the irrigation maintenance framework promotes responsiveness, cost-effectiveness, and shared responsibility, ultimately improving both system performance and farmer satisfaction.

10.2.4 Normal Maintenance Program

A normal maintenance program, sometimes referred to as routine or day-to-day maintenance, forms the foundation of any irrigation system's operations and upkeep. It encompasses the regular, small-scale tasks required to maintain infrastructure functionality and ensure continuous water delivery. These activities are typically conducted by in-house personnel using standard tools and minimal specialized equipment, and they are performed within the constraints of available personnel, equipment, and financing.

Routine maintenance targets frequently used and easily accessible components, including (1) reservoirs, dams, canals, and drainage networks; (2) irrigation channels, structures, and turnouts; (3) canal linings, embankments, and outlet works; (4) communication and control systems; and (5) bridge crossings, irrigation equipment, and ancillary facilities. Tasks often include debris removal, weed control, clearing sediment from minor inlets, lubricating gate mechanisms, and patching small cracks in concrete. Timely attention to these seemingly minor issues helps prevent deterioration and reduces the likelihood of major failures that would require expensive catch-up or emergency repairs. For example, early identification of seepage beneath a minor diversion structure during a routine check may allow for low-cost sealing or compaction before it escalates into structural damage. Similarly, clearing vegetation from canal banks can prevent root intrusion and erosion that could weaken embankments over time. Figure 10.1 illustrates how issues identified during normal maintenance can escalate into preventive, essential, or catch-up maintenance, ultimately leading to failure if not addressed.

Common challenges in implementing normal maintenance programs include physical limitations (e.g., difficult terrain), labor shortages, equipment constraints, budgetary restrictions, and administrative inefficiencies. Addressing these challenges requires strategic planning, prioritization of tasks, and the efficient allocation of available resources. The importance of vigilant observation cannot be overstated. Historical evidence, from ancient Mesopotamian and Roman irrigation systems to modern examples, illustrates that deferred or neglected maintenance

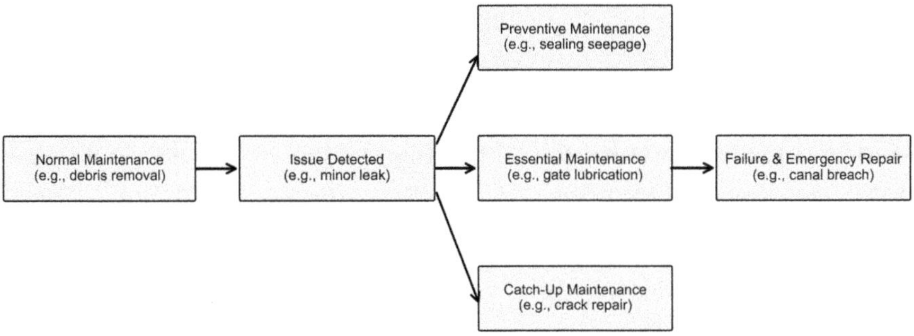

FIGURE 10.1 Maintenance escalation pathway from routine checks to system failure.

often leads to the premature failure of infrastructure that might have otherwise lasted for decades. In contrast, proactive maintenance ensures water users receive reliable service, thereby supporting agricultural productivity and rural livelihoods. For example, in California's Imperial Irrigation District, routine inspections, combined with community-based reporting systems, have led to fewer system failures and lower long-term maintenance costs (https://www.iid.com/customer-service/report-a-problem). A dedicated and systematic maintenance program reinforces irrigation infrastructure resilience and performance, cultivating a culture of regular inspection, prompt intervention, and continuous refinement to safeguard long-term efficiency, equity, and returns.

10.2.5 Catch-up Maintenance Program

A catch-up maintenance program is established to systematically eliminate the backlog of deferred maintenance tasks and restore equipment and infrastructure to their intended performance levels. It begins with a comprehensive inventory of all deferred tasks, each assigned a unique identifier and documented with detailed descriptions of the location, nature of the defect, and criticality. For each task, cost estimates for labor, materials, subcontractor services, and equipment rentals are included (Table 10.2).

To prioritize these tasks, a risk-based matrix is used, scoring each item based on the severity of potential failure and the likelihood of occurrence. Tasks are categorized into three tiers: Tier 1 (Critical), Tier 2 (High Priority), and Tier 3 (Routine). Once prioritized, resource planning outlines the necessary workforce, e.g., internal crews or external contractors, as well as the required tools, spare parts, and consumables. Table 10.2 presents an example from an irrigation district that logged five deferred tasks during April–May 2025. Each task is listed with its ID (CU-01 through CU-05), location, defect description, date logged, risk tier, estimated cost, and required labor hours. The risk matrix identified CU-01 and CU-02 as Tier 1 (Critical), CU-03 and CU-04 as Tier 2 (High Priority), and CU-05 as Tier 3 (Routine). Accordingly, CU-01 and CU-04 are addressed in-house by two masons and laborers. CU-02 is assigned to Contractor A's excavator crew. CU-03 and CU-05 are assigned to Contractor B's mechanic and fitter team. All materials were ordered by June 1,

TABLE 10.2
An Example of Catch-up Maintenance Inventory and Prioritization

Task-ID	Location	Defect Description	Data Logged	Tier	Estimated Cost ($)	Labor (Hour)
CU-01	Main canal (12.5 km)	10 m crack in concrete lining	2025-04-12	Tier 1	5,000	80
CU-02	Branch canal B (3.2 km)	25 m^3 sediment bar obstructing flow	2025-04-15	Tier 1	3,500	60
CU-03	Distributary 3 inlet gate	Corroded spindle and worn sealing gasket	2025-05-01	Tier 2	4,200	70
CU-04	Tertiary inlet C (chainage 45)	Leakage around metal frame	2025-05-10	Tier 2	2,800	50
CU-05	Tertiary inlet D outlet pipe	200 mm pipe wall worn through by abrasion	2025-05-20	Tier 3	1,500	30

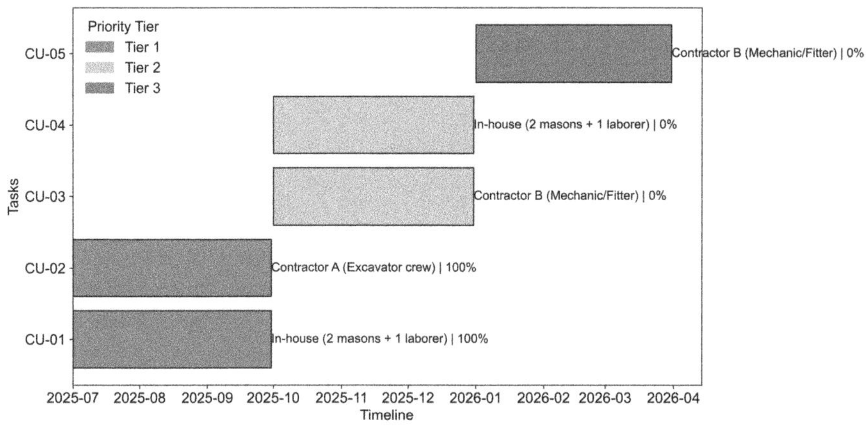

FIGURE 10.2 Catch-up maintenance program Gantt chart.

2025, to support a phased implementation schedule: Tier 1: July–September 2025; Tier 2: October–December 2025; Tier 3: January–March 2026.

Each task is tracked using standardized work orders, which utilize planning tools, such as calendars or Gantt charts to visualize task timelines (Figure 10.2). A monthly backlog reduction report showed that by September 30, 2025, all Tier 1 tasks were completed (100% closure), with actual spending of $8,600, slightly exceeding the estimate of $8,300 (3% over). Tier 2 work was on schedule to begin in October 2025.

10.2.6 Essential Maintenance Program

An essential maintenance program is a structured and time-bound plan that performs recurring maintenance tasks that are critical to the continued performance, reliability, and longevity of irrigation infrastructure. Unlike routine tasks handled daily, essential maintenance activities are scheduled, resource-dependent operations that must occur at regular intervals to prevent functional degradation and reduce the risk of costly repairs.

To ensure the sustained reliability of irrigation infrastructure, a well-structured maintenance program should follow a systematic six-step process. First, a comprehensive inventory of tasks must be compiled. Each task should include a unique ID, description, location, recommended frequency (e.g., quarterly, semi-annual), required labor and materials, cost estimates, and whether the work is performed in-house or contracted. Second, specific responsibilities should be assigned. The management designates the appropriate execution team, such as internal technical staff for routine activities or external contractors for specialized or labor-intensive work. Third, the maintenance schedule should be aligned with operational cycles, such as seasonal cropping patterns, canal downtimes, or low-demand periods, to minimize disruptions to water delivery. Fourth, the plan should be visualized using maintenance calendars or Gantt charts to map task timelines, coordinate resource allocation, and track implementation across time. Fifth, it is essential to document maintenance actions and gather field feedback using standardized inspection forms, logs, and status updates. Finally, the program must include a mechanism to evaluate performance and ensure continuous improvement based on key indicators such as task completion rates, frequency of emergency repairs, and deviations from projected costs.

10.2.7 Preventive Maintenance Program

A preventive maintenance program is designed to identify and address potential issues before they escalate into system failures, structural damage, or costly emergency repairs. Unlike essential maintenance, which follows a routine schedule, preventive maintenance is diagnostic and condition-based, relying on observations, performance indicators, and historical patterns to guide interventions. The goal is to preserve infrastructure integrity, extend service life, and reduce the frequency and severity of corrective actions.

Key elements of a preventive maintenance plan include the following: (1) a detailed inventory of target structures and components (e.g., control gates, embankments, intake screens); (2) description of potential failure mechanisms and associated risk factors (e.g., seepage, sediment buildup, corrosion); (3) recommended inspection methods and frequencies; (4) estimated labor, equipment, and material requirements; (5) documentation protocols for field observations and measured indicators; and (6) a structured timeline and action plan for implementing mitigation or repairs. When properly implemented, preventive maintenance reduces overall lifecycle costs and supports long-term system resilience by targeting small, early-stage issues before they evolve into major failures.

10.2.8 Diagnostic Maintenance Survey

An essential component of any effective maintenance program is the diagnostic maintenance survey, which is a detailed, field-based assessment of the entire irrigation system. This survey involves conducting systematic walk-through inspections of all infrastructure elements, including main canals, branch canals, distributaries, intake and outlet structures, embankments, and access roads. Unlike remote or vehicular assessments, this approach emphasizes on-foot evaluation, ensuring close observation of even subtle or early-stage issues.

During the walk-through, field personnel must document all visible signs of deterioration, regardless of severity. This includes cracks, seepage, sediment buildup, vegetation intrusion, erosion, gate malfunction, and any other maintenance-related concern. Each observation should be georeferenced or associated with chainage and supported by measurements, photographs, or sketches as needed. The collected information forms a foundational inventory of developing issues, critical for early diagnosis, cost estimation, understanding root causes, and designing effective interventions.

By conducting regular diagnostic surveys, irrigation districts can shift from reactive to proactive maintenance, improving long-term infrastructure resilience and operational reliability. These surveys also support evidence-based planning and budgeting, enabling timely and cost-effective resolution of critical issues. Table 10.3 provides an example of a checklist from a diagnostic maintenance survey in an irrigation district. During the post-irrigation season, field staff in the irrigation district conduct a thorough walk-through of the entire system. They record all maintenance issues, ranging from minor surface erosion to critical gate seepage, enabling the district to schedule targeted repairs before the next irrigation cycle.

TABLE 10.3
Sample Field Checklist from a Diagnostic Maintenance Survey

Component	Observed Issue	Location/ Chainage	Severity	Recommended Action	Photo ID
Lined canal	Longitudinal crack (3 m)	Canal A, Ch. 12 + 200	High	Repair with patch lining	IMG_025
Steel gate + concrete frame	Seepage under the frame	Intake Gate B	Medium	Seal joints with epoxy	IMG_029
Gravel road	Erosion and rill formation	Access road (Distributary C)	Low	Grade and compact surface	IMG_041
Outlet structure	Sediment buildup (0.4 m)	Drainage Outfall D	Medium	Dredging recommended	IMG_032
Earth embankment	Gully erosion (2 m wide)	Main canal embankment E	High	Fill and re-compact soil	IMG_034

Note: Ch. 12 + 200 means 12 km + 200 m from a reference starting point.

10.3 TIMELY REPAIR AND REPLACEMENT

Proper maintenance of project facilities requires close and continuous examination of the system, along with timely repair and replacement programs. The ability and efficiency of the personnel assigned to operate and examine the system are of paramount importance. Operators and field personnel must maintain current inventories of infrastructure requiring repair, including detailed descriptions of the problem, its severity, the date of observation, and the recommended action. These inventories should be integrated into the broader maintenance management system and updated regularly through diagnostic surveys (as discussed in Section 10.2.8), walk-throughs, and automated inspections.

A well-designed repair and replacement program should: (1) define repair thresholds and criteria for replacement vs. restoration; (2) include prioritization frameworks to guide decision-making (e.g., risk-based scoring); (3) be incorporated into the annual operation and maintenance budget, with funds allocated for both planned repairs and contingencies; and (4) assign clear roles and responsibilities for execution, quality control, and documentation. In addition, the program must account for the lifespan of each asset, including canals, headworks, gates, pumps, roads, buildings, and other physical infrastructure. As structures age, scheduled replacement cycles should be established based on performance data, material degradation, and environmental exposure. Table 10.4 lists the repair and replacement schedule of major infrastructure components.

TABLE 10.4
Repair and Replacement Schedule and Annual Maintenance Cost of Major Infrastructure Components

Infrastructure Component	Expected Economic Life (Years)	Typical Repair Frequency (Years)	Annual Maintenance as a Percentage of the Initial Investment (%)
	Water Source and Storage		
Dams	50–150	As needed	—
Wells	50–170	Every 10–20	1
Ponds and tanks	20–50	As needed	—
	Water Conveyance		
Main canal lining	15–25	Every 3–5	5
Headworks and control gates	20–30	Annual inspection, repair as needed	1–3
Tertiary inlet structure	15–20	Every 2–3	2–4
	Mechanical and Electrical		
Pump and motors	10–15	Annual	3–6

(Continued)

TABLE 10.4 (*Continued*)
Repair and Replacement Schedule and Annual Maintenance Cost of Major Infrastructure Components

Infrastructure Component	Expected Economic Life (Years)	Typical Repair Frequency (Years)	Annual Maintenance as a Percentage of the Initial Investment (%)
Solar panel/backup generators	10–25	Every 5–10	1–3
	Support Infrastructure		
Inspection/access roads	10–15	Annual (post-irrigation season)	1–2
Steel structures	25–35	Every 5–7	1–2
Maintenance equipment	10–15	Every 2–3	2–5
Safety infrastructure (fencing, signage)	10–30	Every 5–10	1–2
Buildings and ancillary facilities	20–30	Every 5	2–4
	Monitoring and Controls		
Monitoring equipment	5–10	Every 1–2	2–5
Automation and control system (SCADA)	10–20	Every 1–3	3–6
Irrigation management software	5–10	Annual updates	3–5
	Water Application		
Sprinkler systems	10–15	As needed	2–6
Microsystems	1–20	As needed	2–10
Filter station (drip irrigation)	5–10	Every 1–2	2–4
	Drainage and Reuse		
Drainage infrastructure	15–40	Every 10–15	2–4
Water treatment units	10–20	Every 5–10	3–8

Neglecting timely repairs often leads to escalating deterioration, costlier interventions, and service disruptions. Minor cracks left unrepaired may develop into major structural failures; corroded parts may seize during peak demand; and deferred replacements can exacerbate risks to safety and water delivery reliability. In some irrigation governance models, such as those in Japan, routine maintenance of facilities constructed with public funds is delegated to water users (Satoh and Ishii, 2021). However, since major repairs are funded by the government, there is often a financial incentive to postpone minor maintenance until it qualifies as "major". This practice underscores the importance of incentivizing timely and preventive action at all levels of system ownership and operation.

While this discussion primarily focuses on the maintenance of single-purpose irrigation systems, the principles and strategies presented are equally applicable to multipurpose reservoirs or systems involving multiple reservoirs within a shared watershed. In multipurpose systems, the major difference concerns the scale and complexity of operations, while the underlying maintenance concepts remain consistent.

For organizational clarity, the maintenance discussion is structured around key functional elements of the irrigation enterprise. Although some overlap between categories is inevitable, the following thematic arrangement ensures a comprehensive coverage of maintenance considerations: (1) Storage and conveyance systems, including dams, reservoirs, canals, pipelines, gates, and associated control structures; (2) communications and access infrastructure, such as roads, bridges, inspection tracks, and telecommunication systems; (3) buildings and ancillary improvements, including field offices, equipment sheds, fencing, and cattle guards; (4) maintenance equipment and facilities, addressing not only the equipment itself but also its upkeep and storage; (5) weeds management, targeting both terrestrial and aquatic vegetation that may obstruct flow or damage infrastructure; (6) pest control, including rodents, insects, and other biological agents that pose a threat to infrastructure integrity; (7) structural materials, including (a) concrete (e.g., lining, structures), (b) asphalt (e.g., roads, surface), (c) metals (e.g., gates, support), and (d) others (e.g., wood, plastics, composites); and (8) pumps and other special facilities, e.g., maintenance of mechanical and electrical systems critical to water lifting and distribution.

10.4 STORAGE AND CONVEYANCE SYSTEM

The storage and conveyance infrastructure forms the backbone of any irrigation project. Maintaining the full capacity of storage reservoirs is essential. Any loss in storage capacity is unacceptable to the project because the storage facilities are planned and constructed to meet only the irrigation needs, driven by economic considerations. Loss of capacity in storage and re-regulating reservoirs results primarily from three causes: (1) sedimentation (major loss), (2) landslides (less frequent but potentially catastrophic, especially in mountainous terrain), and (3) bank erosion (a gradual process that contributes to basin shallowing and structural degradation).

Sediment accumulation is one of the greatest maintenance challenges faced by storage systems. In many watersheds, particularly in the western United States, streams carry substantial silt loads due to upstream erosion. Unfortunately, no cost-effective method currently exists for large-scale sediment removal from reservoir basins. Consequently, once a reservoir loses a critical percentage of its capacity, continued operation may become uneconomical, leading to abandonment. Until affordable sediment removal solutions are developed, the most effective strategy remains upstream erosion control through the implementation of conservation practices. When a reservoir is located on a silt-carrying stream, periodic surveys must be conducted to determine its capacity, ensuring the accurate measurement of the water actually available in the reservoir. These surveys are referred to as sedimentation surveys (Bureau of Reclamation, 2021). Conducting regular sedimentation surveys is important to understand and quantify storage loss due to various factors, including silt deposition, landslides, and bank erosion.

There are many examples of loss of storage resulting from silt deposits carried into reservoirs from watersheds. Elephant Butte Reservoir, a key reservoir on the

Rio Grande in New Mexico, lost approximately 17% of its storage capacity during the first 32 years of operation (Kelly et al., 2007). The rate of sedimentation lessened considerably later. Sediment surveys on Lake Mead, the storage reservoir, averaged 105,500 acre-feet (AF)/year from 1935 to 1941 (Smith et al., 1960). Due to the closure of Glen Canyon Dam in 1963 on the upstream, the rate of sedimentation greatly decreased. Reservoirs are found throughout the world that have been abandoned as storage facilities because of the accumulation of sediments. Globally, sedimentation has resulted in a 13%–19% reduction in original storage capacity, with future projections indicating even more severe challenges. By 2050, countries such as Seychelles, Japan, Ireland, Panama, and the United Kingdom are expected to lose 35%–50% of their storage capacity (Perera et al., 2023). Landslides can lead to immediate and devastating loss of capacity. The most notable case is Italy's Vajont Dam, where a massive landslide in 1963 displaced nearly the entire reservoir overnight. Bank erosion, as observed at American Falls Reservoir (Idaho), results in a gradual yet significant loss of usable storage and compromises reservoir integrity.

10.5 CASE STUDIES IN IRRIGATION SYSTEM MAINTENANCE

The Imperial Irrigation District (IID), the largest irrigation district in the United States, plays a vital role in delivering water and power across California's Imperial Valley. With a legally protected entitlement to 3.1 million AF of Colorado River water annually, of which approximately 2.6 million AF are delivered in-valley, predominantly to agriculture, IID supports the production of food and fiber for both national and international markets. Ensuring the reliable delivery of this resource requires robust, proactive maintenance strategies that span a vast and aging infrastructure network.

IID owns and operates an expansive infrastructure system that includes 3,000 miles of canals and drains, as well as Imperial Dam. As many of these assets approach or exceed their design life, the district faces growing challenges in maintaining system efficiency, reliability, and safety. In response, IID has launched a comprehensive Asset Management Plan and a 15-year Capital Plan to guide the systematic repair, replacement, and upgrading of its critical infrastructure.

The Asset Management Plan integrates digital tools and structured planning processes. An IT-based inventory has been created, complete with a criticality matrix, lifecycle replacement schedules, and budget forecasts. In support of this effort, IID's Information Technology Department collaborated with infrastructure teams to track and assess all assets, including communication networks and field hardware. The district also invested in state-of-the-art surveying equipment, including drones equipped with lidar, telephoto, and thermal imaging capabilities, as well as a remotely operated submarine device for inspecting submerged infrastructure. These technologies enhance the district's ability to perform remote diagnostics and predictive maintenance, reducing the need for reactive repairs.

Numerous targeted maintenance and infrastructure improvement projects have been completed in recent years (Table 10.5). A major structural rehabilitation effort took place in 2022 at the Gila Gravity Main Canal and Gila Headworks located at Imperial Dam. In partnership with the Bureau of Reclamation's Yuma Area Office

TABLE 10.5
Summary of Key Maintenance Projects in the Imperial Irrigation District

Project Name	Maintenance Focus	Primary Benefits
Dogwood canal relocation	Seepage mitigation	Road safety and future expansion readiness
Baughman drain structure replacement	Bank erosion control	Stabilized flow into the New River
Oak lateral canal lining	Efficiency improvement	Reduced seepage and maintenance costs
Sunbeam drain pipeline installation	Safety near public recreation areas	Enhanced public safety and drainage performance
Gila headworks rehabilitation	Structural gate replacement	Restored operational reliability of Imperial Dam

Imperial Irrigation District (2022–2024).

and technical staff in Denver, the project involved demolishing deteriorated concrete piers and end walls, installing epoxy anchor bars, replacing a failing I-beam trunnion with reinforced concrete corbels, and fabricating and installing three large top seal radial gates. This year-long effort engaged multiple internal departments, including Imperial Dam Construction and Operations, Water Engineering, and Project Management.

IID's maintenance philosophy emphasizes the importance of timely repairs and consistent planning. The district understands that deferring minor issues can lead to major system failures and elevated long-term costs. As a result, the Asset Management Plan includes fiscal and operational safeguards to ensure that necessary maintenance is not delayed due to short-term budget pressures. The results of IID's maintenance planning are significant. Emergency repair incidents have declined, system reliability has improved, and overall customer satisfaction has increased. Furthermore, the district conserves more than 500,000 AF of water annually through both infrastructure modernization and voluntary efforts by growers, which supports the long-term sustainability goals for the Colorado River. Additional conservation commitments are in place through at least 2026.

This case study illustrates how a large, complex irrigation district can effectively manage legacy infrastructure, while adopting modern tools and implementing collaborative strategies. IID's maintenance framework ensures operational resilience and sustainability across one of the most critical water delivery systems in the western United States.

REFERENCES

Bureau of Reclamation (2021). *Guidelines for Developing Reservoir Sedimentation Monitoring Plans*. United States Bureau of Reclamation, Washington, DC.

Imperial Irrigation District (2022–2024). Annual Report. https://www.iid.com/about-iid/mission-vision-statements/annual-reports

Kelly, S., Augusten, I., Mann, J., and Katz, L. (2007). History of the Rio Grande Reservoirs in New Mexico: Legislation and litigation. *Natural Resources Journal*, 47(3), 526–613.

Perera, D., Williams, S., and Smakhtin, V. (2023). Present and future losses of storage in large reservoirs due to sedimentation: A country-wise global assessment. *Sustainability*, 15(1), 219.

Satoh, M. and Ishii, A. (2021). Japanese irrigation management at the crossroads. *Water Alternative*, 14(2), 413–434.

Smith, W.O., Vetter, C.P., and Cummings, G.B. (1960). Comprehensive survey of sedimentation in Lake Mead, 1948–49. Geological Survey Professional Paper 295. United States Department of the Interior, Washington, DC.

TWDB (2016). On farm irrigation strategies for water conservation in the Lower Rio Grande Valley: Final Report, 2016, Agricultural Water Conservation Grant, Texas Water Development Board. https://www.twdb.texas.gov/publications/reports/contracted_reports/doc/1613581997.pdf

11 Operation and Maintenance Equipment

11.1 INTRODUCTION

Maintenance in irrigation systems includes four main components: (1) service, (2) repairs, (3) replacement, and (4) betterment. Service and repairs refer to routine and generally minor actions taken to preserve the functionality and reliability of irrigation equipment. Despite consistent upkeep, equipment will inevitably wear out over time, requiring either replacement or upgrading.

The difference between replacement and betterment should be noted. Replacement refers to replacing old equipment with new one of the same capacity and type, whereas betterment refers to replacing old equipment with new equipment of greater capacity and size. Routine tasks, such as minor repairs and standard replacements, are typically managed by the system's operations team. In contrast, major replacements and betterments require engineering oversight, including detailed design and construction planning, which is often handled by specialized teams or contractors.

The objective of this chapter is to provide a practical overview of how irrigation districts operate and maintain equipment to ensure long-term reliability, serviceability, and cost-efficiency. This chapter examines the decision-making processes involved in selecting, deploying, maintaining, and managing equipment, with real-world examples that illustrate best practices for both small- and large-scale systems.

11.2 FACTORS AFFECTING EQUIPMENT NEEDS

The selection and quantity of equipment required for the operation and maintenance of an irrigation system depend on several interrelated factors. Understanding these influences helps districts optimize their investment and ensure adequate coverage of maintenance needs across their service areas. Key considerations include the following: (1) the areal extent (size of the system); (2) the type of conveyance system, including (a) open channels (lined or unlined canals may require earth-moving equipment for dredging, sediment removal, and weed control) and (b) closed conduit (pressurized pipelines or buried conduits often necessitate excavation machinery, pipe-handling tools, and leak detection technologies); (3) the major items of maintenance in the project, involving (a) sediment (including aggradation and degradation, which requires dredgers, excavators, and sediment transport vehicles), (b) aquatic and terrestrial weeds (may require shredders, aquatic weed harvesters, or herbicide sprayers), (c) age of system and structures (repair tools, replacement equipment, and structural assessment technology), (d) bends and canal cross-sections (may require flexible, compact machines capable of operating in tight or irregular spaces),

DOI: 10.1201/9781003747444-14

(e) erosion control (may require grading equipment, silt traps), and (f) drainage maintenance (may require grading equipment, silt traps, and drainage pipe installation machinery); (4) geographic location of the enterprise, involving (a) isolated (often need to own more equipment due to the lack of nearby contractors or shared resources) or (b) near other districts (may leverage pooled resources or equipment-sharing agreements, reducing the need for full ownership); (5) evolving operational practices and technology, such as (a) chemical weed control for lessening the need for ditchers, crawler tractors, and weed burners, and (b) modernization and system rehabilitation (e.g., converting canals to pipelines) shift equipment needs from earthmoving to pipe installation and pressure regulation; and (6) elimination of outdated practices, such as manual weed burning or soil sterilant application, can reduce health risks and streamline operations.

By evaluating these factors holistically, irrigation districts can make informed decisions about whether to purchase, lease, or share equipment, ensuring cost-effective and efficient maintenance operations across diverse infrastructure types and operational conditions.

11.3 EQUIPMENT TYPES AND APPLICATIONS

The successful operation and maintenance of an irrigation system depend heavily on the availability and suitability of field equipment. Core equipment needs include vehicles for transporting personnel, materials, and tools. Passenger cars and pickup trucks are commonly used to transport staff, while larger trucks and trailer combinations are essential for moving heavy materials and construction equipment. To ensure safe and efficient off-road performance, vehicles should be equipped with positive traction differentials and appropriate tires for rugged terrain. Beyond transportation, the typical inventory for irrigation district maintenance includes a range of heavy and specialized equipment: (1) excavators for trenching, sediment removal, and pipe laying; (2) road patrols (road graders) for maintaining access roads and shaping canal banks; (3) tractors (with attachments such as blades, loaders, mowers etc.); (4) dozers and scrapers (mounted front and rear on crawler tractors for heavy earthwork); (5) mowers or shredders for weed and vegetation management; (6) rubber tired loaders and (7) tractor loaders (require trailer to move) for loading, hauling, or material spreading; (8) pump trucks for dewatering and maintenance of pump stations; (9) cranes for lifting heavy components like gates or concrete panels; (10) other types of construction equipment, such as (a) rollers, (b) drills, and (c) water and fuel tanks; and (11) digital and remote tools, such as lidar-equipped drones and submarine inspection devices for condition assessments of surface and subsurface structures. Table 11.1 provides estimated values for the expected economic life and annual maintenance and repair costs of key equipment commonly used in irrigation districts, based on observed industry standards and manufacturer recommendations (Burton, 2010). These estimates assist in planning for equipment replacement, budgeting, and optimizing the long-term performance of critical assets.

Many excavators and cranes are now designed with modular attachments, which enhance their versatility for various tasks. These attachments may include trenching buckets, dredging buckets, shovels, backhoes, clamshell buckets, augers, hammers,

TABLE 11.1
Estimated Economic Life and Annual Maintenance and Repair Costs of Selected Irrigation Maintenance Equipment

Equipment Type	Primary Application	Expected Economic Life (Years)	Annual Maintenance and Repair Costs (% of Initial Cost)
Dragline excavators	Large canal sediment and weed removal	20–25	4–5
Hydraulic excavators (tracked/wheeled)	Sediment and vegetation removal from canals and drains	15–20	4–5
Graders	Shape canal embankments and maintain access roads	15–20	4–5
Tractor-powered attachments	Vegetation cutting, pumping, and weed control	5–10	5–6
Tractors and chains (vegetation removal)	Manual vegetation clearance using chained systems	10–15	4–5
Front-end loaders/Shovels	Material lifting, loading, and construction support	10–15	4–5
Flat-bed loaders	Transport of heavy tracked equipment	10–15	3
Flat-bed ditchers (Briscoe type)	Cleaning small on-farm channels	10–12	4
Bulldozers	Spoil leveling, embankment shaping, and flood diversion	15–25	5
Scrapers	Earth movement and embankment repair	15–20	4–5
Mowers and shredders	Control vegetation growth along canal banks and right-of-ways	10–12	4–6
Hydraulic backhoes	On-farm channel maintenance	10–15	4
Dredgers	Continuous sediment removal in canals and basins	20–30	5–6
Tipper trucks	Sediment/material transport	10–15	3–4
Cranes	Lifting and positioning of heavy gates, structures, or components in canals and pump stations	30–40	2–4
Vibrating rollers/Sheep's foot	Soil compaction on embankments	10–15	4
Water bowsers	Moisture control for soil compaction	10–12	3–4
Concrete mixers	Portable mixing for repair projects	10–15	3
Concrete vibrators	Concrete compaction for structural integrity	5–10	5–6
Hand-held soil compactors	Finishing compaction near structures	5–8	5–6
Drones (lidar/thermal equipped)	Overhead canal inspection, seepage detection	5–8	6–8 (battery, sensors)
Submarine inspection devices	Underwater structure inspection (e.g., gates, culverts)	5–10	6–8
Concrete mixers	Portable mixing for repair projects	10–15	3
Concrete vibrators	Concrete compaction for structural integrity	5–10	5–6

FIGURE 11.1 Mower used for cleaning drain ditches.

draglines, or pile drivers. The choice to use a particular unit and its attachments depends on the specific requirements of the task at hand. In some cases, more than one of these versatile units may be necessary. For example, the Delta Lake Irrigation District in South Texas spans approximately 25 miles of ditches. The district uses tractors equipped with mowers to clean and drain ditches approximately three times a year (Figure 11.1). The district also has three excavators used to dig ditches, lay pipes, and remove sediment buildup along canal banks and lakeshores (Figure 11.2). It also has a crane (Figure 11.3a) used for heavy lifting and moving large objectives, a grader (Figure 11.3b) used to maintain roads, ditch, and canal banks, and to make embankments, a loader (Figure 11.3c) for moving soil and rock, a bulldozer (Figure 11.3d) for pushing soil, and several small tractors with small mowers.

11.4 EQUIPMENT SELECTION CRITERIA

The equipment must be considered in light of the work to be done. The type and quantity of equipment required vary widely depending on the system's size, infrastructure complexity, geographic conditions, and frequency of maintenance activities. Four key principles can help guide this selection process: (1) Adequate sizing and capability: Equipment must be appropriately sized for the intended tasks. Undersized

FIGURE 11.2 Illustration of an (a) excavator used for digging and laying pipe, (b) excavator with a dredging bucket used to clean sediment from canals, and (c) excavator with a rake used to clean weeds from canals.

machines reduce productivity and increase wear, while oversized machines may be underutilized and uneconomical. The goal is to select equipment that can perform its function efficiently under field conditions without excessive downtime or operational strain. (2) Evaluating used equipment: Although used equipment may offer cost savings upfront, its long-term value depends on condition, remaining service life, and repair needs. Districts should conduct thorough inspections and cost–benefit analyses before purchasing used units. In some cases, the long-term reliability and warranty support of new equipment justify the higher initial investment. (3) Preventive maintenance and readiness: Equipment should be maintained in peak condition and ready for deployment at all times. Preventive maintenance programs, including regular inspections, oil changes, lubrication, and wear-part replacement, can extend equipment lifespan and reduce unexpected failures during critical periods (e.g., canal cleaning before irrigation season). (4) Cost-effective utilization and replacement: Equipment should be used to achieve the lowest cost per unit of work over its lifecycle. As operating costs increase and performance declines with age, timely replacement is often more economical than continual repairs. Districts should

FIGURE 11.3 Illustration of a (a) cane, (b) grader, (c) loader, and (d) bulldozer.

monitor usage hours, repair frequency, and downtime to determine optimal replacement intervals. Strategic equipment trade-ins before major failures can preserve asset value.

The effectiveness of an irrigation project is not solely dependent on the skill of the personnel but also significantly on the quality and capability of the equipment used. For example, the best truck driver or dozer operator in the world cannot move much dirt if all he has to work with is a wheelbarrow; the best ditch rider cannot handle many miles of canal if all he has is a horse; the best bookkeeper cannot keep many records if the only tool he has to work with is a pencil. Therefore, providing the right tools and equipment is crucial for maximizing operational efficiency. This involves an in-depth assessment of the project's requirements and the capabilities of the available equipment.

Modern technology continues to transform equipment functionality. Innovations such as GPS-guided graders, laser-leveling systems, and remote-controlled canal-cleaning attachments enhance both precision and labor efficiency. These advancements should be considered when evaluating equipment purchases or upgrades. Equipment selection is both a technical and financial decision. It must consider long-term operational needs, field conditions, and economic constraints. Strategic selection, combined with regular maintenance and scheduled replacement, ensures that irrigation infrastructure is supported by a reliable and efficient equipment fleet.

11.5 OWNERSHIP OR PURCHASING OF EQUIPMENT

The decision to own, rent, or share irrigation equipment is a fundamental consideration for irrigation districts and depends largely on the size, budget, workload, and location of the enterprise. Smaller districts often face constraints that limit their ability to invest in a full suite of equipment, while larger districts may justify the capital expenditure through the scale and frequency of required maintenance.

For large-scale operations, owning key equipment can enhance response time, increase operational control, and reduce long-term costs. A typical district may invest between $150,000 and $200,000 in maintenance and operations equipment. With a 10-year depreciation schedule, this translates to an annualized cost of $15,000–$20,000. To justify this investment, the equipment should be used frequently and be capable of performing multiple tasks throughout its lifecycle. Ideally, each unit should fulfill more than 90% of its potential use cases to avoid having idle capital tied up in underutilized assets.

A question to be decided by management is the ownership of equipment. By this what is meant is if it is advantageous to (1) own and maintain all needed equipment, (2) rent equipment as needed, (3) pool ownership with adjacent districts or similar organizations, and (4) contract all work and own little, if any, of the equipment.

Districts that own and maintain all necessary equipment gain the greatest control over scheduling and operations. This approach is often necessary in isolated areas where rentals or contractors are not readily available. Ownership requires a robust in-house maintenance program, adequate storage facilities, trained personnel, and sufficient capital reserves for replacement.

Renting equipment as needed can be cost-effective for districts with infrequent or seasonal maintenance demands. It reduces capital investment and depreciation costs but may introduce scheduling challenges during peak demand periods. Rental options are best suited for specialized tasks or when short-term flexibility is more important than long-term ownership.

Neighboring districts with similar needs may form cooperative agreements to share equipment and reduce individual ownership burdens. For example, one district may invest in a crane, while another invests in a dredger, with each party gaining access as needed. This model requires clear agreements regarding scheduling, maintenance responsibilities, and liability.

Some districts outsource most or all of their maintenance work to private contractors. This approach transfers equipment ownership and maintenance responsibilities to the contractor. While this can reduce the need for internal staffing and storage, it also limits control over timing and quality. Contracts must be carefully structured to ensure reliability and accountability.

To manage the financial implications of equipment ownership, districts should plan for systematic replacement and depreciation management. Depreciation represents the reduction in the value of equipment over time and is a critical financial consideration for irrigation districts. Depreciation of equipment can bankrupt an irrigation district. There are different ways to live with it. If nothing is done, it can strike a hard financial blow. One strategy is to establish a sinking fund, in which money is set aside annually, based on the expected lifespans of equipment. This proactive approach ensures funds are available for future replacement needs and avoids sudden budgetary shortfalls. Alternatively, a pay-as-you-go model can be used, where funds are allocated each year, based on current needs and availability. Regardless of the approach, tracking depreciation is essential for accurate budgeting, particularly in the face of inflation and rising equipment costs. Depreciation in 1960 was 70% greater than in 1950. In dollars and cents, this means that in 1960, it accounted for 22 cents of every dollar of farm expenses, compared to 15 cents in 1950. Districts can also benefit from understanding tax depreciation methods, where applicable, to reduce tax burden and recover equipment costs over time through allowable deductions.

In summary, the choice between owning, renting, pooling, and contracting irrigation equipment depends on a district's operational scale, geographic context, and financial structure. Sound economic analysis and long-term planning can optimize equipment strategies and ensure reliable, cost-effective maintenance of irrigation systems.

11.6 COMMUNICATION EQUIPMENT

Effective communication is essential for the efficient operation and maintenance of irrigation systems. Whether responding to emergencies, coordinating water deliveries, or managing maintenance crews, rapid and reliable communication significantly enhances operational efficiency and system safety.

Many irrigation districts have adopted radio communication systems to streamline operations. These systems have proven cost-effective over time by enabling real-time coordination. For instance, one district reported that prompt communication via radio helped prevent a canal breach, potentially saving over $250,000 in repair costs and crop damage. The question of who needs radio communication and how much is needed generally is a project-by-project matter. Basically, the headquarters office, general superintendent, water masters, maintenance superintendent, and ditch riders should be provided with such equipment. On many projects, some radio units are also provided to personnel responsible for the operation of reservoirs, diversions, and other major water control facilities. In many projects, key equipment pieces, such as vehicles and mobile machinery, are also equipped with radio units to ensure seamless communication across all operational sectors.

While radio remains fundamental, commercial telephone services and internet-based tools have become increasingly important. Landline and cellular services, fax, and email support for day-to-day coordination. In remote areas, project staff may also be responsible for maintaining telephone lines and infrastructure, necessitating technical know-how.

With technological advancements, modern irrigation systems are leveraging advanced communication platforms for even greater efficiency: (1) Smartphone applications. Mobile apps now allow field staff to report issues, log maintenance activities, and access real time system updates. Cloud-connected platforms improve data sharing between field and office personnel. (2) Satellite communication. In isolated or mountainous regions where cellular coverage is unreliable, satellite phones provide a dependable communication option, especially critical during emergencies or natural disasters. (3) Internet of Things (IoT) and remote monitoring. Sensors embedded in gates, pipelines, pumps, and canals transmit real-time data to control centers via wireless networks. These systems can monitor flow rates, equipment performance, and gate positions, alerting operators of anomalies without requiring a physical inspection. (4) Supervisory Control and Data Acquisition (SCADA) systems. SCADA integrates sensors, communication devices, and control software to remotely manage water deliveries. Operators can open or close gates, monitor water levels, and dispatch field teams in response to system alerts—all from a centralized location.

Implementing modern communication infrastructure requires staff training, regular system testing, and established protocols to ensure effective use. For example, districts must develop communication hierarchies, emergency contact trees, and procedures for issuing alerts. Periodic drills and updates help maintain readiness and ensure that communication tools are used to their full potential.

As technology evolves, communication systems in irrigation management are expected to become more automated, integrated, and data-driven, reinforcing the need for districts to invest not only in hardware but also in the human capacity to operate these systems effectively.

11.7 MAINTENANCE OF EQUIPMENT

To achieve optimal efficiency from equipment, it must be kept in good repair and regularly updated to remain modern. Properly maintained machinery performs better, lasts longer, and helps prevent unexpected breakdowns that could disrupt water delivery or damage critical infrastructure. To know whether it is worth having and keeping, good records must be kept of its cost and performance. A procedure for maintenance entails (1) preventive maintenance, (2) maintenance facilities and approaches, (3) maintenance records, and (4) a maintenance program.

Preventive maintenance: By systematically identifying and correcting minor problems before they worsen, scheduled preventive maintenance significantly reduces equipment downtime and improves system reliability. This includes regular inspection, lubrication, cleaning, calibration, and timely replacement of worn-out parts. Equipment manuals should be consulted to follow the manufacturer's recommended maintenance intervals.

Maintenance facilities and approaches: Maintenance work can be performed in several ways: (1) In-district shops: Districts with adequate resources may establish on-site workshops for repairs and servicing. (2) Shared central shops: Neighboring districts can pool resources and maintain a centralized facility. (3) External contracts: When internal capacity is limited, hiring local contractors for repairs can be a more cost-effective option. (4) Vendor-supported maintenance: Rental or leased equipment may include service agreements, reducing district responsibilities. The appropriate approach varies with the district's operational scale, financial capacity, and access to nearby service providers, all of which shape the feasibility and cost-effectiveness of different maintenance options.

Maintenance records: Maintaining detailed records of equipment performance, repair history, and costs is crucial for effective asset management. Logs help track usage, identify recurring issues, plan replacements, and justify future capital expenditures. Records should include equipment ID, specifications, and age; maintenance and repair logs; costs of parts, labor, and downtime; and performance evaluations.

Maintenance program: A structured maintenance program must align with the district's operational demands and budget. The program should include (a) an equipment inventory; (b) scheduled service intervals (daily, weekly, seasonal, annual); (c) emergency response protocols; and (d) resource allocation (personnel, tools, replacement parts). A structured program ensures that maintenance is not neglected due to short-term priorities and that emergency response capacity is always available.

The matter of equipment (and its maintenance) is not an easy problem for management to solve. District managers must weigh several factors when determining how to organize maintenance efforts: (1) Workload and equipment use: Assessing the frequency and intensity of equipment usage helps in determining the appropriate maintenance strategy. (2) Cost–benefit analysis: Weighing the costs of different maintenance approaches against their benefits to find the most cost-effective solution. (3) Emergency preparedness: Ensuring that the selected maintenance strategy supports rapid and effective response during emergencies, such as canal breaks or structural failures. (4) Availability of resources: Considering the availability of skilled personnel, facilities, and financial resources to support the chosen maintenance strategy. The decision between ownership and outsourcing often depends on district size. Larger districts may benefit from owning and operating full-service maintenance departments, while smaller ones may find outsourcing more practical.

11.8 EXAMPLE: IMPERIAL IRRIGATION DISTRICT (IID)

As discussed in Chapter 10, the Imperial Irrigation District (IID) in Southern California is used here as a detailed, real-world example, because it exemplifies the challenges and solutions faced by large-scale irrigation agencies. Its operations span vast agricultural lands and incorporate extensive infrastructure, offering a rich context to demonstrate how strategic equipment management contributes to system efficiency, sustainability, and long-term reliability.

IID maintains an extensive inventory of heavy equipment tailored to specific maintenance and operational tasks. The district's water and power departments maintain annual summaries of equipment by class and project proposed additions. Table 11.2 provides an example of the water department's fleet equipment (Engineering Section 703000), which includes seven long-reach excavators, 12 backhoes, five graders, and over 60 pickup trucks in various configurations (Imperial Irrigation District, 2025). For 2025, the district proposed the purchase of one additional angledozer and one long-reach excavator to support increasing operational demands. These additions reflect IID's proactive investment strategy for maintaining infrastructure performance and adapting to future service requirements.

Long-reach and standard excavators are used extensively for dredging, removing sediment, and embankment repairs. Long-reach excavators are especially effective in deep or wide canals where access is limited, while tracked excavators are preferred for unstable terrain. These machines are typically owned when used frequently and depreciated over a 10–15 year period, with associated maintenance costs integrated

TABLE 11.2
Equipment Summary in Imperial Irrigation District (Section Name: Engineering 703000)

Equipment Type	Quantity in 2025	Proposed to Buy in 2025
Angledozer	7	1
Backhoe	12	
Excavator, long reach	7	1
Excavator, track	7	
Excavator, wheel	3	
Grader	5	
1/2 Ton Pickup	43	
1/2 Ton Pickup 4×4	15	
3/4 Ton Pickup 4×4	3	
Semi-trailer lowboy	4	
Water truck	10	
Move truck	5	
Total	132	2

into the district's operating budget. Preventive maintenance routines, such as hydraulic fluid checks, track inspections, and replacement of bucket teeth, are essential to minimize downtime.

Tractors equipped with attachments, such as mowers and sprayers, serve multiple purposes, including canal bank mowing and herbicide application. Many of these models now run on biodiesel or electric power. Graders and bulldozers are employed to maintain canal slopes and access roads and are often shared across water and power departments to optimize capital expenditures. Emergency equipment, such as mobile pumps and portable generators, is stored and deployed during canal breaches or power outages to ensure uninterrupted water delivery.

Preventive maintenance is a cornerstone of IID's asset management approach. Equipment is serviced regularly under a fleet maintenance program, and replacement decisions are based on equipment age (typically over 10 years), usage hours (exceeding 10,000 hours), and repair cost thresholds. These strategies are outlined in the Budget Plan, which also includes a detailed schedule for replacing key fleet assets. For example, in 2025, the district's Capital Plan further supports operation and maintenance efficiency by allocating $41.6 million for water capital projects, including $15.4 million for irrigation and drainage equipment and $23.3 million for water transfer infrastructure (Imperial Irrigation District, 2025). SCADA upgrades, pump station enhancements, and other major investments fall under these allocations. Meanwhile, ongoing challenges such as sediment accumulation, aging infrastructure, and rising operational costs are addressed through IID's broader Asset Management Plan, which emphasizes modernization and lifecycle cost optimization.

REFERENCES

Burton, M.A.S. (2010). *Irrigation Management: Principles and Practices*. CABI, ProQuest Ebook Central. https://ebookcentral.proquest.com/lib/clemson/detail.action?docID=500967

Imperial Irrigation District (2025). 2025–2026 Budget Plan. Finance Office and Staff of Imperial Irrigation District. https://www.iid.com/home/showpublisheddocument/22531/638792712093230000

12 Maintenance of Dams and Conveyance Structures

12.1 INTRODUCTION

The integrity and functionality of dams and conveyance structures are crucial for the effective management of water resources in both agricultural and urban settings. These structures play crucial roles not only in water storage and distribution but also in flood control and environmental management. The maintenance of such infrastructures, therefore, is not just about preserving physical assets; it is also about ensuring the sustainability of a resource that is fundamental to life and livelihoods. Service and repairs constitute the subject matter of this chapter; the maintenance of canals is further discussed in Chapter 13; and replacements and betterments constitute the subject matter of Chapter 16. Service and repairs encompass all components of the irrigation system, from the headworks to the point of water delivery to the farm. This includes the maintenance of (1) dams and headworks, (2) reservoirs, (3) canals, (4) structures, (5) bridges, (6) communication systems, and (7) others, such as guesthouses, inspection houses, roads, and transportation systems.

12.2 RESERVOIRS

Regular monitoring and preventive maintenance are essential for sustaining reservoir integrity, optimizing operational performance, and safeguarding public and downstream safety over the long term. The suggested observation program includes the following: (1) Surveys: Conduct comprehensive surveys every 5 years to assess the condition of the reservoir. In cases of observed degradation, these should be conducted annually. (2) Soil conservation program: A soil conservation program should be implemented in the watershed to reduce erosion and sedimentation, preserving storage capacity and water quality.

12.2.1 Main Storage Dam

In general, except for a large accident, reservoir maintenance is usually limited. The nature of maintenance varies based on the reservoir's design and location. If the reservoir is situated on a small channel, the dam is similar to a diversion dam. When the dam is filled through a feeder canal, the problems of spillway capacity are either eliminated or greatly reduced. The maintenance of the outlet works is similar to that of the headgate of canals. The action of waves on the faces of earth dams may cause

DOI: 10.1201/9781003747444-15

too much trouble. To mitigate this, course material or concrete may be placed on the water side of the dam. In addition, dry rock riprap is often used to mitigate the impact of wave action, reducing erosion and structural wear. However, great care needs to be exercised in the design and emplacement of ripraps if the detrimental impact of water criteria is to be reduced or eliminated. Protecting the downstream slope from rain impacts is also crucial. This can be achieved through natural solutions, such as the growth of vegetation, provided that the soil and moisture conditions are suitable. Alternatively, a layer of gravel or loose rocks can offer the necessary protection against erosion caused by rainwater runoff.

12.2.2 Regular Maintenance

The key to effective dam maintenance lies in addressing several critical aspects: (1) erosion control, (2) leakage management, (3) crack detection, and (4) the upkeep of various appurtenances. Each of these components requires a unique approach and careful attention to detail to maintain the structural integrity and functionality of the dam. The possible causes of dam failure are shown in Figure 12.1.

12.2.2.1 Erosion

Erosion is a major concern for embankment structures. It can occur on both the upstream and downstream faces due to natural elements, such as water flow, wave action, and rainfall.

Erosion caused by wave action is a significant problem for upstream slopes. Protective measures, such as rock riprap over filter material or alternative materials like steel, bituminous, concrete facing, bricks, or concrete blocks, are commonly used. Berm construction can also help dissipate wave energy. When severe erosion occurs on the upstream slope of a dam, prompt repairs should be undertaken. This involves preparing the surface, applying a protective layer, and securing it with a

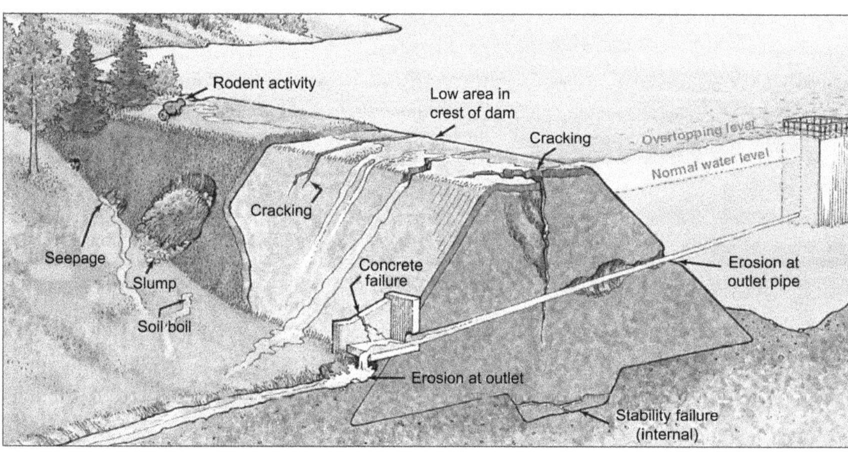

FIGURE 12.1 Illustration of possible causes of dam failure. (Adapted from USDA, 2023.)

bench to ensure that the layer extends below the lowest anticipated pool level. If rock riprap is used, then it should be heterogeneous, and the filter must prevent soil washout. Ripraps require regular maintenance due to natural weathering processes. Regular maintenance, including the replacement of weathered riprap, is crucial. If erosion or riprap deterioration occurs frequently, professional advice should be sought to design a more effective solution.

The downstream face of the embankment is susceptible to erosion due to heavy rainfall. Healthy vegetative cover is crucial for erosion control. Embankments should be designed to promote sheet flow, which distributes water evenly and minimizes erosion. Poorly maintained vegetation or concentrated flows can result in the formation of rills and gullies, which necessitate prompt repair with suitable soil, compaction, and reseeding. Graded rock mixtures on sand filters can be used for severe erosion. At spillway contact points, issues such as poor compaction and settlement can lead to erosion, which may require regrading, resistant surface protection, or the construction of wood beam steps. Consistent maintenance and proactive measures are key in mitigating the impact of erosion on dams. Through careful design, regular inspection, and timely intervention, the longevity of these structures can be significantly enhanced, ensuring that they continue to serve their purpose effectively.

12.2.2.2 Seepage

Seepage naturally occurs in all dams as impounded water seeks pathways of least resistance through the dam and its foundation. However, it is imperative to manage seepage by controlling both its velocity and quantity. Seepage may appear at different locations, including the downstream face, areas beyond the toe of the dam, or downstream abutments at elevations below the normal water level. Typical signs and indicators of seepage include the following: (1) Clear water leaks: Some clear water leaks may be expected in dams due to seepage, which is a natural phenomenon. However, it's essential to closely inspect and measure the discharge of these leaks on a regular basis. An increase in the volume of clear water leaks or the appearance of muddiness in the discharged water can be indicative of trouble. This may suggest an increase in seepage or potential erosion within the dam structure, which requires immediate attention and investigation. (2) Depression or settlement of an earth dam: The presence of depressions or settlements in an earth dam is a cause for concern. These depressions can indicate that the dam structure is experiencing uneven settling, which can compromise its stability. Settlement issues should be thoroughly assessed, and measures should be taken to rectify them to prevent further structural problems or breaches. (3) Piping below the dam: Piping refers to the internal erosion of soil or materials within the dam structure, often caused by seepage or the movement of water through the dam. Piping can weaken a dam's internal structure, potentially leading to breaches or failure. If piping is suspected, immediate action is necessary to investigate the extent of the issue and implement appropriate measures to prevent further erosion and strengthen the dam.

Regular monitoring is crucial for early detection of seepage and preventing potential dam failure. Keeping detailed written records that include the locations where seepage happens, the volume and characteristics of the seepage flow, the size of affected wet areas, and the types of vegetation present is also important.

Photographic documentation can serve as an invaluable record for assessing seepage conditions over time. Inspectors should be particularly alert to any increases in seepage flow and any signs that the flow is carrying soil particles, which can be indicative of erosion. Additionally, routine inspection and maintenance of internal outlets for embankment and foundation drainage help ensure that drainage systems function properly and remain free from blockage.

12.2.2.3 Cracks

Cracks in dam embankments are indicators of potential structural issues and require careful attention. These cracks can be broadly categorized based on their size, orientation, and implications. Short and isolated cracks, often small and inconspicuous, may not immediately signal which is a severe problem but need to be monitored over time for any changes that could signify underlying issues. In contrast, larger, well-defined cracks, particularly those wider than 1/4 in., are more alarming. These cracks indicate serious structural concerns and necessitate prompt and effective action.

There are two primary orientations of cracks: transverse and longitudinal. Transverse cracks, running horizontally across the embankment, often suggest differential settlement within the embankment. This condition can lead to the development of seepage paths and rapid piping, which is a severe problem in dam safety. Longitudinal cracks, aligned parallel to the length of the embankment, may indicate the early stages of a slide or slump on the embankment faces. In newly constructed dams, these cracks may also indicate inadequate compaction during construction, which is a significant issue affecting the structural integrity of the dam.

The first step in managing cracks in dam embankments involves thorough inspection and documentation. Engineers must conduct detailed inspections, categorizing the severity of observed cracks based on established criteria. This process includes measuring the dimensions of the cracks and their locations, and employing photographic evidence and geospatial tagging to create a comprehensive record. Such documentation is vital for both immediate response and long-term monitoring. Once cracks are identified and documented, the next step involves deciding on and implementing repair strategies. For smaller cracks, sealing is often an effective solution. This is typically done by compacting clay into cracks using shovels and compacting tools. After sealing, it is crucial to monitor the area for any movement. Stakes or photographs can be used to document the progression of cracks. Continued movement after sealing indicates more severe issues, such as an impending slide.

12.2.2.4 Appurtenances

In dam maintenance, the care of appurtenances can improve the operational efficiency of the dam. Appurtenances, such as spillways, outlet works, drains, fish facilities, and general equipment, require regular attention and specific maintenance practices (Figure 12.2).

A spillway is an emergency structure of paramount importance that requires regular and diligent maintenance. It must be kept free of debris to function correctly in emergency situations. The intake channel, a critical component of the spillway, should be kept clean with the regular removal of rocks and other obstructions.

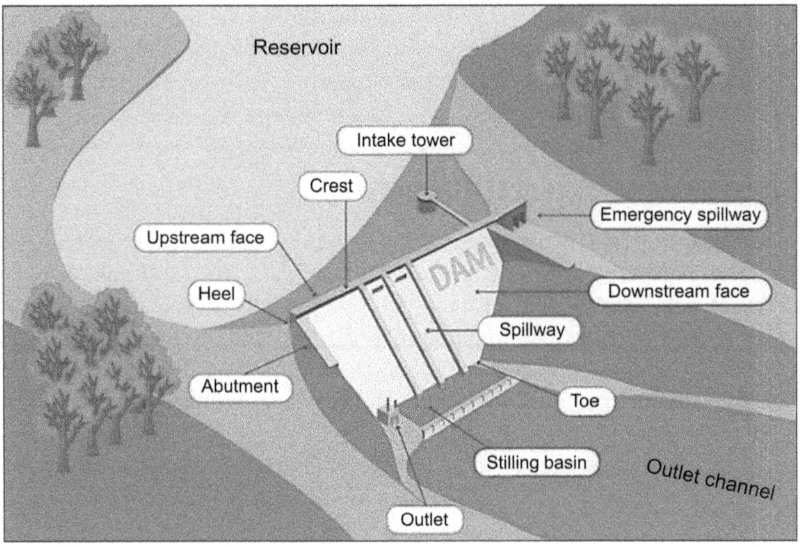

FIGURE 12.2 Major components of a dam. (Adapted from U.S. Army Corps of Engineers, 2023.)

The control section, comprising piers, crests, gates, and entrance and exit structures, requires regular inspections and maintenance to ensure that it remains operational. The chute, whether a tunnel, shaft, or other conveyance, must be meticulously maintained to prevent clogging. This includes ensuring that underdrains are functional and checking slabs for settlement and correcting them as necessary. The stilling basin and return channel also require regular cleaning to prevent wear and tear, and maintaining riprap downstream is essential to ensure the longevity and effectiveness of the basin.

Outlet works, including tubes or tunnels through the dam, are integral to its operation. These structures should be kept clean, and their protective coverings should be kept intact. Controls, such as gates and valves, are central to the operation of outlet works and require consistent maintenance. The stilling basin, similar to the spillway, needs to be kept clean and backfilled to maintain its effectiveness. If a return channel is present, it should be routinely cleaned and maintained.

The maintenance of drains, particularly in and near the dam and its appurtenances, is crucial. Keeping a record of discharge measurements is essential for monitoring the performance and health of a dam. Ensuring that all drains, especially at exits, are open and free from obstructions is a critical part of this maintenance routine.

Fish facilities, depending on their type, require a significant amount of maintenance. This includes the regular removal of trash and debris to ensure that facilities remain functional. The operational status of these facilities is often a legal requirement, making their maintenance not only a matter of efficiency but also of compliance.

General equipment maintenance encompasses a range of activities. Gates and valves should undergo monthly inspections, minor repairs, and adjustments, with testing conducted biannually. Major repairs should be scheduled during off-season periods. Regular painting is also important for the longevity of these components. Trash racks require consistent cleaning and painting, whereas log booms and additional racks require frequent inspections and timely repairs.

12.3 RIVERS OR NATURAL CHANNELS

Rivers or natural channels are important conduits for irrigation. Effective maintenance of these waterways is crucial for reducing water loss and enhancing water conservation. The maintenance of these natural channels is typically shared among a collection of administrative bodies, ranging from local county officials to state authorities, and at times, even federal agencies, such as the U.S. Government. Given the scale of such undertakings, the expenses incurred are typically prorated, reflecting the shared responsibility.

One of the critical aspects of maintenance is the proper management of bridges spanning these waterways. These structures are susceptible to debris accumulation, which, if not promptly addressed, can lead to significant obstruction of water flow. Regular trash removal from bridges is thus essential not only for the unimpeded flow of water but also for the structural integrity of the bridges themselves.

Vegetation along riverbanks and within channels presents both benefits and challenges. While some plant life contributes to bank stabilization and ecosystem balance, excessive vegetation, particularly invasive species, can cause substantial water loss. The roots and biomass of these plants can absorb a large amount of water, reducing the volume available for irrigation purposes. Therefore, controlling vegetation is a necessary maintenance activity to optimize water availability for agricultural purposes.

The accumulation of silt within the channel can alter the river's course, leading to meandering, which not only results in water loss but can also impact land use adjacent to the river. Meandering can encroach on farmlands, causing the loss of fertile soil, thereby reducing the agricultural value of the land. Regular dredging and silt removal can mitigate these effects and maintain the channel's capacity to deliver water effectively.

In colder climates, ice poses a unique challenge. Ice formation can cause damage to bridges and structural components of waterways. The expansion of water as it freezes can exert tremendous force on man-made structures, leading to cracks or even collapse. Moreover, ice jams can lead to sudden flooding upon thawing, which can overwhelm channels and irrigation systems. Preventative measures, timely ice removal, and structural reinforcements are critical to protect the infrastructure from such damage.

12.4 DIVERSION DAM AND HEADWORKS

Diversion dams are special structures whose design depends on local conditions. A diversion dam may not be required in streams that carry large low-water flows almost on a permanent basis in comparison with the amount to be diverted. An example is the case of several canals diverting from the Yellowstone River in the

United States. The headgates in such cases should be planned and constructed such that the flow past them is sufficient for diversion, and no cutting of the canal at the headgate will be executed. For streams with poor foundations, it may be less expensive to use a temporary diversion dam during the low-water period of each year. During high-water periods, securing a diversion is relatively easy. However, one must evaluate the cost of a temporary diversion dam against the annual maintenance costs and the fixed charge of a permanent diversion dam. Streams with torrential or fluctuating flows are not conducive to temporary dams.

In addition to any other component of an irrigation system, headgates should be built more permanently because an injury to the inlet control of the canal may cause a loss of water to the entire irrigation system. When designed and built properly under engineering supervision, these structures should have low maintenance costs. Duplicate control of headgates is desirable for all canals except small canals. This can be achieved by having two sets of gates, so that when one set is damaged, the other set can be used, ensuring the uninterrupted supply of water to the irrigation canals. There should be a provision for making emergency repairs, such as the emplacement of flashboard grooves. To regulate canal flow, a spillway may be constructed immediately downstream of the headgate. Secondary gates located below the spillway can safeguard the canal in the case of upper gate failure and provide better control over the amount of water diverted.

There is a need to prevent the entrance of drifting and fish into the headgates. This may be accomplished using screens across the headgates. Although the entrance of fish may not be harmful to the canal, their entrance should be prevented to safeguard the public interest in fish. In some states in the United States, such prevention may be required by law. The drift occurring in streams consists largely of logs or debris from timber. To prevent this, entry into the canal can be achieved by the proper location and design of headgates. The main idea is to allow clear surface water to be taken into canals. The screens for the prevention of fish entrance should be of much closer mesh than those needed for other purposes. The drift may, however, clog such screens or checks, causing injury to the canals. In the United States, several states require by law the provision of fish ladders at all dams on streams.

12.5 STRUCTURES IN AND ALONG CONVEYANCES

Regardless of the type of material from which the structures are built, they all wear out and deteriorate. Some general types of maintenance that apply to structures in and along waterways include (1) backfill stability, (2) riprap application, and (3) vegetation and sediment management.

Backfill stability: The loss of backfill, including settlement, seepage, piping, and erosive action from both upstream and downstream water, must be minimized. The remedy is to eliminate the cause of the trouble, if possible, and replace the material.

Riprap application: Riprap is used around structures along canals and elsewhere as protection to the earth adjacent to the flow. Flow velocity, wave action, and eddy currents cause erosion, particularly at and near structures, at curves, and in areas where the soil is not resistant to such conditions. Riprap serves best when shaped to fit the section after placement, although much is simply dumped from trucks into distressed areas. The size and density of the material used for riprap should be carefully considered. Small-sized materials or materials of low density are simply carried away by water to form a bar downstream and need to be removed. Extremely large materials, when used in certain locations, may increase the friction factor to a point where the desired discharge cannot be maintained.

For economic reasons, the designer usually includes only the amount of riprap deemed necessary. In many instances, particularly during transitions, checks, and drops, additional tasks must be performed by operation and maintenance personnel. Significant work has been conducted to determine the optimal size and density of materials, as well as the most suitable methods of emplacement, among other factors. Good maintenance demands that this information be used.

Vegetation and sediment management: The accumulation of mosses, weeds, and sediment, including channel erosion and deposition, presents significant maintenance challenges. More will be discussed later in Chapter 15 about weeds and mosses, and much has already been discussed about sediment. Structures exacerbate problems of this type, and the removal of unwanted materials is typically a recurring maintenance task. Weeds and mosses within structures can be controlled to some degree by chemicals, but materials that move with the flow must be removed by hand or mechanically.

Operation and maintenance personnel sometimes upset the planned operational characteristics of the system by modifying structures. At times, this facilitates operation and may be advisable. At other times, however, maintenance problems may increase. For example, modifying an emergency waterway to function as an operational waterway may cause issues in the waterway channel. Some soils can withstand intermittent flows, but when kept constantly wet by operational waste flows, they deteriorate rapidly, and a badly eroded wasteway channel will develop within a short period of time. Much maintenance may be required to prevent the deterioration, and a major construction issue may arise. Safety structures are included in the system to ensure safe operation. When altered or not properly maintained, the safety of the entire system may be jeopardized.

In most irrigation systems, conveyances are aligned across the natural drainage pattern, which is toward the stream from which the water is diverted. These cross-drainage structures are typically a major maintenance item. Floods may wash out part of the system. Flumes may be destroyed, and inverted siphons, if not located below the scour line of the stream, may be uncovered and damaged. Sections of the canal may be washed away or filled with sediments. If stormwater is introduced into the system, sediment and other debris will enter the conveyances and must be removed. In projects, depending on the system's design, climate, soil, and geography, the maintenance of cross drainage may be the most troublesome and expensive item in the system, resulting in a ranking with or above silt and weed problems.

12.6 MEASURING STRUCTURES

In general, measuring devices are empirically developed. Calibrations are obtained for a given set of conditions, which include physical dimensions, velocity distribution in the approach section, the range of head both upstream and downstream, points for making observations, and the quality of the materials and construction. This means that the calibration is valid only if the installation is constructed, maintained, and operated under identical conditions, and only then will reliable results be obtained. The slightest deviation in construction, maintenance, or operational procedures can significantly skew measurements, leading to unreliable data. As such, to ensure the integrity of flow measurements, frequent inspections are essential. Physical dimensions must be routinely checked, and any discrepancies observed must be promptly rectified.

12.7 STRUCTURES

Structures are constructed from a variety of materials, including wood, stone, concrete, asphalt, various metals, and plastics. Regardless of their composition, all structures are subject to deterioration over time. The key to the long-term use of these structures is the maintenance of moving parts, which often require lubrication to operate smoothly.

Structures in and along conveyances, depending on the material of construction, are of three types: (1) wood structures, (2) concrete structures, and (3) steel structures. The selection of a particular structure depends on (1) construction cost, (2) maintenance cost, (3) expected lifespan, (4) type of structure needed, and (5) local conditions related to topography, soil composition, flow characteristics, etc.

Each material has its own life expectancy, which is determined by both its inherent durability and the environmental conditions. In the early years of use, material-related repairs are typically minimal, but erosion and settlement may necessitate significant work. As time passes and usage intensifies, material wear will naturally increase, particularly in wood structures. For example, during the first quarter of their lifespan, wood structures may require minimal repairs, but maintenance costs can increase to as much as 10% of the original cost annually as the structure ages. The specific environment and purpose of the wood structure also affect the frequency and cost of repairs. With increased use, the need for repairs to the structure's material will also increase. Concrete and steel structures generally demand fewer frequent repairs if they are constructed with appropriate materials. However, steel structures require periodic reapplication of protective coatings, and while the repair cost of concrete structures does not necessarily increase with age, it can be affected by factors such as alkaline reactions and temperature fluctuations.

12.7.1 Wood Structures

In general, wood structures decay slowly when they are continuously dry or wet. The least favorable conditions for the life of a wood structure are when it alternately dries and becomes wet. Alternate wetting and drying promotes bacterial activity in wood, leading to rapid decay. The remedial action may include the use of preservatives (such as creosotes) or the replacement or protection of bridge piers, among other measures. Wood decays if vegetative growth occurs.

Wood structures used in irrigation systems can be divided into three classes: (1) wood flumes, (2) wood pipes, and (3) ordinary wood structures. In the case of wood flumes, wood is not in contact with the ground. In wood pipes, water is under more or less pressure, and the wood is not necessarily continuously wet. In ordinary wood structures, the wood is in direct contact with the earth. The lower portions of the trestles and the ends of the wood flumes can be considered to belong to the class of ordinary wood structures.

The life of wood flumes depends on the type of material used, the type of construction, the type of footing, the conditions of use, and care and maintenance. Protective measures such as coatings and flume linings can enhance the lifespan of wood flumes. The serviceable life of flumes is approximately 20 years for redwood or cedar, 12–15 years for flumes of fir, and 8–10 years for pine. Usage patterns also impact lifespan. For example, flumes in continuous use on farms have a shorter life than those on laterals that operate continuously. Bench flumes or flumes set in contact with or near the ground have a shorter life than well-built, higher flumes positioned above ground.

The life of wood-stave pipes greatly depends on the condition of service. In pipes, wood is more in contact with other materials, and its life is highly influenced by environmental conditions, such as full saturation or partial saturation, full or partial burial in tight soil, exposure to the atmosphere, burial in open soil, such as sand and gravel and volcanic ash, light pressure or partially filled pipes, wire-wounds, chemical conditions of the soil, etc., and asphaltum pitch, gas tar, mixtures of tar and asphaltum, etc., have been used for coating wood pipes. Usually, a coating works well for buried pipes.

In the case of wood structures, environmental conditions are not entirely favorable, as parts of the structure may be continuously wet, alternately wet and dry, or continuously dry. Structures built in dense, heavy soils usually have a longer lifespan than those in lighter soils, such as sand and gravel, which allow for more air and moisture penetration. On average, the lifespan of a wood structure ranges around 20 years for more durable woods like redwood and cedar, 15 years for fir, and about 10 years for pine.

12.7.2 Metal Structures and Metals in Structures (Particularly Iron and Steel)

Iron and steel, due to their strength and durability, are materials of choice for irrigation components such as flumes, pipes, and gates. However, these metals are not impervious to the relentless forces of corrosion and rust, particularly where water flow creates conducive conditions for such decay at joints and downstream interfaces.

It is generally accepted that corrosion of iron and steel is an electrochemical action. The protection of metals then may be accomplished by (1) protective coatings or (2) cathodic protection. For protective coatings, insulators, such as paints and enamels, are typically based on vinyl or coal tar. Preparation is the secret of a good protective coating job, including (1) sandblasting down to bare metal, (2) wire brushing, and (3) special scrapers and cleaning and chipping tools. This foundation allows the coating to form an impervious barrier to corrosive elements. Cathodic protection is a technique used to inhibit corrosion on metal surfaces by transforming them into the cathode within an electrochemical cell. This method is particularly effective for iron and steel structures in contact with soil or water, which act as electrolytes that facilitate corrosion. The process can be broken down into several key steps: (1) Electrochemical cell formation: Corrosion occurs when a metal forms an electrochemical cell, akin to a battery. At the anode, metal loss occurs as positive metal ions are released into the electrolyte. These ions travel through the electrolyte and are deposited at the cathode, with the metallic structure completing the circuit as a conductor. (2) Sacrificial anode: To protect the primary metal structure (the cathode), a more reactive metal (the sacrificial anode) is introduced. Metals lower in the electrochemical series, such as magnesium, zinc, or aluminum, are used because they corrode (oxidize) more readily than the protected metal. This induces current from the sacrificial anode to the protected structure prevents the latter from corroding. (3) System design: The design of cathodic protection systems must be specific to the environmental conditions of the installation site. Factors such as the conductivity of the soil or water, the metal requiring protection, and any existing electrical currents must be considered to ensure the system's effectiveness. (4) Application and maintenance: Cathodic protection systems are commonly used for buried pipelines, submerged gates, and other metal installations susceptible to corrosion. It is important that these systems are monitored and maintained appropriately to adapt to any changes in the operational environment.

Another issue arises from the galvanic action that occurs when dissimilar metals are in contact—a phenomenon that can accelerate corrosion. While primarily a concern during the design phase, maintenance teams can address it by employing neutralization techniques or applying insulating barriers to prevent the flow of galvanic current.

The maintenance of steel structures extends beyond surface treatments to include the structural repair of flumes and their supporting trestles. Trestlework, which is often challenging and expensive to repair, is crucial to the overall integrity of the flume system. Similarly, routine checks and upkeep of flume inlets and outlets are necessary to prevent undercutting and structural compromise.

12.7.3 Concrete Structures

Concrete structures are, relatively speaking, supposed to be permanent structures. However, they are harmed by factors, such as alkali reactions, structural undercutting, accidental damage, inadequate drainage, and large temperature fluctuations. Application of linings, waterproof coatings, or paraffin treatments can substantially extend the lifespan of these structures.

One of the advantages of concrete linings is the reduced maintenance expenditure. Lining effectively prevents problems, such as aquatic growth, erosion, silt accumulation, canal breaks due to burrowing animals, and bank softening due to seepage. Nonetheless, concrete linings can be compromised by the failure of earth backing, cracks, and leakage due to faulty construction or extreme temperature fluctuations, misalignment caused by the pressure of the earth during non-operating seasons, and other factors. Regular canal cleaning (e.g., several times a year) is important to maintain flow efficiency (i.e., Manning's n) and control aquatic growth.

12.7.4 Asphalt and Asphalt Structures

It is essential to note that structures have varying useful lifespans, differing initial costs, and varying maintenance costs, depending on the materials used in their construction. Therefore, engineering economy studies must be conducted to determine if annual maintenance costs have reached a point where the structure should be replaced with similar materials, other materials, or possibly completely redesigned, for instance, by replacing a flume with an inverted siphon.

12.7.5 Bridges

Bridges play a crucial role in canal infrastructure, especially when canals intersect with public roads. The responsibility for bridge construction and maintenance is determined by the nature of the crossing and varies according to state laws. In cases where a canal is built across a public road, the canal owner is responsible for constructing the bridge over the canal. Maintenance obligations differ from state to state. For example, in Colorado, Idaho, Montana, New Mexico, and Wyoming, the county will maintain the bridge after its construction and maintenance for 1 year by the canal owner. In some states, it may be optional for the county to maintain the canal, while in others, canal owners are required to maintain it.

On the other hand, if a public road is constructed across a canal, then the public entity, such as the local government or road authority, is required to build the bridge. The laws vary from state to state regarding maintenance, damage, right-of-way, and entities, such as local governments or road authorities. The design parameters for bridges on public roads, including dimensions like width and span, are often dictated by legal requirements.

Regular maintenance primarily involves replacing worn-out flooring, which is a direct result of traffic wear and tear. Ideally, the bridge span should match the full-supply width of the water surface below to ensure a seamless passage and prevent debris accumulation.

12.7.6 Communication Equipment

Effective communication ensures the operational efficiency of irrigation systems. The maintenance of communication equipment is therefore a component of the system's overall functionality. All communication devices within the network must be kept in optimal working conditions at all times. This includes regular checks, repairs, and updates as necessary. Given the specialized nature of this equipment, having skilled personnel within the organization dedicated to its maintenance is often justified. Their expertise ensures that the communication network functions seamlessly. In many cases, local repair facilities are relied upon for general maintenance needs. This approach can be more practical and cost-effective, especially for routine service.

Efficient and reliable communication systems are absolutely vital for the operation of irrigation systems. Each ditch rider's house should be connected to the main office. This setup enables efficient daily reporting and the prompt issuance of instructions for subsequent operations. Having direct telephone lines to headworks and critical waterway junctions is crucial for swiftly addressing urgent situations and coordinating responses. The incorporation of modern communication technologies, such as cellular phones, fax machines, and email capabilities, greatly enhances the system's operational efficacy. These tools facilitate instant communication, which is particularly crucial in emergency situations or for making rapid decisions. A small investment in the beginning will go a long way toward making the operation and management of the system more efficient and productive. It not only streamlines daily operations but also ensures that all parts of the system are promptly responsive to changing conditions and requirements.

12.7.7 Other Maintenance Items

Effective management of irrigation systems extends beyond water channels and mechanical structures; it encompasses a broad spectrum of ancillary infrastructure. This includes roads, buildings, safety devices, fences, and other facilities that are integral to the operation and safety of the system. The upkeep of these components is not merely a seasonal task but a year-round commitment, which is essential for ensuring uninterrupted functionality and user safety.

Roads that provide access to various parts of the irrigation system require regular maintenance. This involves repairing wear and tear, ensuring clear signage, and maintaining roads that are passable in all weather conditions. Well-maintained roads are crucial for the swift movement of maintenance crews and equipment. Operational buildings, such as control centers and storage facilities, demand consistent care. This ranges from structural integrity checks to routine cleaning and repairs, ensuring that these buildings remain safe and functional for staff and equipment. Regular inspections and repairs of safety devices are vital to prevent accidents and injuries. Similarly, fences that secure the premises need consistent maintenance to ensure that they effectively demarcate and protect the area. Given the diverse nature of these maintenance tasks, engaging external contractors can be an economical and efficient approach. Contracting enables the application of specialized expertise where needed, ensuring high-quality maintenance while optimizing resource allocation within the irrigation system.

12.8 CASE STUDIES

This section presents four case studies showing practices, challenges, and outcomes in the maintenance of dams and irrigation conveyance infrastructure.

12.8.1 Mahaweli Irrigation System (Sri Lanka)

The Mahaweli Development Program (MDP) is the largest irrigation-based agricultural development program in Sri Lanka, covering approximately 901,935 acres (365,000 ha). It includes a network of major reservoirs and dams, such as Victoria and Kotmale, as well as hundreds of kilometers of main and distributary canals. Key maintenance strategies include annual desilting of dams, the rehabilitation of canal linings using concrete or PVC sheets, and the periodic geotechnical monitoring of dam structures. A recent study evaluated conveyance losses (2022–2023) in three major irrigation schemes: Maduru Oya, Hurulu Wewa, and the Minipe Left Bank Canal, which ranged from 27% to 33% (Harindu et al., 2025). This study identified canal lining conditions, soil texture, and elevation differences as significant factors influencing water loss. This emphasized that targeted rehabilitation, especially increased lining in high-loss sections and improved maintenance of tail-end reaches, can substantially enhance conveyance efficiency and water use sustainability.

However, evaluations have revealed that the inefficiency of the dam and conveyance system, attributed to technical limitations, such as organizational gaps and water scarcity, may obscure deeper political-economic challenges (Paranage, 2019). These underlying issues include centralized decision-making, limited farmer agency, and restrictive land use policies that demotivate multi-generational farming communities. Addressing these systemic issues, along with enhancing infrastructure upkeep and incorporating farmer knowledge systems, is critical for revitalizing MDP's maintenance and operational effectiveness.

12.8.2 Mwea Irrigation Scheme (Kenya)

The Mwea Irrigation Settlement Scheme (MISS), located in Kirinyaga County, Kenya, spans the Mwea East and Mwea West sub-counties. Originally established in 1954 with just 65 acres under irrigation in the Tebere section, the scheme has since expanded to cover approximately 30,600 acres (12,382 ha). It sources its irrigation water from the Nyamindi and Thiba Rivers and benefits from largely flat terrain with dominant black cotton soils, which are ideal for rice cultivation. The Thiba Dam, which serves as the primary reservoir, is an earth-fill embankment dam with a storage capacity of approximately 15 million m^3 and was constructed to stabilize the water supply to the Mwea Scheme (Kenya National Irrigation Authority, 2025). Its integration into the conveyance system includes modernized intake works and automated flow regulation to optimize seasonal releases. The dam supplies water to a network of lined and unlined canals, where upgrades, such as the concrete lining of primary canals and the installation of reinforced headworks, increase conveyance efficiency and reduce seepage losses. These infrastructure enhancements have doubled rice production, increased the annual production value from KSh 10 billion to KSh 18 billion, and have created 50,000 new jobs, resulting in a total of 175,000. The scheme now supports double cropping at 70,000 acres/year and has stimulated the establishment of over 13 large private rice mills. These advancements have contributed to improved livelihoods, enhanced trade, and the settlement of previously landless families.

The National Irrigation Authority (NIA) is responsible for managing and maintaining primary and secondary infrastructure, while farmers manage tertiary infrastructure. The scheme operates under a Participatory Irrigation Management (PIM) model, involving partnerships between the NIA and farmer organizations, such as the Irrigation Water User Association (IWUA), Mwea Rice Growers Multipurpose (MRGM), and Lainisha Sacco. These institutions share responsibilities in land administration, capacity building, irrigation expansion, and infrastructure rehabilitation. Despite these improvements, the Mwea Scheme faces critical challenges in its long-term sustainability. One major concern is the inadequacy of institutional frameworks for the operation and maintenance (O&M) of irrigation infrastructure. Although project overheads include costs for planning, training, and capacity building in O&M, actual post-project resource allocation has often been insufficient. Without continuous investment in institutional strengthening, these systems risk rapid degradation. Moreover, recent evaluations have raised questions about the cost-effectiveness of large-scale irrigation schemes, such as Mwea, compared with decentralized alternatives in Sub-Saharan Africa (Kikuchi et al., 2021). Unless these systemic issues, particularly those related to governance, long-term O&M funding, and grassroots engagement, are addressed, the return on infrastructure investments may remain suboptimal, resulting in inefficiencies and wasted resources.

12.8.3 Somasila Dam (India)

The Somasila Dam, located across the Penna River in the state of Andhra Pradesh, is a major reservoir that supports irrigation in southern districts. With a storage capacity of 1.6 million acre-feet (AF) (2.0 km^3), the dam supplies the Pennar Delta irrigation network, which covers approximately 145,000 ha. The dam is a 352-m-long earthen structure with a maximum height of 39 m. It features a central gated ogee spillway equipped with 12 radial gates (15.24 × 14.168 m each). Maintenance of the Somasila system involves routine desilting, periodic inspection of radial gates, and reinforcement of spillway training walls. The SCADA-controlled canal system has helped reduce transmission losses. Institutional partnerships with Water User Associations (WUAs) ensure participatory irrigation management, including equitable water delivery to tail-end farmers.

The Central Water Commission (CWC) of India has developed rigorous pre- and post-monsoon maintenance protocols applicable to Somasila and other major reservoirs in India (CWC, 2018). These include daily visual inspections, reservoir level monitoring, and mechanical testing of hoist gates and electrical systems. Maintenance schedules are structured into daily, weekly, monthly, quarterly, semi-annual, and 5-year intervals and include activities such as lubrication, gate testing, sediment clearance, vegetation control, and underwater inspections by divers or remote-operated vehicles.

Despite this robust framework, the Somasila Dam faces challenges due to delayed government funding in 2025. The damage to the dam apron, a critical downstream erosion control feature, following Cyclone Nivar in 2020, remains partially repaired. Project engineers have warned of serious flood risks during the monsoon season if repairs are not completed promptly.

12.8.4 Imperial Dam (California, United States)

The Imperial Dam, completed in 1938, is a reinforced concrete slab-and-buttress ogee weir structure located on the Colorado River at the California–Arizona border, approximately 20 miles northeast of Yuma, Arizona. The dam plays a pivotal role in diverting Colorado River water into two major delivery systems: the All-American Canal, which supplies water to California's Imperial Valley, and the Gila Gravity Main Canal, which serves irrigation districts in Arizona.

The Imperial Dam is part of a broader hydraulic system that includes the Gila Headworks, the All-American Canal headworks, three large desilting basins, and sediment return infrastructure. All these components are operated and maintained by the Imperial Irrigation District (IID) in coordination with the U.S. Bureau of Reclamation, with cost-sharing among state and federal water agencies. Key maintenance and operational strategies include the following: (1) automated SCADA-based control of gate operations, diversion structures, and real-time monitoring of water levels and flows; (2) routine inspection and servicing of gates, spillways, trash racks, canal linings, and desilting equipment; (3) sediment management through continuous scraper operation and periodic desilting basin maintenance; and (4) infrastructure upkeep, including vegetation control, concrete repair, periodic dredging, and preventive maintenance of both irrigation and drainage systems. The Imperial Dam's Operations and Maintenance Section within IID's Water Department is responsible for implementing long-term strategies to ensure asset sustainability, structural integrity, and efficient water delivery. This includes coordinated scheduling of water orders, performance of quality control and reporting, and development of canal water accounting integrated into the Bureau of Reclamation's Colorado River accounting framework (Imperial Irrigation District, 2024).

Notably, the lining of the All-American Canal has significantly reduced seepage losses, contributing to substantial water conservation gains across the Imperial Valley. The Imperial Dam case highlights the importance of sediment control, automation, and inter-agency collaboration in maintaining one of the most productive and technically sophisticated irrigation diversion systems in the western United States.

REFERENCES

CWC (2018). Guidelines for preparing operation and maintenance manual for dams. Doc. No.CDSO_GUD_DS_03_v1.0. Central Water Commission, Government of India.

Imperial Irrigation District (2024). 2024 Annual Report. Imperial Irrigation District. https://www.iid.com/about-iid/news-resources/reports

Kenya National Irrigation Authority (2025). Mwea Irrigation Scheme. https://www.irrigation-authority.go.ke/projects/mwea-irrigation-scheme/#

Kikuchi, M., Mano, Y., Njagi, T.N., Merrey, D., and Otsuka, K. (2021). Economic viability of large-scale irrigation construction in Sub-Saharan Africa: What if Mwea irrigation scheme were constructed as a brand-new scheme? *The Journal of Development Studies*, 57(5), 772–789. https://doi.org/10.1080/00220388.2020.1826443

Paranage, K. (2019). The Mahaweli Development Project and the 'rendering technical' of agrarian development in Sri Lanka. *Heliyon*, 5(6), e01811. https://doi.org/10.1016/j.heliyon.2019.e01811

Sumanasekara, H.S., Jayasinghe, A., Herath, A.B., Siriwardhana, K.D., Imbulana, I.L., Mullegamgoda, K., Herath, R., Paranavithana, G., and Rathnayake, U. (2025). Insights on the irrigation conveyance efficiency of three canal systems for improving agricultural sustainability in Sri Lanka. *Water Supply*, 25(5), 807–824. https://doi.org/10.2166/ws.2025.056

U.S. Army Corps of Engineers (2023). National Inventory of Dams. https://nid.sec.usace.army.mil/#/learn/dams101

USDA (2023). Dam Safety. The U.S. Department of Agriculture. https://www.nrcs.usda.gov/programs-initiatives/watershed-rehabilitation-program/dam-safety-resources-for-dam-owners

13 Maintenance of Canals

13.1 INTRODUCTION

Canals, which serve as critical conduits for water transportation, irrigation, and, in some cases, navigation, require diligent and ongoing maintenance to ensure their effective and efficient operation. The maintenance of canals encompasses a broad array of activities, each targeted at addressing specific challenges posed by both natural forces (e.g., sedimentation, erosion, and vegetation) and human factors (e.g., infrastructure wear and operational errors). This chapter reviews the multifaceted nature of canal maintenance, highlighting how each maintenance activity contributes to preserving the structural integrity and operational performance of canal systems.

From the foundational tasks of priming canals for optimal water flow to the management of aquatic growth and silt accumulation, this chapter covers the strategies and techniques employed to maintain these vital waterways, including (1) priming of canals, (2) mitigating settling or slipping of banks, (3) preventing overtopping of canal banks, (4) protecting against canal erosion, (5) managing silt and sediment in canals, (6) handling aquatic growth, (7) controlling vegetation on canal banks, (8) protecting canals in blowing soils, (9) controlling burrowing animals, and (10) fencing canals right-of-way. These components form the foundation of a sound canal maintenance strategy, ensuring longevity and performance under diverse environmental conditions.

13.2 HYDRAULIC DESIGN OF CANALS

The hydraulic design of conveyances (canals, laterals, ditches, and drains) is dependent on the flow characteristics. The coefficient of hydraulic roughness (i.e., Manning's n) used in the design must be maintained during the operational phases, if the system is to operate satisfactorily and carry the quantity of water needed. Extensive tests over the years have established empirical values of the roughness coefficient to be used by designers under anticipated conditions.

Conveyance may also be designed using stable channel relationships. For this procedure, the designer may adopt a cross-section and slope that results in a velocity and discharge that will not cause aggradation or degradation of the section when the type of material through which the canal is excavated and the sediment load (both quantity and physical properties) are taken into account in the design. This design procedure is being employed at present.

Operating personnel should have a clear understanding of the channel conditions represented by the chosen roughness coefficients or stable channel assumptions and should preserve these conditions by routinely cleaning and reshaping channels and eradicating weeds, willows, moss, and other undesirable growth. If the conveyance was designed for a Manning's n value, which cannot be maintained with normal

DOI: 10.1201/9781003747444-16

operating and maintenance forces at a reasonable cost, major modifications or improvements should be considered.

Maintenance efforts in conveyances, excluding structures, are then primarily directed toward keeping them in such a condition that Manning's n value or the stable channel assumption will remain the same as those used in design. This is not an easy task and requires much attention and expense.

The maintenance of an irrigation system begins the day the system is placed in operation. One of the first problems to be encountered is probably the pudding and priming of the system, if this is not accomplished as part of the construction work, to be certain that the system can be operated as planned without excessive erosion, settlement of canal banks, and structures. This significantly enhances the stability of conveyance system and highlights areas of potential weakness.

Maintaining the cross-section of a conveyance depends on its longitudinal slope. For example, straight reaches require reshaping because of (1) bank sloughing and (2) sediment deposits. Curved reaches may require constant reshaping because of (1) bars (on the inside of the curve) and (2) bank erosion (on the outside). This maintenance may be accomplished during off-season periods when there is no flow in the conveyances. Additionally, much work can be done with flows in the canal via draglines and dredgers. This operation necessitates the use of specialized mechanical equipment, including draglines, backhoes, grade-alls, bulldozers, and similar machinery. Each of these plays a crucial role in the maintenance process, offering distinct capabilities for reshaping, cleaning, and repairing the canal infrastructure.

Canals are generally cleaned and reshaped by machinery, with the most common being the ordinary dragline. Special dredging machines have been developed for use on smaller canals. The dragline is used on larger canals where the banks are strong enough to carry its weight. Endless conveyor types of bucket dredges are used on some canals. Canal cleaning by dragline is most effective when water is present in canals, as it softens the sides and bottoms, facilitating the deposition of excavated material. Canals are cleaned by equipment working from the lower end to the upstream end so that debris, weed seeds, and fine materials are not carried downstream into uncleaned sections where they tend to clog or choke the canal system.

A very important reach of the main canal is the one that is just downstream from the headwork. This reach is very susceptible to sedimentation. The slope of the river is such that existing velocities are adequate to carry considerable sediment. If this material is not removed at the headworks, the flat grade in the canal and consequently lower velocity will cause much of the sediment load to drop in the first few miles of the canal. Where the stream is known to carry high silt loads, the headworks may include devices to remove at least part of the load and return it to the stream. This may include a low-level sluice outlet and a wasteway channel extending back to the river, one to three miles (or farther) downstream from the intake. When water is not needed for irrigation, the reach between the intake and sluice can be scoured out by diverting large flows into the canal. It is often necessary to remove the material from the canal via a dragline.

When fine sediment is removed from the flow in the canal, a settling basin may be built downstream from the headworks. Such a device is a real maintenance problem. If it works as intended, it is filled with sediment, which must be removed, usually mechanically by dredging or draglines. There may not be a right-of-way to store more than a limited accumulation. The silt must then be hauled away for disposal, which is another costly operation.

Bank settlement may cause low plots and distortion of the cross-section. These issues should be corrected as soon as possible because they may be indications of future, more serious problems.

Water weeds in the canal cross-section can significantly change Manning's *n* value in a short period. Many types of weeds infest the system. Some remain completely submerged, and others emerge above the water. Many varieties of moss can be physical and/or chemical in nature. The subject of weeds is covered more thoroughly in Chapter 15. These weeds, including willows and brushes, cause other damage in addition to restricting flow. Weeds growing in water often provide places for silt deposits. The roots may cause leaks in the canal banks, especially after they die and rot out, leaving a natural piping system.

These remarks are primarily applicable to unlined canals. However, conveyances, even those with hard surface linings, suffer from many problems. In addition, many pests thrive on hard surfaces. Among these are freshwater sponges, clams, and many types of marine growth. These deposits tend to increase Manning's *n* value and reduce the carrying capacity unless they are removed. Mechanical removal may be costly and difficult, so chemical treatment may be used.

Land weeds, especially grasses, grow along the berm of the conveyance, where they are well watered and hang over into the flow. They soon collect trash and sediment, radically changing the cross-section. Soil sterilants are used to inhibit growth. This berm may also be bladed off with suitable mechanical equipment.

In earthen and earth-lined conveyances, burrowing animals are a maintenance concern. They dig through banks, and unless a leak is observed, a break may occur.

Notably, one of the objectives of the management of an irrigation enterprise is to deliver water from the source to the irrigator. The major facility for accomplishing this goal is the extensive conveyance system in the project. If these are not maintained to deliver the water in the required quantity at the necessary time, the entire enterprise may fail. The attached sketch shows the consequences of neglected maintenance through three cross-sectional views (Figure 13.1). Figure 13.1a represents a properly maintained canal, with its original cross-section intact. Figure 13.1b depicts a canal with significant silt and sediment accumulation at the bottom, which notably reduced the water flow area. Figure 13.1c shows a canal with eroded and collapsed banks, leading to an unstable, widened, and shallow cross-section that impairs hydraulic efficiency. These examples underscore the importance of routine inspection, sediment removal, bank protection, and other critical maintenance activities throughout the system. Failure to address such issues systematically can result in reduced delivery efficiency, water loss, and even structural failure.

13.3 PRIMING CANALS

Priming is essential, whether the canal is new or old. This process involves gradually introducing water into the canal system to prepare it for full operation. This process serves several important functions. If a canal is old and established, water should be

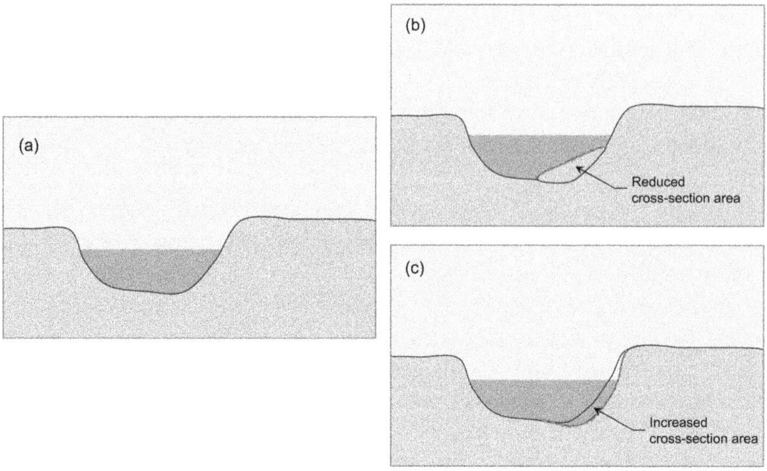

FIGURE 13.1 (a) a properly maintained canal, (b) a canal with significant silt and sediment accumulation, and (c) a canal with eroded and collapsed banks.

turned in at least 2–4 weeks before the anticipated need for water delivery. If two weeks are needed to bring a canal to full depth, priming should be initiated four weeks before sufficient amounts of water are needed to avoid severe interruption of service. The benefits of early priming include (1) the ability to permit sufficient time for repairs if needed and (2) the ability of water released during priming to clear out accumulated weeds and drift, thereby cleaning the canal.

If a canal is new, much care should be exercised when water is run for the first time. This is because the banks of the canal and the backfill around the structures have not yet fully settled and compacted. Thus, the canal should be filled slowly, carrying a shallow depth over an extended period. This gradual process allows banks to absorb moisture, settle properly, and reveal any potential weaknesses for repair. It also helps identify and address issues caused by burrowing animals or around structures. Ideally, priming for a new canal should be conducted in the season prior to its intended operational use.

Abrupt changes in water levels, either shutting off or filling the canal too quickly, can be detrimental. Consider the case in which the canal is running at full depth. The canal banks absorb water, and the portion of the bank adjacent to the water becomes saturated. The plane of saturation depends on the composition of the canal bank and the conditions for water absorption. This plane is held on the inner side of the canal bank. When the water is suddenly lowered, it removes the pressure of the water, and the inner face of the bank may slip if its slope is steeper than the material can withstand in its saturated condition. In light soils, the moisture in banks may drain out bcforc such slipping or sloughing occurs. Sloughing also occurs on canal banks with riprap. In the case of lined canals, the concrete lining may be fenced out of position due to pressure accumulation behind the lining. In heavy soils, sloughing is even more common if canal banks are not protected. This is also true in canals operating at relatively high velocities.

If repairs are necessary during peak water demand, quickly restoring the canal to full discharge can cause erosion. Allowing more time for gradual priming can help prevent such issues.

13.4 MITIGATING SETTLEMENT OR SLIPPING OF BANKS

Canal banks, particularly those on side hills or unstable grounds, are prone to settlement or slippage, largely due to two factors: (1) Seepage-induced settlement. When seepage from the canal reaches uncompacted material, it settles under the influence of seepage water. (2) Slope stability disruption. When the canal is excavated on talus slopes below higher ground, it destroys the natural balance, and the formation may move whether there is water or not.

When a canal passes over or through uncompacted material, such as sand or gravel, seepage from the canal causes the rearrangement of particles in this material, leading to eventual settlement. Where a canal passes through a fill material containing much organic material, such as peat, or where the underlying material is in a semi-liquid condition, the weight of the fill may cause settlement due to the compaction of the lower material.

When a canal is cut into a slope, it destroys the footing of the higher slope. This may cause the upper material to slip either gradually or in larger movements, blocking the upper part of the canal or pushing the foot of the slope out and carrying the canal with it. Such movements are independent of any seepage due to pressure from above the canal and may cause cracks in the lower canal banks.

In the maintenance and mitigation of canal banks, particularly in areas prone to instability, a combination of thoughtful design and responsive maintenance strategies is essential. Designing the canal as a bench flume is often advisable in such regions, as it minimizes disruption to the natural slope and enhances stability. It is better to meet these conditions, as they arise in such cases, rather than trying preventive measures. The strategic placement of wasteways is crucial for quick water diversion, which can significantly limit damage from potential breaks. When excavating material from the canal, a beneficial practice involves using the excavated material as a protective blanket on the lower banks rather than adding it to the bank's crown. This approach helps stabilize the bank. In situations where slipping occurs between different strata, the implementation of piling serves as an effective measure for stabilizing the area. Additionally, seepage-induced sloughing can be addressed through methods such as concrete lining, which is particularly effective in gypsum formations, to prevent saturation of the banks. In instances of breaks, particularly in side-hill canals, it is important to avoid rebuilding along the same alignment to prevent the recurrence of the issue. Overall, maintaining canal banks on unstable grounds or slopes demands an intricate balance of proactive design and agile, responsive maintenance strategies.

13.5 OVERTOPPING OF CANAL BANKS

Overtopping, a critical issue in canal management, occurs when water exceeds the bank level, often leading to potential structural damage and operational disruptions (Figure 13.2a). This overtopping is typically caused by blockages in the canal, such

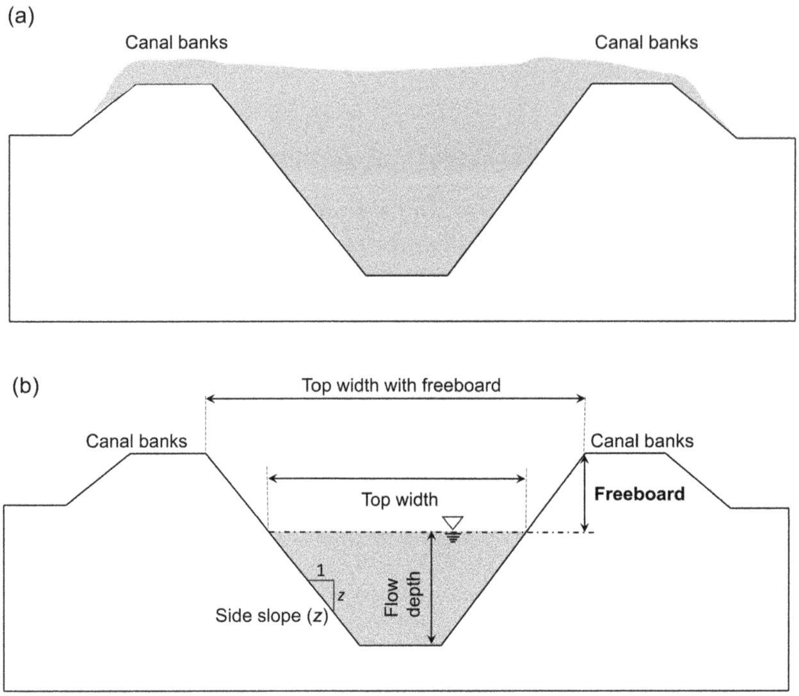

FIGURE 13.2 Illustration of (a) overtopping of canal banks and (b) freeboard in a canal.

as the choking of the flow due to clogged checks or material slippage into the upper slope. To mitigate this risk, maintaining an adequate level of freeboard (Figure 13.2b), i.e., the distance between the water surface and the top of the canal bank, is essential. Overtopping of a canal is usually caused by the choking of the canal due to the clogging of the checks or the slipping of material into the upper slope.

The required freeboard varies with the canal size and capacity. For large and deep canals, a minimum of 3 ft (0.91 m) of freeboard is recommended. Medium-sized laterals, with capacities ranging from 50 to 100 cfs (1.4 to 2.8 m^3/s), should maintain a freeboard of 1.5–2 ft (0.46–0.61 m). For canals with a capacity of less than 50 cfs (1.4 m^3/s), a freeboard of at least 1 ft (0.30 m) is recommended (USBR, 1967).

The width of the canal also impacts the required freeboard. Wider canals may necessitate less freeboard due to their increased capacity to handle water flow. The type of material comprising the canal bank is another crucial factor. For example, small canals in light soils may require freeboard of 2–3 ft (0.61–0.91 m), whereas in heavy soil conditions, a 1-ft (0.30 m) freeboard may suffice for canals carrying 200–300 cfs (5.7–8.5 m^3/s).

Routine inspections of canal banks and structures are vital for identifying potential blockage points or areas of material slippage. Overgrowth of vegetation can lead to reduced canal capacity and increased risk of overtopping. Regular maintenance to control vegetation growth is essential. In areas prone to overtopping, additional structural reinforcements such as riprap or concrete linings may be necessary to strengthen banks and prevent erosion. Having a well-defined emergency response plan for instances of overtopping ensures swift action to minimize damage and restore normal operations.

During the design phase, considerations for future expansions and changes in water demand should be factored in to accommodate increased flow without the risk of overtopping. Periodic assessments of canal capacity and usage patterns can guide decisions to increase canal size or improve flow management strategies.

13.6 PROTECTION AGAINST CANAL EROSION

Erosion control in canal systems is vital for maintaining their structural integrity and operational efficiency. The key to controlling erosion lies in managing the flow velocity within the canal, ensuring that it remains below the maximum threshold tolerable by the materials comprising the canal's structure. The permissible canal velocities for different original materials excavated are shown in Table 13.1 (Scobey, 1939).

Excessive flow velocities are a primary cause of erosion. Keeping the velocity within safe limits for a specific canal material is essential. Linings made from

TABLE 13.1
Maximum Allowable Flow Velocity in the Canal for Different Original Materials Excavated

	Flow Velocity (ft/s)		
Original Material Excavated for Canal	**Clear Water, No Detritus**	**Water Transporting Colloidal Silts**	**Water Transporting Noncolloidal Silts, Sands, Gravels, or Rock Fragments**
Find sand (noncolloidal)	1.50 (0.46 m/s)	2.50 (0.76 m/s)	1.50 (0.46 m/s)
Sandy load (noncolloidal)	1.75 (0.53 m/s)	2.50 (0.76 m/s)	2.00 (0.61 m/s)
Silt loam (noncolloidal)	2.00 (0.61 m/s)	3.00 (0.91 m/s)	2.00 (0.61 m/s)
Alluvial silts (noncolloidal)	2.00 (0.61 m/s)	3.50 (1.07 m/s)	2.00 (0.61 m/s)
Ordinary firm loam	2.50 (0.76 m/s)	3.50 (1.07 m/s)	2.25 (0.69 m/s)
Volcanic ash soil	2.50 (0.76 m/s)	3.50 (1.07 m/s)	2.00 (0.61 m/s)
Fine gravel	2.50 (0.76 m/s)	5.00 (1.52 m/s)	3.75 (1.14 m/s)
Stiff clay (very colloidal)	3.75 (1.14 m/s)	5.00 (1.52 m/s)	3.00 (0.91 m/s)
Graded, loam to cobbles (noncolloidal)	3.75 (1.14 m/s)	5.00 (1.52 m/s)	5.00 (1.52 m/s)
Alluvial silts (colloidal)	3.75 (1.14 m/s)	5.00 (1.52 m/s)	3.00 (0.91 m/s)
Graded, silt to cobbles (colloidal)	4.00 (1.22 m/s)	5.50 (1.68 m/s)	5.00 (1.52 m/s)
Coarse gravel (noncolloidal)	4.00 (1.22 m/s)	6.00 (1.52 m/s)	6.50 (1.98 m/s)
Cobbles and shingle	5.00 (1.52 m/s)	5.50 (1.68 m/s)	6.50 (1.98 m/s)
Shales and hardpans	6.00 (1.83 m/s)	6.00 (1.83 m/s)	5.00 (1.52 m/s)

Adapted from Scobey (1939).

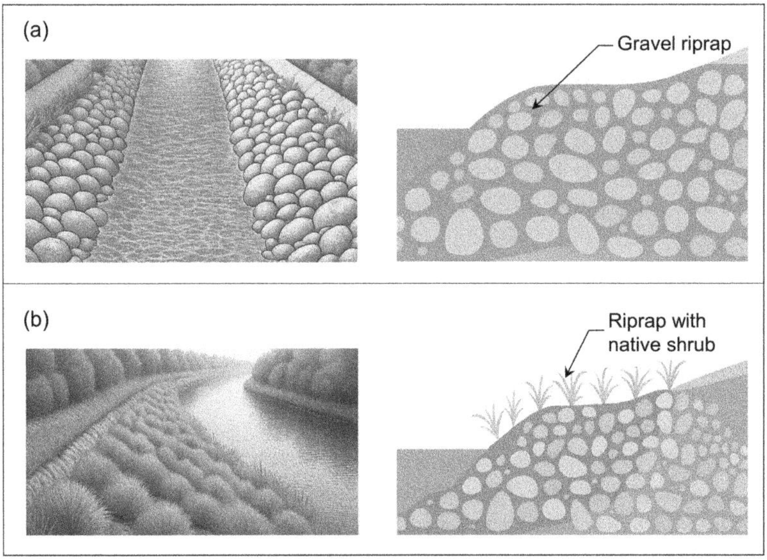

FIGURE 13.3 (a) Gravel riprap and (b) vegetated riprap.

materials, such as concrete, rock, gravel, or brush riprap, can effectively control erosion. Each material has its own advantages and is suitable for different erosion scenarios. Concrete is a popular choice for linings because of its durability and effectiveness in reducing both erosion and seepage. When installed correctly, concrete linings are cost effective, considering their long-term benefits in terms of maintenance reduction. In cases where erosion is primarily due to wave action or high velocities near structures, the gravel lining is beneficial (Figure 13.3a). This is an economical and temporary solution for preventing canal and structural erosion. Although not as long-lasting as concrete, brush riprap can endure for several years and is effective in many situations (Figure 13.3b).

In some canals, it is better to smooth the eroded section rather than to restore the original section. This is because the eroded section more closely conforms to the natural section under actual flow conditions and is more easily maintained.

13.7 MANAGEMENT OF SILT OR SEDIMENT IN CANALS

Silt or sediment generated in canals should be properly managed. Depending on the amount of silt generated, its deposition may be desirable or objectionable. For example, in porous soils, the deposition of sediment in the canal bed and sides reduces seepage losses and is therefore beneficial. The seepage loss, expressed as the percent of discharge lost per kilometer of canal, varies with the size of the cross-section and the material of the canal bed. In canals operating under checks, the deposition of silt is greater. Sometimes, the deposition of silt may be so extensive that it severely affects the canal's capacity.

If the value of water served and the value of preventing possible damage from the waterlogging of lower lands by seepage water are greater than the cost of silting, artificial silting may be considered. Artificial silting may be carried out in different ways: (1) depositing the silt directly at the location desired during the nonoperating season and (2) allowing water containing silt into the canal and allowing it to distribute under the influence of sedimentation. If the entire system is located in sandy land and may require silting, the second method may be preferable. When only short reaches of a canal through porous soils require treatment, the first method is preferable. Neither method is effective in canals where the velocity is sufficiently high to erode the silt. The silt deposited in place during the off-season is referred to as the puddle lining.

If excess silt is deposited, it needs to be removed. The rate of silt deposition is the determining factor in handling silt. Silt can be removed in different ways: (1) design of headworks and canals permitting minimum deposition of silt, (2) removal of silt by large velocities in large canals if sufficient fall is not available, (3) use of an enlarged cross-section at selected locations, and (4) sluicing out of the silt deposits above checks and stirring or loosening the silt with harrows during such sluicing.

13.8 HANDLING OF AQUATIC GROWTH

All types of growth that occur within the water cross-sections of canals are defined as aquatic growth. They include a diverse range of plants, such as mosses, algae, pondweeds, and lichens. A number of methods are employed to handle aquatic growth, including (1) cutting, (2) dragging, (3) harrowing, (4) sun killing, and (5) miscellaneous methods.

Since aquatic plants trail downstream from their place of growth, they can be effectively removed by being cut with a saw or knife. Where applicable, dragging may be preferable to sawing because of greater efficiency and lower cost. Harrowing with disk harrows has been found to produce mixed results. If the water can be drained from the portions of an irrigation system, pond weeds and other aquatic growth can be killed by exposure to the sun for 3–8 days, making this method both inexpensive and effective. Other methods of handling aquatic growth include raking, chemical treatment, oiling canal beds, spraying with arsenic solutions, and cattle grazing.

13.9 HANDLING OF SEMI-AQUATIC PLANTS

A number of plants have roots within the canal water cross-section, but their growth is largely above the water surface. These plants are referred to as semi-aquatic plants. Examples include tules, cat tails, water grass, bamboo, and joint grass. Additionally, species, such as Johnsongrass and Bermudagrass, may extend their roots into the adjoining soil and into the water. Semi-aquatic plant growth is limited to shallower depths. Such growth, where it occurs, may more effectively stop the flow than do aquatic plants.

The presence of semi-aquatic plants in canals is both beneficial and challenging. While they contribute to the canal's ecological balance and can reduce the velocity of water flow, leading to increased silt deposition, they also pose management challenges because of their robust growth. The erect stature of these plants means they are resilient against water currents, making their removal more complex. Semi-aquatic plants can be controlled by (1) cutting and mowing, which are effective for large areas of growth; (2) digging and pulling, which is suitable for smaller infestations or sensitive areas where machinery cannot be used; and (3) grazing, which allows livestock to graze these plants, can be an eco-friendly management option.

13.10 HANDLING OF VEGETATION ON CANAL BANKS

Vegetation growing on canal banks ranges from beneficial to detrimental, impacting both the physical and the ecological aspects of canals. In areas with blowing soils, vegetation on canal banks is advantageous, as it provides stability. However, in other soil types, vegetation growth can be harmful in several ways: (1) plants interfere with the use of banks; (2) they extend into the water and retard the flow; and (3) they are a source of tumbling weeds that clog screens or structures in canals or supply weed seeds, which may foul fields on the farm.

Commonly growing weeds on canal banks vary with the soil and climate. Common growing weeds include tumbling weeds, sweet clover, bronco grass or cheatgrass, Johnsongrass, bermudagrass, watergrass or barnyard millet, willows, etc. Tumbling weeds found along canals are three kinds: Russian thistle, tumbling mustard, and ordinary tumbleweed. A survey of the weeds on the irrigation canal system maintained by the Elephant Butte Irrigation District in New Mexico and Texas identified 76 plant species, which were grouped into 13 categories (Schroeder et al., 2012). Common species included Bermudagrass, scouring rush, rice cutgrass, Johnsongrass, curly dock, *Cyperus* species, *Echinochloa* species, and *Plantago* species. Bermudagrass was the most common and was found at more than half the sampled locations, and its presence was influenced by the soil potassium and phosphorus levels. Scouring rush was more common in smaller, intermittent canals and was affected by salinity and soil organic matter. Rice cutgrass was more prevalent in areas with higher soil organic matter. The distributions of Johnsongrass and curly dock were influenced by the soil texture and chemical properties. The occurrence of *Cyperus* and *Echinochloa* species varied with soil phosphorus level, whereas *Plantago* species were less common under highly saline or sodic conditions.

There are many methods for controlling vegetation on canal banks. These include grazing, mowing, chemical treatment, seeded grasses, and others. In the case of canal banks with heavy vegetation, grazing by sheep has been found to be useful, especially when it is nutritive. Sheep pack the banks without wearing them down, graze more closely, and will themselves make some growth on much leaner pasturage; they are more beneficial than other stock. However, mowing is not very effective for the control of the Russian thistle because of its proximity to the ground. Chemical treatment includes controlling vegetation with the use of chemical solutions that are available in different forms. Two types of treatments are generally used: (1) the root absorption method and (2) the leaf absorption method. For the root absorption method, herbicides are added to the soil to prevent plant growth and seed germination. The absorption of chemicals in the soil can kill plants or prevent seeds from growing. This approach may be effective for canal banks. For the leaf-absorption method, the chemical is sprayed directly onto the plant to kill it or inhibit seed maturation.

Efforts to establish specific types of grasses on canal banks, such as western wheatgrass, brome grass, alfalfa, bluegrass, and clover, have yielded mixed results, often falling short of expectations. Other control methods, such as the harrowing and oiling of canal banks, are also practiced, although their specifics are not detailed in the passage. The ongoing challenge is to find a plant that can be easily established on canal banks, maintain itself with minimal care, and have an unobjectionable impact on the canal environment.

13.11 PROTECTION OF CANALS IN BLOWING SOILS

It is always difficult to maintain canals in light soils because of wind erosion or bank blowing. The difficulty is twofold. First, the drifting soil tends to fill the canal cross-section. Second, erosion weakens the structural integrity of canal banks. To address these issues, various protective measures have been implemented. These include cultivating rye to stabilize the soil, fencing to protect against range stock, applying stable manure as a coating, covering banks with gravel, growing Russian thistle and securing it with woven wire, and using cement coating, calcium chloride, crude oil, sugar syrup, coal tar, wind breaks, and sand fencing.

13.11.1 Canal Screens

Canal screens play a crucial role in weed management. They are typically installed at the inlets of siphons and other feasible locations, although some systems do not use them at all. While these screens are effective in preventing the clogging of inlets and siphons by weeds, they themselves can become clogged and require frequent cleaning. In main canals where the water velocity is high, screens may not be necessary, as the flow is sufficient to carry weeds through the siphons. However, in laterals, especially where siphons are smaller, the risk of clogging is greater, making screens a more common feature.

13.11.2 Controlling Burrowing Animals

Canal maintenance is further complicated by the presence of burrowing animals. These animals make holes that can become weak points, leading to canal breaks. When the water surface is raised above these holes, breaks are likely to occur. Such holes are also troublesome and dangerous to fill. The prevention or repair of breaks caused by such animals may be the most significant component of canal maintenance costs.

There are different kinds of animals that damage canals. These include pocket gophers, ground squirrels, muskrats, meadow mice or rats, kangaroo rats, badgers, etc. Pocket gophers cause the most trouble and are the most common. The natural enemies that kill gophers include hawks, owls, coyotes, badgers, and weasels.

Ground squirrels are quite common and are diurnal in nature. Muskrats burrow into banks from the inside of the canal below the water surface. Prairie dogs seldom infest canal banks, but they are destructive to crops. Meadow mice or rats may be harmful to smaller laterals, especially around structures. Kangaroo rats sometimes make burrows in ditch banks. Badgers live on various burrowing animals. Therefore, they do more good than harm. The same is true of weasels, for they kill gophers and mice.

There are a multitude of methods for controlling burrow animals. These include poisoning, trapping, suffocation, natural, and some special methods. Strychnine forms the basis of most of the poisons used. It can be used in a variety of baits, depending on the feeding habits of the target animals. Traps have been used for gophers and squirrels. Suffocation of animals with the exhaust fumes of automobiles or the fumes of carbon bisulfide is also an effective method of killing animals. Because certain burrowing animals may be part of the food of other animals, larger animals should be protected. For example, gophers and squirrels are preyed upon by badgers, weasels, skunks, hawks, and owls. With badgers, the harm caused may exceed the benefit. Other control methods may include concrete lining, the pasturing of sheep, the use of oil in holes placed along banks, and the use of sand deposits.

13.11.3 Fencing Canal Rights-of-Way

The issue of fencing canals and the legal implications involved depend on the nature of the right-of-way. Generally, there are two types of rights-of-way: (1) ownership by the canal company, similar to a farmer's ownership of land, and (2) mere easement, where the company is granted the right to use the land solely for canal-related purposes. In cases where land adjacent to the canal is privately owned, the landowner must take reasonable measures to prevent any damage to the canal by their livestock and is liable for compensating for any damage incurred. Furthermore, they must obtain permission from the canal company before erecting fences across the right-of-way.

When the right-of-way is held as an easement, the canal owner is referred to as the dominant estate, whereas the original grantor of the easement is the servient estate. Here, the landowner maintains the right-of-way, provided that it does not interfere with canal operations. The canal owner is responsible for safeguarding the canal against the landowner's ordinary use of the land. Challenges in this context include issues related to the erection of fences along canal rights-of-way, the potential damage to crops by livestock, the liability for such damage, and the costs associated with fencing.

13.12 CASE STUDIES

13.12.1 Canal Seepage and Conveyance Loss Reduction in the Lower Rio Grande Valley, Texas

The Lower Rio Grande Valley (LRGV) in Texas is one of the most productive agricultural regions in the United States and is supported by a complex system of irrigation canals that divert water from the Rio Grande. This system, which is managed by

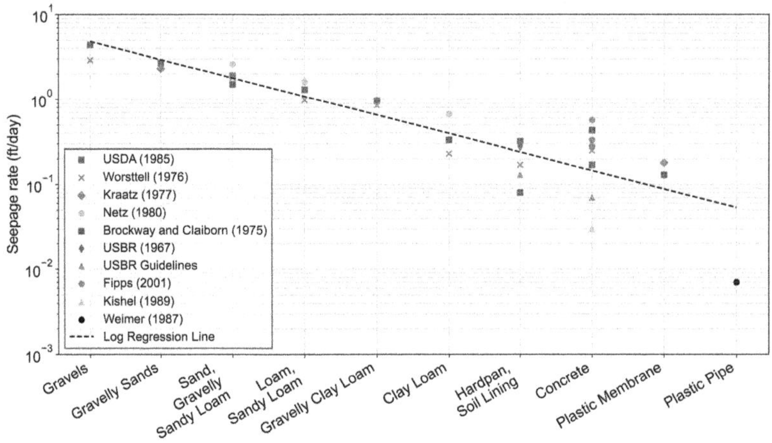

FIGURE 13.4 Canal seepage rates measured for unlined canals with different soil types and for canals with typical lining materials. (Some of the data used in the figures are adapted from Sonnichsen, 1993.)

more than 20 irrigation districts, many of which were established in the early 20th century, has suffered from aging infrastructure. Many networks, including earthen canals and aging concrete linings, have experienced significant seepage and conveyance losses over time. These inefficiencies, combined with recurrent droughts, rapid urban growth, and international water-sharing requirements, have intensified the urgency of improving water use efficiency.

To address these challenges, a study conducted in 1998 across 28 irrigation districts of the LRGV quantified two major sources of water loss: seepage losses and conveyance losses (Fipps, 2001). Seepage losses were measured via the standard ponding method (Kraatz, 1977), where water level declines were recorded in isolated canal segments under nonflowing conditions. The results revealed seepage rates ranging from 0.2 to 0.6 ft/day, even in some concrete-lined canals (0.3 and 0.6 ft/day), indicating deterioration in construction or maintenance. As shown in Figure 13.4, typical seepage rates for concrete-lined canals ranged from 0.1 to 0.3 ft/day, whereas unlined canals varied widely depending on the soil type.

In response, several irrigation districts have initiated targeted canal rehabilitation programs. These efforts included (1) replacing or reinforcing old linings with new reinforced concrete, (2) reshaping and compacting earthen canals, (3) installing pipeline segments, and (4) implementing flow monitoring and automation systems. Rehabilitation efforts were prioritized via detailed seepage maps and cost–benefit analyses to address the most inefficient sections.

Water management strategies (WMSs) submitted by districts in three planning cycles were synthesized to create a general conservation framework. These include (1) canal lining and pipeline installation (including interdistrict interconnects); (2) repairs and improvements (automated gates, new meters, and SCADA); and (3) metering at farm gates and within canal networks. The total estimated water savings from all submitted WMSs in 2080 is 151,233 acre-feet (AF)/year, based on current and potential conveyance efficiencies (targeting 90%). For districts without specific seepage data, average values from Fipps (2001) were applied to estimate water conserved per mile of canal lining or pipeline replacement, assuming negligible evaporation losses.

13.12.2 The Maintenance of the All-American Canal in California

The All-American Canal (AAC) is an 82 miles (132 km) long water delivery system located in southeastern California, which is the largest irrigation canal in the world. Completed in 1942 under the Boulder Canyon Project Act, it conveys water from the Colorado River to the Imperial and Coachella Valleys, enabling irrigation across more than 630,000 acres of farmland and serving nine cities (USBR, 2025). The canal is managed by the Imperial Irrigation District (IID) under contract with the U.S. Bureau of Reclamation. Due to its scale, age, and desert location, the AAC faces multiple maintenance challenges, including seepage losses, sedimentation, bank erosion, vegetation encroachment, and structural degradation. To address this, the IID performs regular activities such as canal cleaning, debris and vegetation removal, and bank shaping to maintain flow efficiency and structural integrity.

Routine cleaning and debris removal are fundamental maintenance activities on the AAC. Desert winds frequently blow sand into the canal, and runoff from nearby fields contributes silt and organic matter, which must be cleared to preserve conveyance capacity and prevent overtopping. Cleaning is typically performed annually or biannually, using canal beds shaped with templates to maintain the designed cross-section. Similarly, vegetation management is conducted seasonally. Deep-rooted bushes and trees on embankments are cleared to prevent cracks and seepage, improve visibility for inspection, and reduce interference with hydraulic structures. Prior to the major canal lining project in the 2000s, vegetation was systematically removed to prepare for construction and reduce risk to the concrete structure.

Historically, one of the most pressing issues affecting the ACC was seepage, particularly in the Algodones Dunes region where sandy soils led to substantial water loss. The estimated water loss was up to 67,700 AF annually, much of which seeped across the U.S.-Mexico border (IID, 2025). To address this, the All-American Canal Lining Project (AACLP) was implemented between 2007 and 2009. The $200 million project constructed a 23-mile concrete-lined canal parallel to the original earthen section from Pilot Knob to Drop 3 (as shown in Figure 13.5). The lined canal significantly reduced seepage and maintenance needs and improved long-term delivery reliability. In addition to reducing water loss, the new lined section avoided areas prone to sand invasion, thus lowering future maintenance costs.

Environmental mitigation was a significant part of the AACLP. The Chanan Remington Memorial Wetland, located between Drops 3 and 4 (Figure 13.5a), was developed as a mitigation site, restoring 44 acres of riparian habitat, including marshes, cottonwood-willow groves, and mesquite stands, highlighting how infrastructure upgrades can be aligned with ecological goals.

Ongoing erosion control includes reshaping, compacting, and stabilizing canal banks using methods such as rock riprap and grass sod placement. Drop structures and emergency spillways are used to dissipate flow energy during peak irrigation.

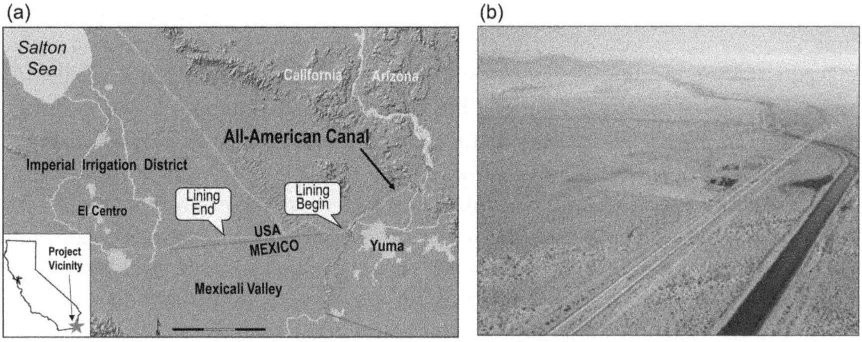

FIGURE 13.5 (a) Location of the All-American Canal and (b) a section of the canal showing the new concrete-lined canal parallel to the original unlined canal. (Adapted from USBR, 2006.)

Structural maintenance, such as sealing cracks in the concrete lining, repairing embankments, and upgrading weirs and gates, is scheduled based on seasonal inspections or initiated in response to observed damage. Notably, riprap was first installed in 1950 (USBR, 1950), with major repairs carried out later, reflecting a long-term pattern of post-season assessments followed by targeted interventions.

Maintenance schedules are strategically aligned with irrigation seasons, with major work planned during off-peak or post-harvest periods. Long-term capital planning has also been emphasized, particularly in the context of the AACLP and future modernization efforts. Looking ahead, IID has proposed an ambitious set of modernization goals between 2025 and 2028. These include the creation of a canal lining and sediment management plan by 2026, a weir repair and replacement strategy by 2027, and a full fleet and equipment assessment by 2028. To support operational efficiency, IID is investing in remote-controlled gates, high-water alert systems, and smart dashboards that enable real-time decision-making. Additionally, the Water Conservation Program provides funding to farmers for on-farm improvement projects, with deadlines scheduled for March and August every year (IID, 2025).

The maintenance of the All-American Canal illustrates how a complex, large-scale irrigation system must adapt to both engineering and environmental challenges over time. Through a combination of routine maintenance, major upgrades like the AACLP, and forward-looking modernization strategies, the IID continues to ensure the long-term functionality, efficiency, and resilience of this critical water delivery system.

13.12.3 New Method for Canal Seepage and Void Detection

In 2015, the U.S. Bureau of Reclamation (USBR) investigated a range of innovative methods for canal inspection with the goal of developing faster, more cost-effective, and reliable techniques to detect seepage and voids (Reclamation, 2015). Early detection of seepage and structural voids is essential for preventing costly failures and maintaining the reliable delivery of irrigation water.

Traditional field methods, such as the ponding method, dye tracer studies, and inflow–outflow measurements remain widely used. The ponding method, which involves isolating a section of canal and observing the rate of water level decline, is among the most accurate methods but requires canal shutdowns and controlled conditions. Dye tracer methods are helpful for confirming specific seepage pathways once problem areas have been identified. Inflow–outflow tests, which use flow measurement devices, such as weirs or flumes, are more suitable for evaluating longer canal reaches but are limited by the precision of measuring equipment. Additional field-based techniques include the use of seepage meters—such as constant-head permeameters—to measure infiltration directly at specific points and the estimation of losses on the basis of soil and lining characteristics.

To enhance large-scale assessments, several remote and geophysical methods have emerged. Geologic and hydrogeologic investigations using well logs, groundwater level data, and new boreholes can help determine whether elevated groundwater near a canal is due to seepage or other recharge sources. Remote sensing technologies have gained attention for their ability to rapidly screen long canal sections. Thermal imaging can reveal temperature anomalies linked to moisture presence, whereas ground-penetrating radar (GPR), electrical resistivity, self-potential methods, and microwave radiometry provide insights into subsurface water content and soil conditions. However, these methods may be influenced by nearby structures, utilities, and surface materials, and they typically indicate the presence of moisture without showing precise flow paths.

Other advanced methods include hydrochemical tracer studies, which use isotopes or chemical tracers to detect canal seepage into surrounding soils or aquifers. While informative, these techniques are often expensive and require specialized expertise, and the short residence time of canal water can limit detection accuracy. Fiber optic distributed temperature sensing (FO-DTS) systems offer a highly sensitive method for detecting leaks over long distances by tracking temperature changes along the canal lining. Although effective, the high installation and maintenance costs restrict their application to high-risk canal segments where failure poses serious hazards.

In summary, no single method is universally applicable. Instead, these technologies should be viewed as complementary tools that can be selectively deployed on the basis of canal type, soil conditions, operational constraints, and risk level. A combined approach that integrates traditional field methods with modern remote sensing and geophysical tools provides the most robust framework for effective canal seepage detection and long-term maintenance planning.

REFERENCES

Fipps, G. (2001). Potential Water Savings in Irrigated Agriculture in the Lower Rio Grande Valley of Texas. Final Report. The Texas Water Resources Institute, TWRI TR-183.

IID (2025). All-American Canal Lining Project. Imperial Irrigation District https://www.iid.com/water/library/all-american-canal-lining-project (Access on August 1, 2025).

Kraatz, D.B. (1977). *Irrigation Canal Lining*. United Nations Food and Agriculture Organization, Rome.

Reclamation (2015). Inspecting Reclamation Canals in New Ways: Accurately Predicting Canal Seepage. https://www.usbr.gov/research/docs/updates/ 2015–06-canal

Schroeder, J., Murray, L., Ulery, A., Fiore, C., Nguyen, H., and Liu, X. (2012). Identification and Detection of Weeds on Irrigation Canals: Survey of the Vegetation and Soils of the Leasburg Canal System, 2002–2006. New Mexico State University. https://pubs.nmsu.edu/research/water/RR777/index.html

Scobey, F.C. (1939). *Flow of Water in Irrigation and Similar Canals*. Technical Bulletin No. 652. United States Department of Agriculture, Washington, DC.

Sonnichsen, R.P. (1993). Seepage Rates from Irrigation Canals. Water Resources Program, Open-File Technical Report, OFTR 93-3. Washington State Department of Ecology.

USBR (1950). Annual Project History, Boulder Canyon Project–All-American Canal–Imperial Division, Interior Bureau of Reclamation. Washington, DC.

USBR (1967). Design Standards No. 3. Canals and Related Structures. U.S. Department of the Interior Bureau of Reclamation. Washington, DC.

USBR (2006). All-American Canal lining project. Supplemental Information Report. U.S. Bureau of Reclamation.

USBR (2025). Boulder Canyon Project – All-American Canal System. https://www.usbr.gov/projects/index.php?id=514 (Access on August 1, 2025).

14 Water Quality

NOTATION

TDS total dissolved solids (mg/L)
EC conductivity of water (dS/m)
SAR sodium adsorption ratio
EC_e electrical conductivity of the saturated soil extract (dS/m)
LF leaching fraction
I required irrigation depth (mm)
ET_c crop evapotranspiration (mm)

14.1 INTRODUCTION

Water quality is an important consideration for the operation and maintenance of irrigation systems. Poor-quality water can lead to numerous issues, including emitter clogging, soil degradation, crop toxicity, and reduced system lifespan. In addition to technical implications, water quality is increasingly governed by federal and state regulations that place responsibility on water users to prevent environmental contamination. In the United States, the Clean Water Act (CWA) is the primary federal law regulating water pollution. Originally enacted as the Federal Water Pollution Control Act of 1948, the legislation was reorganized and comprehensively expanded in 1972. The term "Clean Water Act" became the common name following further amendments in 1977. The Act defines the fundamental framework for regulating pollutant discharges and setting water quality standards for surface waters in the United States. Under CWA, irrigation districts and water delivery entities may be subject to regulation if their activities result in the discharge of pollutants, such as return flows, tailwater, or maintenance-related effluents, into streams, rivers, or other regulated surface waters. While agricultural return flows from irrigated fields are generally exempt under Section 402 of the Act, point source discharges from pumping stations, construction activities, or canal maintenance operations may require permits under the National Pollutant Discharge Elimination System (NPDES).

The U.S. Environmental Protection Agency (EPA) classifies water pollution sources into two main categories: point sources and nonpoint sources. Point sources refer to discrete, identifiable conveyances, such as pipes, channels, or outfalls from facilities, such as wastewater treatment plants, industrial operations, or pump stations. Nonpoint sources are diffuse in nature and include agricultural runoff, stormwater, erosion from unlined canals, drainage from irrigated fields, and the leaching of nutrients or pesticides into surface or groundwater.

It is important for irrigation district managers and engineers to familiarize themselves with the regulatory framework of the CWA, particularly when engaging in activities that may affect nearby water bodies. By complying with state and federal water quality standards, irrigation managers can minimize legal exposure and contribute

 DOI: 10.1201/9781003747444-17

to sustainable water management and environmental stewardship. Where necessary, consultation with state environmental agencies or legal counsel may be warranted to ensure that appropriate permitting and reporting procedures are followed.

The responsibilities associated with maintaining water quality vary significantly from one project to another. In smaller projects with high-quality water sources, drainage water and return flows to streams are minimal and generally of good quality, posing little concern. However, larger projects, particularly those involving sewage treatment plants or the return of large quantities of poor-quality water to streams, may find pollution control to be a major responsibility. Additionally, the impact of water quality on the operation and management of both the project and individual farm units within the project is a critical consideration.

Key considerations for operation and maintenance personnel regarding water quality include the following: (1) Irrigation water must meet specific standards for chemical and sediment content, and thermal conditions, which vary depending on cropping patterns and soil types; (2) addressing water quality issues requires (a) measuring flow at various locations, (b) sampling flow at numerous sites, (c) analyzing these samples, and (d) maintaining accurate records; and (3) conducting thorough inspections and observations across the watershed, up to the point of water release from the project, is essential. These topics are explored in the following sections of the chapter, which provide guidance on assessing and managing irrigation water quality.

14.2 IRRIGATION WATER QUALITY AND MANAGEMENT

14.2.1 Water Quality Standards for Agriculture

Water quality for irrigation should meet certain standards to produce an adequate environment for crop growth, achieve good yields, sustain soil health, and ensure the long-term sustainability of the irrigation system. The quality varies, depending on the dissolved salts present, which originate from the natural weathering of rocks and minerals like lime and gypsum. When irrigation water evaporates or is taken up by plants, these salts remain in the soil and may accumulate over time. The suitability of water for irrigation depends on both the total concentration of salts and the types of salts present. Poor-quality water can lead to reduced crop yields, soil structure degradation, and management challenges. Therefore, assessing water quality requires consideration of local soil properties, climatic conditions, crop sensitivity, and the experience of water users.

Factors that are commonly used to evaluate water quality include the following: (1) Salinity: High salt concentrations reduce the availability of water to plants, impairing their uptake and growth; (2) Water infiltration rate: Excess sodium or insufficient calcium can slow down the rate at which water penetrates the soil, potentially leading to an insufficient water supply for crops between irrigations; (3) Specific ion toxicity: Accumulation of certain ions like sodium (Na^+), chloride (Cl^-), or boron (B) in sensitive crops can reach toxic levels, harming crops and reducing yields; (4) Miscellaneous issues: Excess nutrients can decrease yield or quality; deposits on crops can affect marketability; and high corrosion levels can increase equipment maintenance and repair costs. Guidelines for evaluating irrigation water quality are shown in Table 14.1 (Ayers and Westcot, 1994). The guidelines presented here focus

TABLE 14.1
Guidelines for Irrigation Water Quality

Potential Irrigation Water Quality Problem	Describing Parameter	Degree of Restriction on Use		
		None	Slight to Moderate	Severe
Salinity (affects crop water availability)				
EC[a]	dS/m	<0.7	0.7–3.0	>3.0
TDS[b]	mg/L	<450	450–2,000	>2,000
Infiltration (affects water infiltration rate, which is evaluated using EC and SAR[c] together				
	SAR		EC, dS/m	
	0–3	>0.7	0.2–0.7	<0.2
	3–6	>1.2	0.3–1.2	<0.3
	6–12	>1.9	0.5–1.9	<0.5
	12–20	>2.9	1.3–2.9	<1.3
	20–40	>5.0	2.9–5.0	<2.9
Specific Ion Toxicity (affects sensitive crops)				
Sodium (Na^+)[d]				
Surface irrigation	SAR	<3	3–9	>9
Sprinkler irrigation	meq/L	<3	>3	
Chloride (Cl^-)				
Surface irrigation	meq/L	<4	4–10	>10
Sprinkler irrigation	meq/L	<3	>3	
Boron (B)	meq/L	<0.7	0.7–3.0	>3.0
Miscellaneous Effects (affects sensitive crops)				
Nitrogen (NO_3^-N)[e]	meq/L	<5	5–30	>30
Bicarbonate $(HCO_3)^-$ overhead sprinkling only	meq/L	<1.5	1.5–8.5	>8.5
pH			Normal range 6.5–8.4	

[a] EC is the electrical conductivity of the irrigation water in deciSiemens per meter at 25°C (dS/m) or in millimhos per centimeter (mmh/cm). 1 dS/m = 1 mmho/cm = 1000 μmho/cm.

[b] TDS is the total dissolved solids in mg/L. Generally, the following equations can be used:

$$\text{TDS}(\text{mg/L}) = \text{EC}(\text{dS/m}) \times 640 \text{ for EC} < 5 \text{ dS/m}$$

$$\text{TDS}(\text{mg/L}) = \text{EC}(\text{dS/m}) \times 800 \text{ for EC} > 5 \text{ dS/m}$$

[c] SAR is the sodium adsorption ratio.

[d] For surface irrigation, most tree crops and woody species are sensitive to sodium and chloride; refer to the values in Table 14.1. Most annual crops are not sensitive; refer to the values in Table 14.3. For chloride tolerance of selected fruit crops, see Table 14.4. Under overhead sprinkler irrigation and low relative humidity (<30%), sodium and chloride can be absorbed through the leaves of sensitive crops, increasing the risk of foliar injury

[e] NO_3^-N denotes nitrate-nitrogen reported in terms of elemental nitrogen (NH_4^-N and Organic-N should be included when wastewater is being tested).

TABLE 14.2
Terms, Units, and Useful Conversions for Understanding Water Quality Analysis

Parameter	Symbol	Atomic Weight	Units	Usual Range in Irrigation Water
		Constituents		
(1) Cations				
Calcium	Ca^{2+}	40.1	meq/L	0–20
Magnesium	Mg^{2+}	24.3	meq/L	0–5
Sodium	Na^{+}	23.0	meq/L	0–40
Potassium	K^{+}	39.1	meq/L	0–2
(2) Anions				
Bicarbonate	HCO_3^-	61.0	meq/L	0–10
Sulfate	SO_4^{2-}	96.1	meq/L	0–20
Chloride	Cl^-	35.3	meq/L	0–30
Carbonate	CO_3^{2-}	60.0	meq/L	0-0.1
Nitrate	NO_3^-	62.0	meq/L	0–10
		Trace Elements		
Boron	B	10.8	mg/L	0–2
Acid/basic	pH	1–14	—	6.0–8.5
		Total Salinity		
Electric conductivity	EC	—	dS/m mmho/cm	0–3
Total dissolved solids	TDS	—	mg/L	0–2,000
		Sodium Hazard		
Sodium adsorption ratio	SAR	—	—	0–15

Source: Adapted from Ayers and Westcot (1994).

Note: meq/L = millequivalent per liter $\left(\frac{\text{mg/L}}{\text{equivalent weight}} \text{ or } \frac{\frac{\text{mg}}{\text{L}} \times \text{Valence}}{\text{Atomic weight}}\right)$

1 mg/L (milligrams per liter) ≈ 1 ppm (parts per million).

on water quality parameters that are commonly observed and have a significant impact on crop production. The laboratory tests and calculations required to apply these guidelines are shown in Table 14.2.

14.2.2 Frequency of Water Testing

The frequency of irrigation water quality testing is dependent on the type and source of the water being used, as well as its variability over time. Surface water, including rivers, reservoirs, and canals, is affected by seasonal and weather-related fluctuations in flow, temperature, runoff, and sediment load. These variations can affect salinity, nutrient levels, and contamination risk. As a result, surface water sources typically require more frequent monitoring, such as seasonally or annually, to ensure continued suitability for irrigation and to detect emerging concerns. In contrast, groundwater sources generally exhibit greater chemical stability, particularly when drawn from deep, confined aquifers, such as basalt formation. These sources may only require testing every 3–5 years. However, shallow aquifers, such as those in alluvial deposits, are more vulnerable to changes due to land use activities, return flows, and surface infiltration. These sources should be tested more frequently, especially in regions with intensive agriculture or groundwater overdraft. Monitoring becomes increasingly important as groundwater levels decline, which can lead to increasing salinity or concentrations of other dissolved solids. In such cases, adjustments to irrigation management and treatment strategies may be necessary. Water analysis is also strongly recommended when new agricultural land is purchased or leased or when a new well or water source is developed to establish a baseline for long-term monitoring.

For wastewater sources, such as dairy lagoon effluent, food processing discharge, or treated municipal wastewater, the testing frequency should be tailored to site-specific applications and any applicable regulatory requirements. While agricultural wastewater is typically subject to less stringent oversight than municipal or industrial effluent, routine testing is still recommended to ensure effective nutrient management, crop safety, and environmental protection. Testing helps prevent overapplication of nutrients, accumulation of toxic elements, and potential contamination of surface or groundwater.

14.2.3 Salinity and Sodicity

Irrigation water naturally contains various salts, which can accumulate in soils at levels that exceed their original concentrations in the applied water. The degree of salt buildup is affected by the quality of water, irrigation methods, drainage efficiency, and evapotranspiration rates. If the salt concentration becomes too high, it can reduce crop yield and impair soil health. To avoid this, salt levels in the root zone must be maintained below thresholds that would impact yield. Generally, high-quality irrigation water poses minimal salinity risks. However, as water quality declines, managing salt accumulation becomes more critical. With increased water salinity, more attention is needed to flush salts from the root zone to prevent yield-impacting concentrations. In some cases, it may be necessary to cultivate crops that can withstand the expected salinity levels in the root zone. The required leaching fraction depends on both the salinity of the irrigation water and the crop's tolerance level.

The two main problems related to salts in irrigation water are salinity and sodality. Salinity can be measured by determining the electrical conductivity of water (EC) or its total dissolved solids (TDS). Electrical conductivity of water is generally measured in deciSiemens per meter (dS/m) or millimhos per centimeter (mmho/cm). TDS are expressed in milligrams per liter (mg/L), a unit equivalent to parts per million (ppm). Sodicity, or sodium hazard, is generally expressed as sodium adsorption ratio (SAR), which reflects the relative proportion of sodium ions to calcium and magnesium in water. High sodium levels can deteriorate the soil structure by causing clay particles to disperse, which impairs infiltration and reduces soil permeability.

Calcium and magnesium help counteract these effects, and bicarbonate concentrations can also influence SAR through their impact on calcium availability.

The SAR is calculated using Equation (14.1), where all ion concentrations are in meq/L:

$$\text{SAR} = \frac{[\text{Na}^+]}{\sqrt{\frac{[\text{Ca}^{2+}]+[\text{Mg}^{2+}]}{2}}} \tag{14.1}$$

A practical example of SAR and TDS calculations based on laboratory water analysis is provided in Example 14.1. More detailed information about the SAR interpretation can be found in Singh and Su (2022).

Example 14.1

A farmer receives a laboratory analysis of water intended for irrigation. The report provides the following measurements: electrical conductivity (EC): 2.2 dS/m, sodium (Na^+): 276 mg/L, calcium (Ca^{2+}): 80 mg/L, and magnesium (Mg^{2+}): 48.6 mg/L. Estimate the total dissolved solids (TDS) and the sodium adsorption ratio (SAR), and evaluate water quality hazards.

Solution:

Convert mg/L to meq/L using $\frac{\frac{\text{mg}}{\text{L}} \times \text{Valence}}{\text{Atomic weight}}$

$$\text{Na}^+: \quad \frac{276 \times 1}{23.0} = 12.0 \text{ meq/L}$$

$$\text{Ca}^{2+}: \quad \frac{80 \times 2}{40.1} \approx 4.0 \text{ meq/L}$$

$$\text{Mg}^{2+}: \quad \frac{48.6 \times 2}{24.3} = 4.0 \text{ meq/L}$$

Since EC = 2.2 dS/m < 5 dS/m, the total dissolve solids

$$\text{TDS} = \text{EC} \times 640 = 2.2 \times 640 = 1{,}408 \text{ mg/L}$$

The sodium adsorption ratio (SAR) is

$$\text{SAR} = \frac{[\text{Na}^+]}{\sqrt{\frac{[\text{Ca}^{2+}]+[\text{Mg}^{2+}]}{2}}} = \frac{12.0}{\sqrt{\frac{4.0+4.0}{2}}} = 6.0$$

The EC value of 2.2 dS/m falls within the "slight to moderate restriction" category for salinity, according to Table 14.1. This suggests that while water is usable for most crops, some degree of salinity management is required to avoid long-term soil degradation. Additionally, the SAR value of 6.0, in combination with an EC of 2.2 dS/m, indicates a slight to moderate risk of infiltration problems, particularly in fine-textured soils. To mitigate these risks, appropriate leaching practices should be applied to prevent the buildup of salts in the root zone. Regular monitoring of soil infiltration rates is also recommended, and gypsum application may be beneficial if reduced permeability is observed. Where possible, growers should avoid cultivating highly sensitive crops.

14.2.3.1 Crop Tolerance to Salinity

Crop tolerance to salinity varies significantly across species and varieties. Some crops maintain high yields in moderately to strongly saline environments due to their ability to adjust osmotically, maintain turgor pressure, and extract water despite reduced water potential in saline soils. This physiological adaptability is especially valuable in regions where soil or water salinity cannot be sufficiently reduced, allowing producers to shift toward salt-tolerant and economically viable alternatives.

Some salt-tolerant crops enable the use of moderate saline water for irrigation, which was once considered unsuitable. Over the past several decades, the relative salt tolerances of many crops have been empirically determined, providing essential guidance for irrigation planning and salinity management. The impacts of different values of electrical conductivity of the irrigation water (EC) and different values of soil salinity, as measured from the electrical conductivity of the saturated extract of the soil (EC_e), are shown in Table 14.3. Forage crops generally exhibit the highest salinity tolerance, followed by field crops, vegetables, and fruits (the most sensitive).

TABLE 14.3
Salt Tolerance for Different Crops

Crop	Botanical Name	Salt Tolerance Threshold[a] (EC_e), dS/m	Yield Decline Rate[b] (Y_d), % per dS/m	Maximum EC_e, dS/m	Tolerance rating[c]
Field Crops					
Barley[d]	*Hordeum vulgare*	8.0	5.0	28	T
Bean	*Phaseolus vulgaris*	1.0	19	6.3	S
Broad bean	*Vicia faba*	1.6	9.6	12	MS
Corn	*Zea mays*	1.7	12	10	MS
Cotton	*Gossypium hirsutum*	7.7	5.2	27	T
Cowpea	*Vigna unguiculata*	4.9	12	13	MT
Flax	*Linum usitatissimum*	1.7	12	10	MS
Guar	*Cyamopsis tetragonoloba*	8.8	17.0	15	T
Millet, foxtail	*Setaria italica*	—	—	—	MS
Oats	*Avena sativa*	—	—	—	MT
Peanut	*Arachis hypogaea*	3.2	29	6.6	MS
Rice, paddy[e]	*Oryza sativa*	3.0	12	11	S
Rye	*Secale cereale*	11.4	10.8	21	T

(*Continued*)

TABLE 14.3 (*Continued*)
Salt Tolerance for Different Crops

Crop	Botanical Name	Salt Tolerance Threshold[a] (EC_e), dS/m	Yield Decline Rate[b] (Y_d), % per dS/m	Maximum EC_e, dS/m	Tolerance rating[c]
Safflower	*Carthamus tinctorius*	—	—	—	MT
Sesame	*Sesamum indicum*	—	—	—	S
Sorghum	*Sorghum bicolor*	6.8	16	13	MT
Soybean	*Glycine max*	5.0	20	10	MT
Sugar beet[f]	*Beta vulgaris*	7.0	5.9	24	T
Sugarcane	*Saccharum officinarum*	1.7	5.9	19	MS
Sunflower	*Helianthus annuus*	—	—	—	MS
Triticale	× *Triticosecale*	6.1	2.5	46	T
Wheat[d]	*Triticum aestivum*	6.0	7.1	20	MT
Wheat (semi-dwarf)	*Triticum aestivum*	8.6	3.0	42	T
Wheat, durum	*Triticum turgidum*	5.9	3.8	32	T
Grasses and Forage Crops					
Alfalfa	*Medicago sativa*	2.0	7.3	16	MS
Barley (forage)[d]	*Hordeum vulgare*	6.0	7.1	20	MT
Bermuda grass	*Cynodon dactylon*	6.9	6.4	23	T
Clover, alsike	*Trifolium hybridum*	1.5	12	9.8	MS
Clover, berseem	*Trifolium alexandrinum*	1.5	5.7	19	MS
Clover, ladino	*Trifolium repens*	1.5	12	9.8	MS
Clover, red	*Trifolium pratense*	1.5	12	9.8	MS
Clover, strawberry	*Trifolium fragiferum*	1.5	12	9.8	MS
Corn (forage)	*Zea mays*	1.8	7.4	15	MS
Cowpea (forage)	*Vigna unguiculata*	2.5	11	12	MS
Fescue, tall	*Festuca elatior*	3.9	5.3	20	MT
Foxtail, meadow	*Alopecurus pratensis*	1.5	9.6	12	MS
Harding grass	*Phalaris tuberosa*	4.6	7.6	18	MT
Lovegrass	*Eragrostis* sp.	2.0	8.4	14	MS
Orchard grass	*Dactylis glomerata*	1.5	6.2	18	MS
Ryegrass, perennial	*Lolium perenne*	5.6	7.6	19	MT
Sphaerophysa	*Sphaerophysa salsula*	2.2	7.0	16	MS
Sudangrass	*Sorghum sudanense*	2.8	4.3	26	MT

(*Continued*)

TABLE 14.3 (*Continued*)
Salt Tolerance for Different Crops

Crop	Botanical Name	Salt Tolerance Threshold[a] (EC_e), dS/m	Yield Decline Rate[b] (Y_d), % per dS/m	Maximum EC_e, dS/m	Tolerance rating[c]
Trefoil, big	*Lotus uliginosus*	2.3	19	7.6	MS
Trefoil, narrowleaf birdsfoot	*Lotus corniculatus* var. *tenuifolius*	5.0	10	15	MT
Vetch, common	*Vicia angustifolia*	4.5	2.6	26	MT
Wheat (forage)	*Triticum aestivum*	4.5	2.6		MT
Wheat, durum (forage)	*Triticum turgidum*	2.1	2.5	24	MT
Wheatgrass, standard crested	*Agropyron sibiricum*	3.5	4.0	28	MT
Wheatgrass, fairway crested	*Agropyron cristatum*	7.5	6.9	22	T
Wheatgrass, tall	*Agropyron elongatum*	7.5	6.9	31	T
Wildrye, beardless	*Elymus triticoides*	2.7	6.0	19	MT
Vegetable and Fruit Crops					
Artichoke	*Helianthus tuberosus*	—	—	—	MT
Asparagus	*Asparagus officinalis*	4.1	2.0	54	T
Bean	*Phaseolus vulgaris*	1.0	19	6.3	S
Beet, red[f]	*Beta vulgaris*	4.0	9.0	15	MT
Broccoli	*Brassica oleracea* var. *botrytis*	2.8	9.2	14	MS
Brussels sprouts	*B. oleracea* var. *gemmifera*	—	—	—	MS
Cabbage	*B. oleracea* var. *capitata*	1.8	9.7	12	MS
Carrot	*Daucus carota*	1.0	14	8.1	S
Cauliflower	*B. oleracea* var. *botrytis*	—	—	—	MS
Celery	*Apium graveolens*	1.8	6.2	18	MS
Corn, sweet	*Zea mays*	1.7	12	10	MS
Cucumber	*Cucumis sativus*	2.5	13	10	MS
Eggplant	*Solanum melongena* var. *esculentum*	1.1	6.9	16	MS
Kale	*Brassica oleracea* var. *acephala*	—	—	—	MS
Lettuce	*Lactuca sativa*	1.3	13	9.0	MS

(*Continued*)

TABLE 14.3 (*Continued*)
Salt Tolerance for Different Crops

Crop	Botanical Name	Salt Tolerance Threshold[a] (EC_e), dS/m	Yield Decline Rate[b] (Y_d), % per dS/m	Maximum EC_e, dS/m	Tolerance rating[c]
Onion	*Allium cepa*	1.2	16	7.4	S
Pepper	*Capsicum annuum*	1.5	14	8.6	MS
Potato	*Solanum tuberosum*	1.7	12	10	MS
Pumpkin	*Cucurbita pepo* var. *pepo*	—	—	—	MS
Radish	*Raphanus sativus*	1.2	13	8.9	MS
Spinach	*Spinacia oleracea*	2.0	7.6	15	MS
Squash, scallop	*Cucurbita pepo* var. *melopepo*	3.2	16	9.4	MS
Squash, zucchini	*Cucurbita pepo* var. *melopepo*	4.7	9.4	15	MT
Strawberry	*Fragaria* sp.	1.0	33	4	S
Sweet potato	*Ipomoea batatas*	1.5	11	11	MS
Tomato	*Lycopersicon esculentum*	2.5	9.9	13	MS
Turnip	*Brassica rapa*	0.9	9.0	12	MS
Watermelon	*Citrullus lanatus*	—	—	—	MS
Woody Crops					
Almond	*Prunus dulcis*	1.5	19	6.8	S
Apple	*Malus sylvestris*	—	—	—	S
Apricot	*Prunus armeniaca*	1.6	24	5.8	S
Avocado	*Persea americana*	—	—	—	S
Blackberry	*Rubus* sp.	1.5	22	6.0	S
Boysenberry	*Rubus ursinus*	1.5	22	6.0	S
Date palm	*Phoenix dactylifera*	4.0	3.6	32	T
Fig	*Ficus carica*	—	—	—	MT
Gooseberry	*Ribes* sp.	—	—	—	S
Grape	*Vitis* sp.	1.5	9.6	12	MS
Grapefruit	*Citrus paradisi*	1.8	16	8.0	S
Guayule	*Parthenium argentatum*	8.7	11.6	17	T
Jujube	*Ziziphus jujuba*	—	—	—	MT
Lemon	*Citrus limon*	—	—	—	S
Lime	*Citrus aurantiifolia*	—	—	—	S
Loquat	*Eriobotrya japonica*	—	—	—	S
Mango	*Mangifera indica*	—	—	—	S

(*Continued*)

TABLE 14.3 (*Continued*)
Salt Tolerance for Different Crops

Crop	Botanical Name	Salt Tolerance Threshold[a] (EC_e), dS/m	Yield Decline Rate[b] (Y_d), % per dS/m	Maximum EC_e, dS/m	Tolerance rating[c]
Olive	*Olea europaea*	—	—	—	S
Orange	*Citrus sinensis*	1.7	16	8.0	S
Papaya	*Carica papaya*	—	—	—	MT
Passion fruit	*Passiflora edulis*	—	—	—	S
Peach	*Prunus persica*	1.7	21	6.5	S
Pear	*Pyrus communis*	—	—	—	S
Pineapple	*Ananas comosus*	—	—	—	MT
Plum, prune	*Prunus domestica*	1.5	18	7.1	S
Pomegranate	*Punica granatum*	—	—	—	MT
Pummelo	*Citrus maxima*	—	—	—	S
Raspberry	*Rubus idaeus*	—	—	—	S
Rose apple	*Syzygium jambos*	—	—	—	S
Sapote, white	*Casimiroa edulis*	—	—	—	S
Tangerine	*Citrus reticulata*	—	—	—	S

Source: Adapted from Maas and Hoffman (1977).

[a] Salt tolerance threshold (EC_e) is the electrical conductivity of saturation extract of soil in dS/m, which is 1.5 times that of the electrical conductivity of irrigation water, i.e., $EC_e = 1.5$ EC.

[b] Percent yield decline (Y_d) is the yield reduction rate per unit increase in salinity beyond the threshold. The rates of EC_e and EC are the same.

[c] Salt tolerance ratings are sensitive (S), moderately sensitive (MS), moderately tolerant (MT), and tolerant (T) in Figure 14.1.

[d] During germination and seedling stage, EC_e should not exceed 4 or 5 dS/m. These data may not be applicable to recently developed semi-dwarf wheat varieties.

[e] Values are for soil water while plants are submerged. Less tolerant during the seedling stage.

[f] During germination, EC_e should not exceed 3 dS/m.

Full yield potential is achievable for most crops with water salinity below 0.7 dS/m, hence no restriction on use is indicated. For water in the slight to moderate salinity range, achieving full yield is still possible but requires careful management to maintain soil salinity within crop tolerance range. For water with relatively high salinity levels or for sensitive crops, increased leaching may be impractical due to the excessive water needed, and switching to a more tolerant crop might be advisable. With water salinity in the slight to moderate range, sensitive crop production may be limited, particularly in clay-rich soils. Water with salinities exceeding 3.0 dS/m may still be usable but should be limited to salt-tolerant crops and more permeable soils where high leaching is feasible. Table 14.3 can be used to evaluate the economic impact of installing on-farm drainage systems. The divisions for classifying crop tolerance to salinity, including sensitive (S), moderately sensitive (MS), moderately tolerant (MT), and tolerant (T), are shown in Figure 14.1.

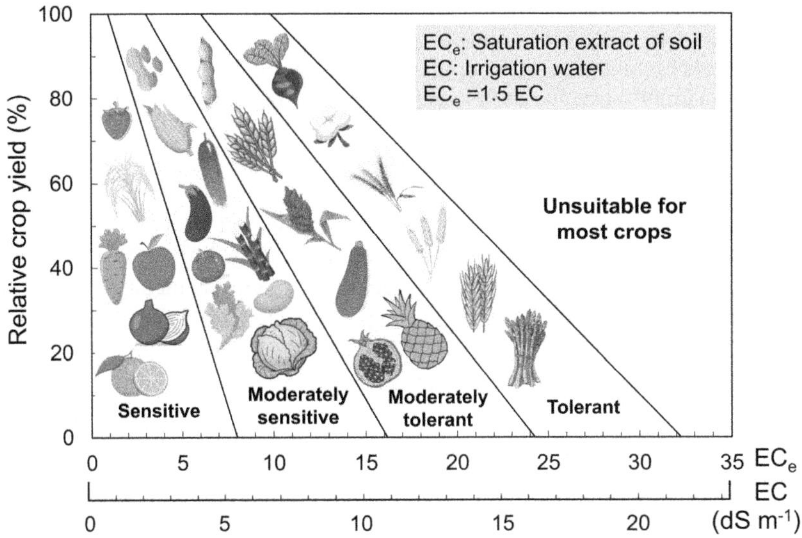

FIGURE 14.1 Divisions for classifying crop tolerance to salinity and representing crops in different categories. (Adapted from Su et al., 2025.)

Example 14.2

A farmer has irrigation water with an electrical conductivity (EC) of 2.5 dS/m. They are considering planting either tomato or cucumber. Based on the salt tolerance thresholds and yield decline rates provided in Table 14.3, which crop is more suitable? What yield reduction should be expected for each crop?

Solution:

The electrical conductivity of the saturated soil extract (EC_e) is approximately 1.5 times the irrigation water EC:

$$EC_e = 1.5EC = 1.5 \times 2.5 = 3.75 \text{ dS/m}$$

From Table 14.3, the salt tolerance thresholds and yield decline rates for tomato and cucumber are:

Crop	Salt Tolerance Threshold (EC_e), dS/m	Yield Decline Rate (Y_d), % per dS/m
Tomato	2.5	9.9
Cucumber	2.5	13.0

The yield loss of tomato:

$$\text{Yield loss} = (3.75 - 2.5) \times 9.9 \approx 12.4\%$$

The yield loss of cucumber:

$$\text{Yield loss} = (3.75 - 2.5) \times 13.0 = 16.25\%$$

Both tomato and cucumber are moderately sensitive to salinity, and both would experience yield reductions at this irrigation salinity level. However, tomato performs slightly better than cucumber under the given conditions, with a lower yield loss. If the farmer must choose between these two crops, tomato is the more suitable option.

14.2.3.2 Salinity Monitoring

Effective salinity monitoring is crucial for managing salt accumulation and mitigating its adverse effects on crop productivity and soil health. A robust monitoring strategy integrates both direct sampling techniques and advanced remote sensing technologies to assess salinity levels in irrigation water, soil, and the vadose zone over time (Figure 14.2).

The traditional approach involves direct sampling of soil and water, followed by laboratory analysis to quantify the concentration of dissolved salts. This method provides precise and reliable measurements and is especially useful in establishing a baseline understanding of salinity levels in a given area. Regular sampling is recommended to track spatial and temporal changes in salinity, thereby enabling early detection of emerging problems and the implementation of timely mitigation strategies. For soil analysis, it is recommended that samples be collected at multiple representative points across the field, specifically from two depths: 0–15 cm (surface layer) and 15–30 cm (root zone) (Gibson et al., 2021). These depth intervals provide insight into salt accumulation patterns that may affect both seed germination and root development. By analyzing both soil and water samples over time, farmers and land managers can develop informed strategies to maintain soil health and protect crop productivity.

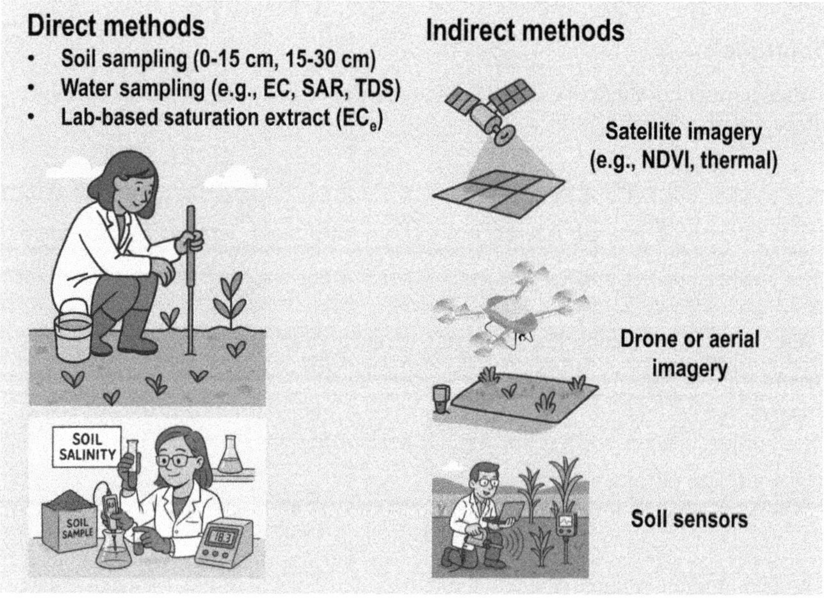

FIGURE 14.2 Different salinity monitoring methods.

Indirect monitoring methods, enhanced by recent technological advances, provide cost-effective scalable solutions for salinity monitoring over large agricultural landscapes. These include the following:

1. **Satellite remote sensing**: Multispectral and hyperspectral imagery enables detection of vegetation indices, e.g., normalized difference vegetation index (NDVI), which reflect plant health and stress due to salinity. Changes in canopy temperature (from thermal imagery) and soil reflectance can also serve as early warning indicators of salinity. These spatial datasets are invaluable for identifying salinity hotspots, tracking land degradation trends, and guiding targeted interventions.
2. **Photogrammetry and photointerpretation**: High-resolution drone or aerial imagery supports 3D mapping of micro-topography, vegetation patchiness, and soil surface features such as crusting, common symptoms of localized salinity. These techniques enhance the detection of subtle field-scale salinity variations.
3. **Electromagnetic induction (EMI) and soil sensors**: EMI sensors and in situ salinity probes provide rapid, non-destructive mapping of soil salinity variability across fields. These tools are widely adopted in precision agriculture to create salinity management zones and optimize irrigation practices.

In regions affected by shallow saline water tables or saltwater intrusion, monitoring groundwater dynamics is critical. As introduced in Chapter 9 on drainage, tools such as piezometers and observation wells can be installed to monitor seasonal fluctuations in groundwater depth and pressure. Maintaining the water table below critical thresholds reduces the risk of upward salt migration via capillary rise, thereby preventing secondary salinization. By integrating satellite observations, geospatial tools, field-based sensors, and laboratory diagnostics, irrigation managers can implement proactive, data-driven strategies to monitor and manage salinity. This integrated approach supports more resilient irrigation scheduling, sustainable land management, and long-term soil productivity.

14.2.3.3 Salinity Control

The goal of salinity control is to sustain acceptable crop yields by managing the accumulation of salts in the root zone. Effective salinity management often requires a combination of strategies tailored to local soil, crop, climate, and water quality conditions. There are four common practices:

1. **Leaching**: Applying extra irrigation water to flush salts below the root zone is a critical method of controlling salinity. The amount of water needed for leaching varies, based on the crop's salt sensitivity and the salinity of the irrigation water. In cases where excessive leaching becomes impractical, switching to salt-tolerant crops may be necessary.
2. **Adequate drainage**: Proper drainage is essential to prevent water table rise and salt accumulation. Without sufficient drainage, leached salts can return to the root zone through capillary rise.

3. **Crop selection**: When salinity levels cannot be economically controlled, cultivating salt-tolerant crops becomes a viable option.
4. **Cultural practices**: Adjustments in irrigation frequency, land grading, fertilization timing, and seeding methods can also help manage salinity, especially in dealing with short-term or temporary salinity increases.

14.2.3.3.1 Leaching Water Requirements

Leaching refers to the practice of applying excess irrigation water to dissolve and flush salts below the crop root zone, helping to maintain soil salinity at levels that do not hinder plant growth. This technique is only effective when adequate drainage allows the salts to be removed from the soil profile. Without proper drainage, salts may reaccumulate through capillary rise, particularly in areas with shallow water tables or high evaporation rates. While leaching effectively reduces salinity, it may also mobilize other soluble substances, including nutrients, fertilizers, and pesticides, potentially affecting groundwater quality. Therefore, leaching should be implemented carefully, considering both agronomic benefits and environmental concerns. Ayers and Wescot (1976) suggested an equation to calculate the leaching fraction (LF), the minimum proportion of applied water required to maintain root zone salinity within a crop's tolerance range. This equation accounts for the variability in salt concentrations throughout the soil profile and their temporal dynamics, as discussed by Rhoades (1974) and Rhoades and Merrill (1976):

$$\mathrm{LF} = \frac{\mathrm{EC}}{5\mathrm{EC}_e - \mathrm{EC}} \tag{14.2}$$

where LF is the leaching fraction (dimensionless), EC is the electrical conductivity of the applied irrigation water in dS/m, and EC_e is the maximum average soil salinity tolerated by the crop as measured on a soil saturation extract in dS/m. Table 14.3 can be used to obtain the EC_e value for a given crop based on the appropriate acceptable yield. For general salinity management, it is advisable to select an EC_e value associated with at least 90% yield potential, ensuring that leaching maintains near-optimal productivity.

Once the leaching fraction is determined, the total irrigation requirement depth (I) that meets both crop evapotranspiration and salinity control needs can be calculated using the following equation:

$$I = \frac{\mathrm{ET}_c}{1 - \mathrm{LF}} \tag{14.3}$$

where I is the required irrigation depth (mm) and ET_c is the seasonal crop evapotranspiration (mm). This equation ensures that sufficient water is applied to meet both the crop's consumptive use and leach excess salts below the root zone. In practice, the actual leaching requirement may vary based on soil texture, rainfall, irrigation method, and climate. Moreover, uniform water distribution and appropriate timing of leaching relative to crop growth stages are essential for effective salt management.

Example 14.3

Estimate the seasonal irrigation depth for corn irrigated with a surface irrigation system and an application efficiency of 70%, if the annual evapotranspiration is 700 mm. The electrical conductivity of the water is 1.3 dS/m, and a 90% yield potential is expected.

Solution:

Since a 90% yield potential is expected, the salt tolerance threshold of corn is 1.7 dS/m and the yield decline rate (Y_d) is 12% per dS/m, as shown in Table 14.3. If a 90% yield potential is expected, then EC_e at the 90% yield potential is

$$EC_e = 1.7\ \text{dS/m} + \frac{(100\% - 90\%)}{12\%\ \text{per}\ \frac{\text{dS}}{\text{m}}} \approx 2.5\ \text{dS/m}$$

Given the EC = 1.3 dS/m, the leaching fraction is

$$LF = \frac{EC}{5EC_e - EC} = \frac{1.3}{5 \times 2.5 - 1.3} \approx 0.12$$

The seasonal irrigation depth is

$$I = \frac{ET_c}{1 - LF} = \frac{700\ \text{mm}}{1 - 0.12} \approx 795.5\ \text{mm}$$

Considering that the application depth is 1,000 mm (700 mm/0.7 = 1,000 mm) and if all water losses were due to deep percolation (not due to runoff), the application depth is greater than the seasonal irrigation depth adjusted for leaching of 795.5 mm. Thus, extra leaching will not be necessary.

14.2.3.3.2 Adequate Drainage

In irrigated agriculture, salinity issues are often linked to inadequate drainage and shallow water tables, typically within 1–2 m of the soil surface (Figure 14.3). In such scenarios, the upward movement of water through capillary action brings salts to the root zone. Over time, especially in hot climates or fallow fields, this process can lead to significant salt accumulation near the surface, adversely affecting soil health and crop productivity. The rate at which salinity builds up in these circumstances is influenced by several factors, including depth and salinity of groundwater, frequency and efficiency of irrigation, soil texture and structure (affecting capillary rise), and climatic conditions (temperature, humidity, and rainfall).

In arid and semi-arid regions, effectively addressing salinity issues that arise from or are exacerbated by inadequate drainage requires stabilizing the water table at a safe depth, typically around two meters or more. This often involves the installation of open or tile drains, or drainage wells, which help remove saltwater from the subsurface and transport it to a designated area for safe disposal. Once drainage is properly established, salinity issues primarily arise from the quality of irrigation water and the management of salts introduced through irrigation. Therefore, an integral part of effective salinity control is ensuring adequate drainage to manage the water

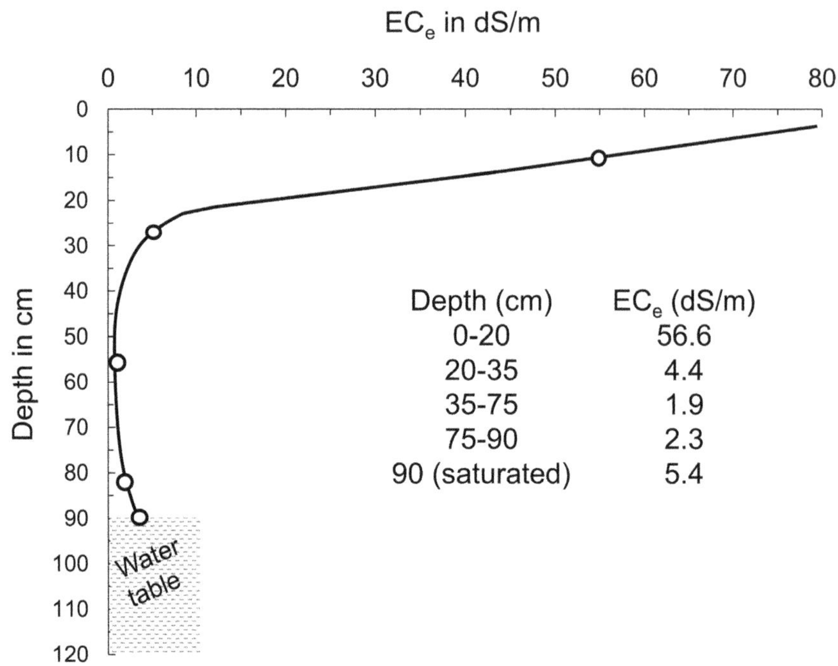

FIGURE 14.3 Salinity profile in the soil with a shallow water table. (Adapted from Mohamed and Amer, 1972.)

table, coupled with periodic leaching to remove accumulated salts. Maintaining a net downward movement of water applied to the surface is key to achieving the necessary leaching and controlling salinity. The guidance provided in Table 14.1 is based on the assumption that all salt accumulation in the crop root zone originates from the irrigation water, indicating that drainage is sufficient. In the absence of proper drainage, additional salinity control strategies are required.

14.2.3.3.3 Salt-Tolerant Crops

Crops vary widely in their ability to tolerate salinity, with significant differences not only among species but also among cultivars within the same species. The most notable variations are found in salt-tolerant crops such as barley, cotton, and sugar beet. However, even within these species, certain cultivars may be more or less tolerant than the general levels indicated in Table 14.3. Therefore, selecting appropriate salt-tolerant varieties is essential, particularly when irrigation water has moderate to high salinity or when soil salinity is difficult to manage through leaching and drainage alone.

In recent years, plant breeding and genetic research have focused on improving salt tolerance, especially among closely related genotypes. These efforts aim to develop cultivars that can maintain high yields and quality under saline conditions, making more effective use of saline or marginal-quality water resources. However, most genetically enhanced or selectively bred salt-tolerant varieties take 5–15 years to reach commercial viability and require thorough field evaluation to confirm their adaptability, yield stability, and economic competitiveness under real-world conditions. Until these new cultivars become widely available, the tolerance thresholds provided in Table 14.3 remain the most reliable guide for salinity management. These thresholds indicate the average soil salinity (EC_e in dS/m) at which specific crops begin to experience yield reductions, assuming ideal conditions for other growth factors.

Environmental conditions, particularly climate and season, significantly influence how crops respond to salinity stress. In cooler climates or during cooler seasons, crops tend to tolerate higher salinity. This is due to reduced evapotranspiration rates, lower water uptake demands, and greater opportunity for natural or irrigation-induced leaching. In hot, dry conditions, especially during peak summer months, salinity stress is amplified due to accelerated evaporation from the soil surface, rapid soil moisture depletion, increased salt concentrations near the root zone, and limited ability of plant roots to absorb water under osmotic stress. As a result, the same level of soil salinity may have more severe effects on crop growth during high ET periods than during cooler, wetter periods. Therefore, both crop selection and planting date should consider the seasonal variability in salinity response.

14.2.3.3.4 Cultural Practices

Cultural practices play a critical role in managing salinity, especially when physical or chemical control measures (e.g., gypsum application) are limited. When thoughtfully implemented, these practices can significantly reduce salt accumulation and improve crop establishment and yield under saline conditions. One of the most fundamental practices for salinity control in surface-irrigated fields is land leveling. Uneven fields lead to non-uniform water application, causing salt accumulation in elevated areas and waterlogging in depressions. Over time, this results in uneven crop growth and increased salinity stress. Laser or precision leveling is recommended to maintain uniform slopes and improve irrigation efficiency. Periodic regrading helps correct micro-topographic variations that develop due to tillage, erosion, or settling.

Another practice is the timing and frequency of irrigation. When saline water is used, it is important to avoid long intervals between irrigations, especially during hot and dry periods, as this can lead to excessive salt concentrations near the root zone. Increasing irrigation frequency and scheduling irrigation prior to anticipated rainfall events can help leach accumulated salts below the root zone. Additionally, pre-plant irrigation is often used to flush salts from the germination zone, improving seedling establishment and early crop growth.

Seed placement is another key consideration under saline conditions. Salinity can significantly hinder germination and seedling emergence, particularly when salts accumulate near the soil surface. Adjusting planting depth and bed shape can improve growing conditions for young plants. For example, placing seeds on the sides of beds or ridges rather than on the tops, where salt tends to concentrate, can reduce salt exposure. Using flat or broad beds also helps promote more uniform water distribution and salt leaching around the seed zone.

Fertilizer management must also be adjusted in saline environments. Because fertilizers are composed of soluble salts, excessive or poorly timed applications can exacerbate salt stress. Proper fertilization can improve crop yield when nutrient availability is limited, but it does not increase a crop's inherent salt tolerance. Therefore, the type, amount, timing, and placement of fertilizers should be carefully managed to avoid salt injury, especially during early growth stages. Strategies, such as banding fertilizers below or to the side of the seed, avoiding large applications of chloride- or ammonium-based fertilizers, and splitting applications based on crop demand, can reduce the risk of salt buildup.

The choice of irrigation method influences both water use efficiency and the pattern of salt accumulation in the soil. Flood irrigation can promote salt leaching but requires well-leveled fields and adequate drainage. Sprinkler irrigation provides uniform water application but may be less effective for salt removal and can cause foliar salt damage if saltwater contacts leaves. Drip irrigation is highly efficient and maintains consistent soil moisture near the root zone, but salts tend to accumulate at the edges of the wetted area, requiring periodic leaching irrigation to prevent long-term surface salinity.

Effective salinity control involves integrating multiple cultural practices, including land leveling, irrigation scheduling, proper seed placement, careful fertilizer management, and the selection of appropriate irrigation methods. These low-cost, field-level adjustments complement engineering and chemical solutions, offering farmers a practical and sustainable approach to managing salinity under diverse agricultural conditions.

14.2.4 Infiltration Problems

Permeability issues in soil arise when there is a high sodium content and low calcium levels, either in the soil itself or in the irrigation water. This imbalance can lead to a significant decrease in the infiltration rate of soil, making it challenging to supply enough water for crops. Sodium disperses soil particles, breaking down aggregates and leading to the formation of a compact, dense surface layer that significantly reduces the soil's ability to absorb and transmit water. This reduction in permeability restricts water movement to the root zone and limits the effectiveness of irrigation, making it difficult to supply adequate moisture for crop growth. These issues are particularly severe in soils with high levels of exchangeable sodium and low calcium availability, and they can be further aggravated by additional factors, such as low salinity and elevated soil pH.

To evaluate and diagnose these infiltration-related problems, the sodium adsorption ratio (SAR) is commonly used. SAR quantifies the relative concentration of sodium to calcium and magnesium in irrigation water and provides insight into the potential for soil dispersion. However, SAR alone may not provide a complete picture. In many cases, it is more informative to consider both SAR and the electrical conductivity of the irrigation water together. While high SAR increases the risk of soil structure breakdown, adequate EC can help maintain flocculation of soil particles, thereby preserving permeability. For example, high SAR water with low EC is more damaging than high SAR water with moderate or high EC. Table 14.1 and Figure 14.4 provide guidance on the threshold values of SAR and EC that indicate infiltration and permeability hazards.

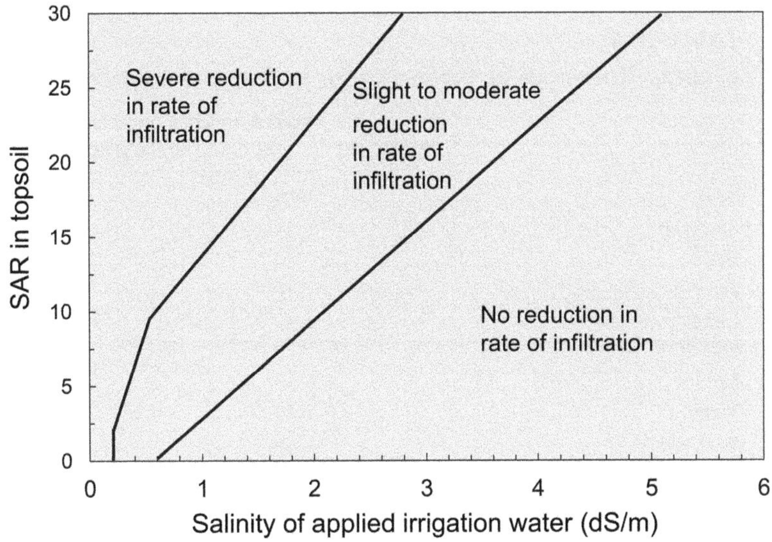

FIGURE 14.4 Water infiltration rate affected by salinity and sodium adsorption ratio (SAR). (Adapted from Rhoades, 1977; Oster and Schroer, 1979.)

In soils where bicarbonates are present, calcium can precipitate out as calcium carbonate, reducing the availability of calcium in the soil solution. In such cases, an adjusted SAR value is used to better reflect the actual impact on soil structure and infiltration potential. One of the most widely used management practices to mitigate these issues is the application of gypsum (calcium sulfate). Gypsum provides a soluble source of calcium, which replaces sodium on the soil's cation exchange sites, thereby improving soil aggregation and restoring infiltration and permeability. In some situations, especially where bicarbonates are prevalent, additional chemical amendments or acidifying agents may be used to enhance calcium availability and maintain a favorable soil structure. Managing infiltration problems requires an integrated approach that considers irrigation water quality, soil chemistry, and soil physical properties. Regular monitoring of SAR and EC, appropriate amendment applications, and careful water management are essential for maintaining soil health and ensuring efficient water use in agricultural systems.

14.2.5 Toxicity

Crop toxicity issues arise when specific elements present in the soil-water solution are absorbed by plants. These elements can build up to levels that detrimentally affect crop yield or quality. Typically, toxicity occurs as plants absorb ions like chloride, boron, and sodium with soil water. These ions then travel through the plant's transpiration stream and accumulate in the leaves at harmful concentrations, potentially causing plant damage. However, it is important to note that not all crops are equally sensitive to these ions. The chloride tolerance levels for a variety of crops are shown in Table 14.4. Crops such as strawberry, beans, and onion are among the most

TABLE 14.4
Chloride Tolerance of Various Crops in Order of Tolerance

Crop	Maximum Cl^{-}[a] Concentration without Loss in Yield[b]	
	meq/L	ppm
Strawberry	10	350
Bean	10	350
Onion	10	350
Carrot	10	350
Radish	10	350
Lettuce	10	350
Turnip	10	350
Rice, paddy[c]	30[d]	1,050
Pepper	15	525
Clover, strawberry	15	525
Clover, red	15	525
Clover, alsike	15	525
Clover, ladino	15	525
Corn	15	525
Flax	15	525
Potato	15	525
Sweet potato	15	525
Broad bean	15	525
Cabbage	15	525
Foxtail, meadow	15	525
Celery	15	525
Clover, Berseem	15	525
Orchard grass	15	525
Sugarcane	15	525
Trefoil, big	20	700
Lovegrass	20	700
Spinach	20	700
Alfalfa	20	700
Sesbania[c]	20	700
Cucumber	25	875
Tomato	25	875
Broccoli	25	875
Squash, scallop	30	1,050
Vetch, common	30	1,050
Wild rye, beardless	30	1,050
Sudan grass	30	1,050
Wheat grass, standard crested	35	1,225
Beet, red[c]	40	1,400
Fescue, tall	40	1,400

(Continued)

TABLE 14.4 (*Continued*)
Chloride Tolerance of Various Crops in Order of Tolerance

	Maximum Cl^{-a} Concentration without Loss in Yield[b]	
Crop	**meq/L**	**ppm**
Squash, zucchini	45	1,575
Harding grass	45	1,575
Cowpea	50	1,750
Trefoil, narrow-leaf bird's foot	50	1,750
Ryegrass, perennial	55	1,925
Wheat, Durum	55	1,925
Barley (forage)[c]	60	2,100
Wheat[c]	60	2,100
Sorghum	70	2,450
Bermudagrass	70	2,450
Sugar beet[c]	70	2,450
Wheatgrass, fairway crested	75	2,625
Cotton	75	1,625
Wheatgrass, tall	75	2,625
Barley	80	2,800

Source: Adapted from Tanji and Wallender (1990).

[a] These data serve only as guidelines for determining relative tolerances among crops. Absolute tolerances vary, depending on climate, soil conditions, and cultural practices.

[b] Cl^- concentrations in saturated extracts sampled in the root zone.

[c] Less tolerant during the seeding stage and emergence.

[d] Values for paddy rice refer to the Cl^- concentration in the soil water during flooded cultivation conditions.

sensitive to chloride, whereas barley, cotton, and wheatgrass show high tolerance. This information is essential for selecting appropriate crops or adapting management strategies when chloride levels in irrigation water are elevated.

Similarly, boron toxicity is a widespread concern, especially in arid and semi-arid regions where groundwater or drainage return flows may contain elevated B concentrations. Boron toxicity symptoms typically appear as leaf tip necrosis or marginal chlorosis. Table 14.5 categorizes crops into three broad groups based on boron tolerance: sensitive (e.g., pecan, walnut, grapes), semi-tolerant (e.g., tomato, barley, corn), and tolerant (e.g., asparagus, date palm, alfalfa) (Wilcox, 1960). Knowing a crop's boron tolerance range and comparing it with irrigation water quality are essential to avoid long-term toxicity buildup in soils.

Sodium toxicity, distinct from sodium-induced soil structure issues, occurs when high levels of exchangeable sodium are taken up by a plant. This can result in leaf burn or stunted growth, particularly in sensitive crops. Table 14.6 shows crop tolerance levels based on exchangeable sodium percentage (ESP), ranging from extremely sensitive species, such as citrus and avocado (ESP < 10), to highly tolerant forage grasses

TABLE 14.5
Limits of Boron in Irrigation Water for Different Degrees of Boron Tolerance by Crops

Sensitive (0.3–1.0 ppm)	Semi-tolerant (1.0–2.0 ppm)	Tolerant (2.0–4.0 ppm)
Pecan	Sunflower (native)	Athel (*Tamarix aphylla*)
Walnut (black or Persian or English)	Potato	Asparagus
Jerusalem artichoke	Cotton (Acala and Pina)	Palm (*Phoenix canariensis*)
Navy bean	Tomato	Date palm (*Phoenix dactylifera*)
American elm	Sweetpea	Sugar beet
Plum	Radish	Mangel
Apple	Field pea	Garden beet
Grape (Sultanina and Malaga)	Ragged-robin rose	Alfalfa
Kadota fig	Olive	Gladiolus
Persimmon	Barley	Broadbean
Chery	Wheat	Onion
Peach	Corn	Turnip
Apricot	Milo	Cabbage
Thornless blackberry	Oat	Lettuce
Orange	Zinnia	Carrot
Avocado	Pumpkin	
Grapefruit	Bell pepper	
Lemon	Sweet potato	
	Lima bean	

Adapted from Wilcox (1960).

TABLE 14.6
Tolerance of Crops to Exchangeable Sodium Percentage (ESP)

Tolerance to ESP	Influencing Range of ESP	Crop	Growth Responsible under Field Conditions
Extremely sensitive	2–10	Deciduous fruits Nuts Citrus Avocados	Sodium toxicity symptoms even at low ESP values
Sensitive	10–20	Beans	Stunned growth at low ESP values even though the physical condition of the soil may be good
Moderately sensitive	20–40	Clover Oats Tall fescue Rice Dallisgrass	Stunned growth due to both nutritional factors and adverse soil conditions

(Continued)

TABLE 14.6 (*Continued*)
Tolerance of Crops to Exchangeable Sodium Percentage (ESP)

Tolerance to ESP	Influencing Range of ESP	Crop	Growth Responsible under Field Conditions
Tolerant	40–60	Wheat Cotton Alfalfa Barley Tomatoes Beets	Stunned growth usually due to adverse physical conditions
Most tolerant	>60	Crested and fairway wheatgrass Tall wheatgrass Rhodes grass	Stunned growth usually due to adverse physical conditions of soil

Source: Adapted from Pearson (1960).

and wheatgrass cultivars (ESP > 60). In many cases, sodium toxicity occurs alongside infiltration and permeability problems, complicating diagnosis and management.

Visual indicators of toxicity can assist in field diagnosis. One such indicator is the presence of white salt deposits on leaves or fruit surfaces, particularly under sprinkler irrigation using water with high bicarbonate content. While these deposits may not always indicate harmful toxicity levels, they often signal the beginning of ion accumulation and prompt further investigation. In some cases, visible injury may occur even when the overall EC of the water or soil is within acceptable limits, underscoring the importance of monitoring individual ion concentrations.

In addition to chloride, boron, and sodium, trace elements, such as selenium (Se), arsenic (As), and mercury (Hg), may be present in some irrigation water sources, particularly in return flows or drainage effluent from areas with naturally occurring geochemical deposits. While these elements may not always be phytotoxic at low concentrations, they can accumulate in edible plant parts and pose serious risks to animal and human health. This highlights the need for comprehensive monitoring of irrigation water quality, not only in terms of salinity but also in terms of elemental composition, to ensure safe and sustainable crop production.

Overall, effective management of ion toxicity involves selecting tolerant crops, adjusting irrigation and leaching practices, and periodically analyzing both irrigation water and plant tissue. Tables 14.4–14.6 serve as critical references for diagnosing and managing toxicity-related risks, enabling informed decisions that protect crop yield, product quality, and environmental health.

Example 14.4

A farmer receives the results of an irrigation water quality test, which shows the following concentrations of potentially harmful elements: chloride (Cl^-) at 45 meq/L (approximately 1,575 ppm), boron (B) at 1.8 ppm, a sodium adsorption ratio (SAR) of 12, and an exchangeable sodium percentage (ESP) of 18 in the soil. The farmer is considering planting barley and alfalfa and wants to determine whether either crop is at risk of toxicity under these conditions.

Solution:

From Table 14.4, the maximum chloride tolerance is approximately 2,800 ppm for barley and 700 ppm for alfalfa. Since the measured chloride level is 1,575 ppm, it is well below the tolerance level for barley but more than double the threshold for alfalfa. Therefore, barley is suitable with respect to chloride concentration, whereas alfalfa is at a high risk of chloride toxicity, which may result in symptoms, such as leaf burn, marginal necrosis, and significant yield reduction.

With respect to boron, Table 14.5 indicates that barley is semi-tolerant (1.0–2.0 ppm), whereas alfalfa is considered tolerant (2.0–4.0 ppm). Given that the measured boron concentration in the water is 1.8 ppm, both crops fall within their respective tolerance ranges, and no immediate boron toxicity is expected.

Finally, based on Table 14.6, both barley and alfalfa are considered tolerant to exchangeable sodium, with tolerances extending up to ESP values of 40. The current ESP of 18 is well below this limit, indicating no significant risk of sodium toxicity for either crop. While boron and sodium levels are acceptable for both crops, the chloride concentration in the irrigation water exceeds the safe limit for alfalfa. Therefore, the farmer should avoid planting alfalfa on this field under current conditions. Barley, however, is within safe thresholds for all measured parameters and is suitable for cultivation.

14.2.6 Miscellaneous Problems

In addition to common water quality challenges, such as salinity, sodicity, and specific ion toxicity, several other issues, often referred to as miscellaneous problems, can significantly affect crop performance, soil structure, and the durability of irrigation systems. These include excess nitrogen, abnormal pH levels, scale formation, elevated magnesium concentrations, and the accumulation of trace elements. Although often overlooked, these factors can influence both short- and long-term agricultural productivity and must be monitored and managed carefully.

Effective management of these problems involves a multifaceted approach: (1) controlling excess nitrogen; (2) maintaining an appropriate pH range; (3) preventing scale buildup from slightly soluble salts like calcium and bicarbonates, which can affect crop appearance and irrigation efficiency; (4) managing the calcium-to-magnesium ratio to avoid infiltration and nutrient uptake issues; and (5) monitoring trace elements to prevent their accumulation in soil and crops. Each of these factors, although secondary to salinity and sodicity in many regions, plays a critical role in ensuring sustainable, high-quality crop production, particularly where marginal or reclaimed water sources are used. Proactive water testing, periodic system maintenance, and adaptive nutrient management are essential components of an integrated water quality strategy that addresses these miscellaneous hazards.

Excess nitrogen in irrigation water, primarily in the form of nitrate (NO_3^-), can act similarly to over-fertilization, promoting excessive vegetative growth, delaying crop maturity, and reducing yield quality. For instance, excessive nitrogen in water used for irrigation can increase the size of sugar beets but decrease their sugar content.

In grapes, excessive nitrogen can prolong vegetative growth at the expense of fruit set and ripening, ultimately reducing yield and sugar concentration. These effects are particularly problematic during the flowering and fruiting stages when nutrient balance is critical. In addition to crop effects, high nitrogen levels also promote algae growth in canals and reservoirs, increasing maintenance requirements. Thus, regular monitoring of nitrogen content in irrigation water and appropriate adjustments to fertilizer applications are essential for both agronomic and operational efficiency.

Irrigation water pH typically ranges from 6.5 to 8.4. Values outside this range may indicate chemical imbalances or potential for toxic ion availability. For example, a low pH can increase the solubility of toxic metals such as aluminum, whereas a high pH may reduce micronutrient availability. Although soil has a buffering capacity that resists rapid pH changes, long-term application of high- or low-pH water can alter soil chemistry. In general, adjusting the soil pH directly, such as by using lime to increase the pH or sulfuric acid to lower it, is more practical than attempting to modify the pH of irrigation water. Nonetheless, attention must also be given to the corrosive effects of low-pH water on pipelines and sprinkler systems, which can lead to infrastructure damage.

Scale formation is another common issue, particularly in sprinkler systems that use water high in slightly soluble salts, such as calcium, bicarbonate, and sulfate. As water droplets evaporate from leaf and fruit surfaces, salts precipitate and form white scale deposits. While not toxic, these deposits can reduce the marketability of produce due to unsightly residue and may interfere with photosynthesis if deposition is severe. To mitigate this issue, growers may acidify irrigation water to prevent salt precipitation or adjust sprinkler scheduling to reduce evaporation losses during application.

Magnesium-related issues arise when water or soil contains high levels of exchangeable magnesium relative to calcium. Although magnesium behaves more like calcium than sodium, excessive magnesium can contribute to infiltration problems and reduce calcium availability to plants, especially in soils with low total calcium. A low calcium-to-magnesium ratio in irrigation water may warrant closer monitoring and, in some cases, the application of gypsum to supplement calcium and improve soil structure.

Trace elements are required by plants only in small quantities, but many can become toxic at relatively high concentrations. While most irrigation water contains trace elements at safe levels, long-term use, especially wastewater or drainage water, can lead to accumulation in soil. This accumulation is often irreversible and may lead to chronic soil contamination or food safety concerns. Common trace elements of concern include arsenic (As), cadmium (Cd), selenium (Se), and zinc (Zn). Table 14.7 presents the recommended maximum concentrations for trace elements in irrigation water, along with notes on their toxicity thresholds and behavior in soil systems (National Academy of Sciences and National Academy of Engineering,1972; Pratt, 1972). For instance, arsenic is highly toxic to rice at concentrations below 0.05 mg/L, while selenium, though essential in trace amounts for animals, becomes toxic to plants and livestock at slightly higher levels.

Trace elements, while essential for plant growth in small quantities, can become toxic at higher concentrations. These elements are generally present at low levels

TABLE 14.7
Recommended Maximum Concentrations of Trace Elements in Irrigation Water

Element	Recommended Maximum Concentration (mg/L)	Remarks
Al (aluminum)	5.0	Can cause non-productivity in acid soils (pH < 5), but in more alkaline soils (pH > 5.5), the ion precipitates, and any toxicity is effectively eliminated
As (arsenic)	0.1	Toxicity to plants varies widely, ranging from 12 mg/L for Sudan grass to less than 0.05 mg/L for rice
Be (beryllium)	0.1	Toxicity to plants varies widely, ranging from 5 mg/L for kale to 0.5 mg/L for bush beans
Cd (cadmium)	0.01	Toxic to beans, beets, and turnips at concentrations as low as 0.1 mg/L in nutrient solution. Conservative limits recommended because of its potential to accumulate in plants and soils to levels that may be harmful to humans
Co (cobalt)	0.05	Toxic to tomato plants at 0.1 mg/L in nutrient solution. Tends to be inactivated in neutral and alkaline soils
Cr (chromium)	0.1	Not generally recognized as an essential growth element. Conservative limits are recommended because of limited information on plant toxicity
Cu (copper)	0.2	Toxic to a number of plants at concentrations between 0.1 and 1.0 mg/L in nutrient solution
F (fluoride)	1.0	Generally inactivated in neutral and alkaline soils
Fe (iron)	5.0	Not toxic to plants in aerated soils, but can contribute to soil acidification and reduced availability of essential nutrients such as phosphorus and molybdenum. Overhead sprinkling may cause unsightly deposits on plants, equipment, and buildings
Li (lithium)	2.5	Tolerated by most crops up to about 5 mg/L and is relatively mobile in soil. Toxic to citrus at low concentrations; a recommended limit for citrus is 0.075 mg/L. Acts similarly to boron
Mn (manganese)	0.2	Toxic to many crops at concentrations ranging from a few tenths of a mg/L to a few mg/L in acid soils
Mo (molybdenum)	0.01	Generally nontoxic to plants at normal concentrations in soil and water, but can be toxic to livestock if forage is grown in soils with high levels of available molybdenum
Ni (nickel)	0.2	Toxic to many plants at concentrations of about 0.5–1.0 mg/L; reduced toxicity at neutral or alkaline soil
Pb (lead)	5.0	Can inhibit plant cell growth at very high concentrations
Se (selenium)	0.02	Toxic to plants at concentrations as low as 0.025 mg/L and toxic to livestock if forage is grown in soils with relatively high levels of added selenium. An essential element for animals, but only in very low concentrations

(Continued)

TABLE 14.7 (*Continued*)
Recommended Maximum Concentrations of Trace Elements in Irrigation Water

Element	Recommended Maximum Concentration (mg/L)	Remarks
Sn (tin)	—	These elements are effectively excluded by plants; specific tolerance levels are unknown
Ti (titanium)		
W (tungsten)		
V (vanadium)	0.1	Toxic to many plants at relatively low concentrations
Zn (zinc)	2.0	Toxic to many plants at widely varying concentrations. Toxicity is reduced at pH $\geq$ 6 and in fine-textured or organic soils

in most water sources but can accumulate in soil over time. This accumulation is mostly irreversible, and repeated application can lead to soil contamination. Most trace elements in irrigation water originate from human activities, such as wastewater disposal. Monitoring for trace elements is essential, particularly in projects using wastewater for irrigation. Recommended maximum concentrations for trace elements in irrigation water are shown in Table 14.7.

14.3 WATER SOURCES

Agricultural irrigation systems rely on a variety of water sources, including surface water from watersheds, return flows, and groundwater. Each water source presents unique challenges regarding quality, sustainability, legal access, and environmental impact. Understanding the characteristics, limitations, and proper management of each source is essential for maintaining both irrigation performance and environmental integrity.

14.3.1 Surface Supply from Watershed

Surface water supplies, derived from rivers, streams, reservoirs, and natural impoundments, are a major source for irrigation. However, their diversion and use must be balanced against ecological concerns. Altering stream flows can reduce the ability of aquatic ecosystems to dilute pollutants, support aquatic life, and sustain habitats for fish and wildlife. Reduced flows can disrupt natural seasonal patterns and threaten biodiversity.

To address these concerns, guidelines should be developed in collaboration with aquatic biologists and conservationists. These guidelines should focus on determining optimal stream flows to maintain ecological balance. They should include strategies for preserving necessary flushing flows throughout the year and ensuring consistent daily flows to prevent the cessation of stream movement. Maintaining this delicate balance between fulfilling human irrigation needs and preserving stream integrity for water quality and aquatic life is essential.

In addition to ensuring ecological stream flow, maintaining the water quality of surface supplies is also important. This goal can be accomplished through several key strategies: (1) Watershed improvement programs. These programs aim to reduce sediment and potential chemical pollution in water sources. (2) Isolation and control flows of poor-quality water from sources. These include identifying and managing low water quality sources, such as effluent from mines, springs, and seeps. Control measures should be implemented to prevent these sources from degrading the overall water quality. (3) Restrictions on urban and industrial wastewater discharge. Regulating the discharge of wastewater from urban and industrial activities into water supplies is critical. Implementing stringent restrictions and monitoring systems can prevent pollution and maintain the quality of water for irrigation purposes.

14.3.2 Source from Return Flows

Return flows are water discharges that re-enter rivers and streams after use in municipal, industrial, or agricultural applications. These include irrigation tailwater, urban runoff, wastewater effluent, and recirculated industrial water. Return flows represent an important supplemental source of water, especially in water-scarce regions. The use of return flows is governed by water rights frameworks, permits, and state or federal agency approvals. These permits specify allowable volumes, usage conditions, and treatment requirements. Water reuse strategies increasingly include the use of treated municipal wastewater for irrigation. Wastewater typically undergoes primary, secondary, and tertiary treatments. Primary treatment involves removing settleable and floating materials, whereas secondary treatment stabilizes most organic materials and reduces nutrient levels, often incorporating the use of lagoons and constructed wetlands. Tertiary treatments, such as filtration and disinfection, produce high-quality reclaimed water suitable for irrigation and sometimes similar in quality to potable water.

Reclaimed water use must comply with trace element concentration limits, such as those outlined in Table 14.7, and consideration must be given to long-term accumulation in soil. For example, Mexico's Irrigation District No. 003 in Tula Hidalgo has used treated wastewater from Mexico City for decades to irrigate crops. In some cases, drainage water, which often contains high salt concentrations, is mixed with fresher water from rainfall or overflow to irrigate pastures outside designated irrigation districts. If return flows are of acceptable quality, they may also be used for groundwater recharge or sold for use in downstream projects.

Maintaining a quality source may comprise (1) a joint effort with upstream users to alleviate pollution (sediment and chemicals) and (2) the development of a program for mixing water of different qualities to obtain the best quality water for the project.

14.3.3 Groundwater Supply

Groundwater is a critical irrigation source, particularly during droughts or periods of low surface flow. The amount of groundwater that can be sustainably withdrawn depends on the aquifer's natural recharge rate, storage capacity, and local hydrological conditions. In some aquifers, full extraction may be feasible without degrading quality, while others require adherence to safe yield or extraction limits, which are often determined by hydrologists and regional water agencies.

Groundwater can be managed conjunctively with surface water to optimize cost and reliability. For example, in El Paso, Texas, surface and groundwater supplies are used in tandem: groundwater is drawn during droughts for urban use, while excess surface water during wet years is used for irrigation and aquifer recharge.

If the quality of groundwater differs significantly from that of surface water, a blending strategy may be required. Groundwater is often cleaner during low-flow periods when surface water is heavily polluted. In these cases, expanding groundwater use may be preferable. However, if groundwater is the sole source and its quality deteriorates over time, adjustments such as changing cropping patterns or implementing advanced treatment may be necessary.

14.4 WATER IN STORAGE

Water stored in reservoirs, tanks, or holding ponds may experience quality deterioration due to evaporation, sediment accumulation, or increased salinity from contact with surrounding substrates. Thermal stratification can also alter water chemistry and contribute to oxygen depletion or thermal pollution. To address these issues, it is important to develop operational strategies that preserve stored water quality. These methods may include (1) selective withdrawal methods to access the best-quality water layer; (2) bypass or off-channel storage options to reduce sedimentation; and (3) flushing and sediment management procedures to maintain storage capacity. In some regions, such as Texas, water storage is managed by regulatory zoning. The storage capacity is divided into three zones. Zone 1 (storage capacity is greater than 80%): In-stream flows can be passed to the project up to the monthly median historic flow; Zone 2 (storage capacity is between 50% and 80%): In-stream flows can be passed up to the 25th monthly percent stream flow; and Zone 3 (storage capacity falls below 50% storage capacity): In-stream flow is passed in amounts established by a water quality standard published by the Texas Commission on Environmental Quality.

Evaporation losses from storage systems must also be accounted for in water planning. Factors such as surface area, wind speed, air temperature, and humidity influence evaporation rates and must be considered in reservoir design and irrigation scheduling.

14.5 WATER IN THE PROJECT

14.5.1 Sources of Pollution

Within irrigation projects, water quality can be compromised by multiple sources of pollution that originate from both agricultural and non-agricultural activities. These pollutants may include (1) wastewater from stagnant pools and seepage zones, (2) chemical residues from herbicides and insecticides, (3) eroded sediment from poorly managed fields, and (4) organic waste from livestock operations, such as feedlots. Additionally, industrial and municipal wastewater discharges can contribute contaminants, while stagnant pools within distribution networks may encourage algal growth and microbial proliferation. Among these, eroded soil is one of the most significant carriers of agricultural pollutants, as it often transports fertilizers, pesticides, and other agrochemicals bound to fine particles. Soluble substances, such as nitrates and certain herbicides, can bypass soil adsorption and move freely with irrigation runoff or percolate into groundwater, posing long-term risks to both water bodies and drinking water supplies.

14.5.2 Action against Pollution

To mitigate pollution within irrigation projects, implementing a comprehensive management approach that includes surveillance, regulatory enforcement, and the adoption of BMPs is essential. Project managers must regularly inspect irrigation systems and enforce water quality regulations, taking legal action when necessary to prevent unlawful discharge. Key strategies to reduce pollutant transport include minimizing runoff and soil erosion through improved field-level practices, such as irrigation scheduling, land leveling, and the adoption of efficient irrigation technologies like drip or sprinkler systems. Installing grass buffer strips between fields and drainage ditches can help trap sediment and agrochemicals before they enter the drainage network. Similarly, settling basins and dikes placed in drainage channels can slow water flow, encouraging sedimentation and reducing pollutant loads. Constructed wetlands, which act as natural biofilters, are highly effective at removing nutrients, pathogens, and suspended solids from return flows.

Promoting environmental stewardship through farmer education and capacity building is also critical. The emphasis should be placed on water measurement, proper chemical application, and irrigation practices that reduce both surface runoff and deep percolation. These efforts are especially important when using marginal-quality water or irrigating near sensitive water bodies. Ultimately, pollution control within irrigation projects is a shared responsibility that ensures not only the protection of soil and water resources but also the long-term sustainability and productivity of irrigated agriculture.

14.6 RETURN FLOWS FROM THE PROJECT

Return flows generated within an irrigation district re-enter nearby water bodies, such as rivers or streams, and, in some cases, may ultimately discharge into estuaries or the ocean. These return flows carry residual water from irrigated fields, which may contain nutrients, sediments, pesticides, or other contaminants. Monitoring and managing the quality of these return flows is essential to protect downstream aquatic and riparian ecosystems. To this end, a comprehensive water quality management plan should be developed, beginning with the identification of pollution sources within the district. The plan should include the assessment of receiving water bodies, strategies for reducing nitrogen, phosphorus, and pesticide loads, and the establishment of a clear implementation timeline. Key components should include public education, BMP adoption, criteria for total maximum daily load (TMDL) compliance, and performance evaluation based on pollutant load reduction.

BMPs play a central role in minimizing pollution from return flows. These may include transitioning to more efficient irrigation systems to reduce runoff, implementing precise irrigation scheduling and fertigation practices, and establishing vegetative buffer strips between fields and main drainage channels. Advanced watershed modeling tools can support the development and evaluation of BMPs. One of the computer programs commonly used to predict environmental effects at the watershed level is the Soil and Water Assessment Tool (SWAT) (Arnold et al. 2012). The Hydrologic and Water Quality System (HAWQS) offers a web-based Decision Support System (DSS) that uses SWAT as its primary modeling engine, which is used for simulating hydrological and water quality processes in watersheds (HAWQS, 2020). It provides an intuitive interface for creating SWAT projects encompassing the contiguous United States (CONUS, 48 states), which can be easily used by non-experts.

To evaluate the environmental impact of return flows, several water quality parameters are commonly measured.

Dissolved oxygen (DO): DO is a key indicator of water quality and aquatic ecosystem health. It represents the concentration of oxygen that is dissolved in water and available to aquatic organisms. Low DO levels, typically below 2 mg/L, often lead to fish kills and severely stressed aquatic life, while concentrations above 10 mg/L during daylight can be indicative of an unhealthy aquatic environment with excessive algal growth due to high photosynthetic activity. Algal blooms can cause significant diurnal fluctuations in DO: they generate oxygen through photosynthesis during the day but consume it at night through respiration and decomposition, potentially leading to hypoxic (low oxygen) conditions. DO levels are influenced by factors, such as temperature, salinity, the presence of organic matter and oxygen-demanding pollutants, and biological activity. Accurate DO measurements are commonly obtained using dissolved oxygen probes or optical sensors in field or laboratory settings.

Biochemical oxygen demand (BOD): BOD is an essential indicator of organic pollution in water. It quantifies the amount of dissolved oxygen required by aerobic microorganisms to decompose the organic matter in a water sample over a specified incubation period, typically 5 days at 20°C (BOD_5). A higher BOD indicates a greater load of biodegradable organic material, which can lead to oxygen depletion in receiving waters and stress on aquatic life. BOD encompasses both dissolved and suspended oxygen-demanding substances. In some analyses, samples may be filtered to isolate dissolved components. Microorganisms metabolize organic compounds through respiration, consuming oxygen to break complex molecules into simpler forms. This process links BOD to ecosystem health, particularly in rivers, lakes, or effluent-dominated streams where excessive organic loading can result in hypoxic conditions and ecological degradation.

Total suspended solids (TSS): TSS refers to the concentration of undissolved particles, both mineral and organic, suspended in water. It is commonly measured as the dry weight of material retained on a filter, expressed in milligrams per liter (mg/L) or parts per million (ppm). Elevated TSS levels can result from soil erosion, algal blooms, stormwater runoff, or effluent discharge. High algal concentrations, in particular, can significantly raise TSS values. In addition to turbidity and light attenuation, suspended solids can act as carriers for nutrients (e.g., phosphorus), heavy metals, and other pollutants. These attached contaminants may settle into sediments or be transported downstream. Furthermore, organic matter in TSS contributes to sediment oxygen demand (SOD), the microbial consumption of oxygen at the sediment–water interface, which can reduce DO levels and degrade aquatic habitats. Monitoring TSS is therefore essential for evaluating water quality impacts on ecosystems, irrigation infrastructure, and downstream users.

Nitrogen: Nitrogen is used as fertilizer for plant nutrition. In aquatic systems, small amounts of nitrogen can promote the healthy growth of algae and phytoplankton. However, excessive nitrogen, particularly from return flows containing fertilizers or wastewater, can result in nutrient enrichment (eutrophication), leading to algal blooms, oxygen depletion, and ecosystem imbalance. Key nitrogen parameters used in water quality monitoring include nitrate-nitrite nitrogen, total Kjeldahl nitrogen (TKN), and ammonia nitrogen, all typically reported in milligrams per liter (mg/L). Elevated nitrate and nitrite concentrations are especially problematic in coastal and estuarine environments, where they contribute to hypoxia and fish kills. According to Texas state screening criteria, nitrate-nitrite levels in tidal waters should not exceed 2.76 mg/L, while drinking water aquifers must maintain nitrate levels below 10 mg/L to protect human health. Regular monitoring of nitrogen forms is essential in return flow management to prevent downstream pollution and comply with water quality regulations.

Phosphorus: Phosphorus is an essential nutrient required in small amounts for the growth of crops, aquatic algae, and phytoplankton. However, elevated phosphorus levels can contribute to the occurrence of algal blooms, which can cause daily swings in dissolved oxygen. Phosphorus is typically measured in two forms for water quality assessment: orthophosphate phosphorus, which is immediately bioavailable, and total phosphorus, which includes all phosphorus forms (dissolved and particulate). Both are reported in milligrams per liter (mg/L). Monitoring phosphorus levels in return flows is critical for controlling nutrient loading into downstream systems, especially in regions with sensitive estuarine or lake environments where phosphorus is the limiting nutrient for algal growth.

Proper management of return flows is not only critical for maintaining water quality but also essential for meeting regulatory requirements and protecting ecosystem services. Proactive monitoring, data-informed modeling, and farmer engagement in BMP implementation are the foundation of a sustainable return flow strategy.

REFERENCES

Arnold, J.G., Moriasi, D.N, Gassman, P.W., Abbaspour, K.C., White, M.J., and Srinivasan, R. (2012). SWAT: Model use, calibration, and validation. *Transactions of the ASABE*, 55, 1508.

Ayers, R.S. and Westcot, D.W. (1976). *Water Quality for Agriculture*. FAO. Irrigation and Drainage Paper No. 29. Food and Agriculture Organization of the United Nations, Rome.

Ayers, R.S. and Westcot, D.W. (1994). Water Quality for Agriculture, Revised. FAO Irrigation and Drainage Paper 29 Rev.1. Food and Agriculture Organization of the United Nations, Rome.

Gibson, N., McNulty, S., Miller, C., Gavazzi, M., Worley, E., Keesee, D., and Hollinger, D. (2021). *Identification, Mitigation, and Adaptation to Salinization on Working Lands in the U.S. Southeast*. Forest Service, U.S. Department of Agriculture, Washington, DC.

HAWQS. (2020). *HAWQS System and Data to Model the Lower 48 Conterminous U.S using the SWAT Model.* Texas Data Repository Dataverse, V1. https://doi.org/10.18738/T8/GDOPBA.

Maas, E.V. and Hoffman, G.J. (1977) Crop salt tolerance: Current assessment. *Journal of Irrigation & Drainage Engineering, ASCE*, 103(IRZ), 115–134. Proceeding Paper 12993.

Mohamed, N.A. and Amer, F. (1972). Sodium carbonate formation in Ferhash area and possibility of biological dealkalization. *Proceedings of the International Symposium on New Developments in the Field of Salt Affected Soils*, 4–9 December 1972, Ministry of Agriculture, Cairo. p. 346.

National Academy of Sciences and National Academy of Engineering (1972). Water quality criteria. United States Environmental Protection Agency, Washington DC. Report No. EPA-R373-033. 592 p.

Oster, J.D. and Schroer, F.W. (1979). Infiltration as influenced by irrigation water quality. *Soil Science Society of America Journal*, 43, 444–447.

Pearson, G.A. (1960). *Tolerance of Crop to Exchangeable Sodium.* Agriculture Information Bulletin No. 216. Agricultural Research Service, United States Department of Agriculture, Washington, DC.

Pratt, P.F. (1972). *Quality Criteria for Trace Elements in Irrigation Waters.* California Agricultural Experiment Station, Riverside, CA, 46 p.

Rhoades, J.D. (1974). Drainage for salinity control. In: J. Van Schilfgaarde, Ed., *Drainage for Agriculture.* American Society of Agronomy Monograph No. 17. American Society of Agronomy, Madison, WI, pp. 433–462.

Rhoades, J.D. (1977). Potential for using saline agricultural drainage waters for irrigation. *Proceedings of the Water Management for Irrigation and Drainage*, 20–22 July 1977, ASCE, Reno, NV. pp. 85–116.

Rhoades, J.D. and Merrill, S.D. (1976). Assessing the suitability of water for irrigation: Theoretical and empirical approaches. In: *Prognosis of Salinity and Alkalinity.* FAO Soils Bulletin 31. FAO, Rome, pp. 69–110. https://www.fao.org/4/x5870e/x5870e00.htm#Contents

Singh, V.P. and Su, Q. (2022). *Irrigation Engineering: Principles, Processes, Procedures, Design, and Management.* Cambridge University Press, Cambridge.

Su, Q., Kambale, R.D., Tzeng, J., Amy, G.L., Ladner, D.A., and Karthikeyan, R. (2025). The growing trend of saltwater intrusion and its impact on coastal agriculture: Challenges and opportunities. *Science of the Total Environment*, 966, 178701.

Tanji, K.K. and Wallender, W.W. (1990). The nature and extent of agricultural salinity problems. In: K.K. Tanji, Ed., *Agricultural Salinity Assessment and Management.* ASCE Manual Reports on Engineering Practices, 71. American Society of Civil Engineers, Reston, VA, pp. 1–41. Reprinted with permission from ASCE.

Wilcox, L.V. (1960). *Boron Injury to Plants.* USDA Bulletin No 211. Agricultural Research Service, U.S. Department of Agriculture, Washington, DC, p. 7.

15 Weed Control

15.1 INTRODUCTION

One of the major problems involved in the maintenance of irrigation systems is the control of weeds and aquatic vegetation. This is particularly important in canals, drains, and reservoirs to ensure that water flow remains within the design parameters. The maintenance level to be achieved determines these limits. In the western United States alone, weed control in irrigation systems incurs costs running hundreds of thousands of dollars annually. This cost does not even include the financial impact of water loss, seepage, and reduced operational efficiency.

Weeds are a serious problem from an operation-and-maintenance viewpoint because preventing the inspection of ditch banks and their root systems provides habitat and food for rodents, such as gophers and ground squirrels. Decaying roots can create seepage problems, while growing plants can transpire significant amounts of water. Weeds on canal banks, especially those overhanging into the water, can trap other debris, reducing the system's capacity and water supply to farmers. This often necessitates costly and frequent mechanical cleaning. Similarly, weeds in water can cause silt accumulation, reduce water flow, increase water levels, and increase seepage. Drains are affected by weeds in the same way as canals. Given the economic losses due to weed proliferation, significant research is being conducted to find cost-effective control or elimination methods that are more affordable than periodic mechanical cleaning and canal rebuilding.

This chapter provides a comprehensive overview of weed control in irrigation systems, emphasizing the challenges and solutions in managing various types of weeds that threaten the efficiency of these systems. It categorizes weeds into land, water, waterline, and transient types, each requiring specific control strategies ranging from manual and mechanical methods to chemical and biological approaches.

15.2 TYPES OF WEEDS

Weeds in irrigation systems are generally categorized into two main types: land weeds and water weeds. Land weeds are those that grow on farms and range areas, deserts, and canal systems, e.g., on the tops of banks, outer slopes, or downhill slopes as far as the water line (Figure 15.1). Examples include bindweeds, Canada thistle (*Cirsium arvense*), foxtails, goosegrass, and Russian-thistle. Common plants, such as sweet clover (*Melilotus alba, Melissa officinalis*), willow (*Salix* spp.), and salt cedar (*Tamarix* spp.), become weeds when they are allowed to grow uncontrolled on canal banks. Watery weeds require considerable moisture and usually grow in water. Examples of watery weeds include tules, cattails (*Typha* spp.), pondweeds (*Potamogeton* spp.), and water plants, described as mosses. Algae are usually considered in the latter class.

DOI: 10.1201/9781003747444-18

FIGURE 15.1 Examples of land weeds (top panel) and water weeds (bottom panel).

However, there are two additional classes that might be added. The third category, which, as a practical matter, fits neither of the two main classes, might be called "waterline" weeds because of the lack of a better term. These are the weeds that grow on ditch banks at the water surface and are dependent on the water in the canal for continuous growth. They cannot survive when completely submerged, nor can they survive or propagate on dry land or on soil that is saturated with water. This group includes plants, such as sedges, barnyard grass, willows, and cottonwoods. These plants cause silt berms to develop and are difficult to eradicate.

The fourth category, if it is a category, might be termed "transient weeds". These are the weeds that mature away from the conveyances and are blown into the canal prism or into the water by wind action. This category includes Russian thistles, seep weed (plants of the genus Suaeda), ragweeds (*Ambrosia artemisiifolia*), and other types that grow dense foliage and break or are blown loose from the ground when mature. In some areas, these moving weed skeletons cause many difficulties in operation and maintenance. Several online resources assist in identifying aquatic weeds and algae, such as the University of Florida Center for Aquatic and Invasive Plants (https://plants.ifas.ufl.edu/plant-directory/) (University of Florida, 2025), the Virginia Tech Weed Identification Website and APP (https://agweedsci.spes.vt.edu/extension/weedid.html) (Virginia Tech, 2025), Aquaplant from the Texas A&M AgriLife Extension (https://aquaplant.tamu.edu/) (Texas A&M AgriLife Extension, 2025), and the National Invasive Species Information Center of the USDA (https://www.invasivespeciesinfo.gov/) (NISIC, 2025).

15.3 WEED CONTROL

Each category of weeds is a separate problem, but there is some relationship among different categories. Each requires different methods of control. There are different methods for weed control, including:

1. **Manual control**: Historically, in regions with lower labor costs, manual weeding has involved hand removal or simple tools. While effective in small-scale or hard-to-reach areas, it is increasingly impractical because of rising labor costs and the mechanization of agriculture.
2. **Mechanical control**: Machinery, such as mowers, weed harvesters, and ditch-cleaning equipment, is used to manage large infestations efficiently. Mechanical methods are well suited for extensive canal networks but may be limited by terrain or water conditions.
3. **Chemical control**: Herbicides offer a targeted means of weed suppression. However, their use must account for potential environmental impacts, herbicide resistance, and regulatory restrictions. Selective herbicides are often preferred to minimize harm to nontarget species.
4. **Biological control**: This method involves the introduction of natural predators or competitive organisms, such as insects or fish, that reduce weed populations sustainably. Biological control aligns with environmentally conscious practices and is especially effective in aquatic environments.

Integrated Weed Management (IWM) offers a comprehensive and adaptive framework that combines multiple control strategies to improve the long-term sustainability and effectiveness of weed control efforts. Rather than relying on a single method, IWM promotes the use of tailored tactics based on specific weed species, scale of infestation, and local environmental conditions. Preventive measures, such as establishing dense, sod-forming grasses, can suppress weed emergence by outcompeting invasive species. Cultural practices, such as crop rotation, canal bank stabilization, and timely mowing create conditions that are unfavorable to weed growth. In addition, regular monitoring enables early detection and rapid response, preventing minor infestations from developing into costly problems. By integrating mechanical, chemical, biological, and cultural approaches, IWM reduces dependency on herbicides, mitigates environmental risk, and supports the long-term ecological health of irrigation infrastructure.

Successful weed control programs must be adaptive and grounded in a local context. Strategies should be reassessed regularly to address emerging challenges, such as herbicide resistance or changes in weed behavior caused by climate variability. The involvement of experts, including agronomists, weed scientists, and irrigation specialists, is essential for selecting appropriate tools and technologies that are both effective and environmentally friendly. Economic considerations are equally important. Evaluating the cost-effectiveness of weed control measures can help decision-makers identify approaches that not only reduce maintenance costs but also increase water delivery efficiency and crop yields. Weed management in irrigation systems must maintain a balance between operational performance, environmental protection, and long-term sustainability.

15.4 LAND WEEDS AND THEIR CONTROL

The banks of irrigation canals and adjacent lands, if not properly managed, are highly susceptible to invasion by a variety of terrestrial weed species. Commonly encountered weeds in canal systems include Russian-thistle, morning glory, bindweed,

TABLE 15.1
Comparison of Land Weed Control Methods

Method	Effectiveness	Selectivity	Environmental Impact	Cost	Ideal Use Case
Pasturing	Moderate	Low	Low	Low	Gently sloped, accessible areas with available livestock
Mowing	Moderate to high	Low	Low	Low to medium	Routine maintenance of easily accessible canal banks
Cutting	Moderate	Low	Low	Low	Periodic cutting of annuals or shallow-rooted perennials
Burning	High (temporary)	Low	Medium to high	Low	Large, dry infested areas away from sensitive infrastructure
Flame throwers	High (with repetition)	Medium	Medium	Medium	Spot treatment in fire-prone or burn-restricted zones
Chemical herbicides	High (selective)	High	Variable	Medium	Targeted removal of invasive species
Spraying machines	High (with precision)	Medium to high	Variable	Medium	Large-scale treatment where labor or terrain limits manual work
Soil sterilants	Very high (long-term)	None	High	High	Remote areas requiring long-term control and no regrowth
Grass seeding	High (preventive)	High	Low	Medium	Erosion-prone slopes or areas needing vegetative cover

sunflowers, and sweet clover. If left unchecked, these weeds can obstruct inspection routes, impede water flow, promote rodent activity, and degrade bank stability. The most effective long-term solution is the establishment of a dense, sod-forming grass cover that competes with and suppresses weed germination and growth. However, this approach can be challenging in arid or semi-arid regions, where rough terrain, low soil moisture, and saline conditions complicate vegetative establishment. Effective control of land weeds typically requires an integrated approach that combines multiple control methods on the basis of site-specific conditions, weed types, labor availability, and long-term management goals. Table 15.1 provides a summary of different land weed control methods, which are discussed in detail in this section.

15.4.1 Use of Pasturing and Mowing

Two widely practiced methods for controlling terrestrial weeds along irrigation canals are pasturing and mowing. These approaches are often used in conjunction with grass seeding and other mechanical or chemical controls as part of an integrated weed management strategy. Pasturing involves the controlled grazing of livestock, such as sheep, goats, or cattle, along canal banks. This method reduces weed biomass naturally and can be cost-effective for large or inaccessible areas. However, careful management is necessary to prevent overgrazing, soil compaction, or damage to desirable vegetation. Mowing involves the mechanical cutting of weeds via powered equipment, such as rotary or flail mowers. Regular mowing helps suppress weed regrowth, prevents seed production, and improves access for inspection and maintenance. It is especially effective for flat or gently sloping rights-of-way where machinery can operate safely. While neither method provides complete eradication, both are valuable tools for suppressing weed growth and maintaining clear, stable canal banks. Their effectiveness is enhanced when timed appropriately and integrated with reseeding or other control measures.

15.4.2 Cutting

Manual or mechanical cutting remains a widely practiced and traditional weed control method (FAO, 1961). When timed appropriately, typically before flowering, cutting effectively reduces weed seed production. However, it rarely eliminates established weeds outright, especially those with deep root systems. Persistent control requires repeated cutting cycles throughout the growing season. A variety of tools can be employed for cutting, including sickles, scythes, and different types of mowing machines. The choice of tool depends on the specific type of weed present and the nature of the terrain being treated.

15.4.3 Burning and Flame Throwers

Controlled burning is widely used to destroy unwanted vegetation. In some climates, weeds die off or dry out during certain seasons of the year, and at this stage, they are combustible and can then burn. A well-controlled fire can destroy green growth and viable seeds, providing an inexpensive solution for extensive infestations. This method is especially valuable in areas where mechanical or manual removal is impractical because of access limitations or labor constraints. However, open burning must be approached with caution. It requires close attention to weather conditions, fuel moisture, wind speed, and the proximity of sensitive infrastructure, such as wooden bridges, culverts, power poles, and nearby croplands. Without proper planning and containment measures, fires can spread rapidly, causing unintended damage to irrigation infrastructure, adjacent property, or ecosystems. Burn permits and adherence to local fire regulations are essential prerequisites before undertaking such operations.

In areas where open burning is restricted or unsafe, flame throwers provide a more controlled mechanical alternative. These devices are designed to project a concentrated, high-temperature flame onto target vegetation. A typical flame thrower unit consists of a pressurized tank (commonly filled with kerosene or other flammable liquids), a pump and power source, and a boom outfitted with multiple nozzles. The fuel is atomized and ignited, producing flame jets that can be directed at the base of weeds to desiccate and scorch plant tissues. Flame throwers are often mounted on wheeled carriages, trailers, or utility vehicles for mobility. The boom can be adjusted vertically and horizontally to accommodate different terrain conditions and plant heights. Operators use hand or hydraulic controls to regulate flame intensity and direction, allowing for the precise treatment of infested areas without affecting adjacent vegetation.

While initial flaming may not kill all weeds, especially deep-rooted perennials in moist soils, repeated flaming can significantly weaken root systems and suppress regrowth over time. This method is particularly effective for managing young or shallow-rooted weed species, as well as preventing seed production in annuals. Nevertheless, flame treatment carries inherent fire risks, especially in dry environments. Operators must use proper personal protective equipment (PPE), establish firebreaks, and have fire suppression tools readily available during operation. Additional safety measures include scheduling applications during low-wind conditions, moistening adjacent areas, and ensuring that the surrounding vegetation is green and nonflammable. When used appropriately, burning and flame throwers can serve as valuable tools within an integrated land weed management program, particularly for spot treatment and difficult-to-access areas. The effectiveness of these methods can be increased when they are combined with mowing, chemical control, or reseeding with competitive grasses to prevent reestablishment.

15.4.4 Chemical Weed Killers

Chemical herbicides have long been a cornerstone of weed management, particularly when labor or access limitations restrict manual or mechanical approaches. Recent advancements have led to the development of selective herbicides that target specific weed species, while sparing beneficial vegetation that stabilizes soil and prevents erosion. Selective herbicides are particularly useful in irrigation systems, where vegetation such as sod-forming grasses may serve to stabilize canal banks and reduce erosion. By selectively suppressing invasive or noxious weeds, these products help maintain both hydraulic efficiency and ecological balance.

(1) The choice of herbicide should be guided by a detailed understanding of the target weed species, their growth stages, and local environmental conditions. Important factors include the soil type, proximity to water sources, rainfall patterns, and sensitivity of the surrounding flora and fauna. Equally critical is compliance with local, state, and federal regulations, as certain chemicals may be restricted or banned owing to their toxicity, persistence in the environment, or impact on human and aquatic health. Because of these complexities, it is strongly recommended that landowners and irrigation managers consult with licensed applicators, agronomists, or weed scientists before initiating a chemical weed control program. Expert guidance ensures proper product selection, application timing, dosage, and safety protocols, which maximize effectiveness while minimizing unintended consequences. Chemical control should also be considered within the broader framework of the IWM. When combined with mechanical, biological, or cultural practices, herbicides can play a strategic role in long-term, sustainable vegetation management. Regular monitoring and reassessment of chemical use are essential to avoid the development of herbicide resistance and reduce the environmental impact over time.

In the western regions of the United States, the herbicides most commonly used include glyphosate (known commercially as Roundup), 2,4-Dichlorophenoxyacetic acid (2,4-D), and 2,4,5-Trichlorophenoxyacetic acid (2,4,5-T). These herbicides can be applied individually or in combination, based on the user's experience and specific requirements.

Glyphosate, chemically known as *N*-(phosphonomethyl)glycine, is a broad-spectrum systemic herbicide primarily used to eliminate perennial weeds. It is absorbed by the plant's foliage and moves to growth points, making it effective only for actively growing plants and not as a pre-emergent herbicide. This herbicide is commonly applied by spraying onto leaves, injecting into tree trunks, or applying directly to stumps and is extensively used in forestry. It is the most frequently used herbicide in the United States. Glyphosate is particularly favored in areas where crops genetically engineered to be glyphosate resistant (Roundup Ready) are cultivated, allowing its use as a postemergence herbicide against a wide range of weeds.

2,4-D is highly effective against most terrestrial weeds, especially when they are young and actively growing. It is a systemic herbicide predominantly used for controlling broadleaf weeds and ranks as the most extensively used herbicide globally and third in North America. Sometimes, areas might require a second application if the initial treatment does not fully eradicate the weeds or if new growth occurs.

For aquatic environments, the herbicide Komeen is frequently employed to control species such as Hydrilla (*Hydrilla verticillata*), Water hyacinth (*Eichhornia crassipes*), Elodea, Naiads, Coontail, Eurasian Watermilfoil, Sago Pondweed, American Pondweed, and Water Lettuce. Komeen, a copper-based herbicide, is generally applied in amounts ranging from 1.7 to 3.3 gallons/acre-foot (16 to 31 L/ha), depending on the specific species targeted. In cases where the treated water is also a source of drinking water, ensuring that copper levels do not exceed 1 mg/L is crucial.

It is essential to adhere strictly to the recommended application rates on herbicide labels to prevent harm to aquatic life, as many herbicides can be toxic to fish. Here is an example of weed control cost. In the 1960s, the expenses for weed spraying along conveyance systems in the Colorado Big Thompson Project in Northern Colorado, United States, provided an insightful benchmark. The costs at that time were approximately \$2.00/acre or ranged from \$8.00 to \$20.00/mile. An analysis of the cost distribution revealed that labor constituted approximately 40% of the total expense, equipment accounted for 25%, and the remaining 35% was attributed to the materials used in the process.

15.4.5 Spraying Machines

The application of chemical herbicides in irrigation systems typically relies on a range of spraying equipment designed to ensure uniform coverage and minimize labor. These machines vary in size and sophistication, from small, portable knapsack sprayers to large, power-driven systems mounted on tractors or utility vehicles. Knapsack sprayers, carried on the operator's back, are commonly used for targeted, small-scale applications along ditch banks or in hard-to-reach areas. For larger treatment areas, power sprayers are more efficient. These typically consist of a tank, a pump, and an adjustable boom with multiple nozzles, allowing the herbicide solution to be delivered evenly across a wide swath of land. Compared with manual methods, such systems can treat extensive areas with minimal manpower and in a shorter time.

Effective herbicide application requires careful attention to several key factors, including (1) Mixing and dosage: Proper mixing and dosing are critical to ensure that the herbicide performs as intended while minimizing environmental impact.

The application rates must align with label recommendations, as underdosing can lead to poor control and resistance, whereas overdosing may damage nontarget vegetation or contaminate water sources. (2) Preventing spray drift: Careful management is also essential, especially near crops, trees, or open water bodies. The wind speed, nozzle type, and boom height must be adjusted to prevent off-target movement of chemicals. (3) Maintenance: Regular maintenance of spraying equipment—including cleaning, calibration, and nozzle inspection—is necessary to ensure operational accuracy and equipment longevity.

Because most herbicides are toxic, operators must take appropriate safety precautions. Protective clothing such as rubber gloves, goggles, respirators, and long-sleeved garments should always be worn during mixing and application to prevent dermal absorption or inhalation. In addition to liquid formulations, some herbicides are available as powders or granules and can be applied via hand-cranked dusters or engine-powered applicators, especially in areas where liquid spraying is not feasible.

15.4.6 Soil Sterilants

For long-term control, especially in areas where complete vegetation removal is desired, soil sterilants may be used. Commonly used sterilants include sodium chlorate, bromacil, and carbon bisulfide. These substances function by penetrating the soil and rendering it inhospitable to plant life for extended periods. The duration of sterility depends on both the concentration of the chemical and the depth of soil penetration, making careful calibration essential for predictable results.

Sterilants can be applied in liquid or granular form, either broadcast on the surface or injected into the soil via mechanical equipment. In areas that are difficult to reach, such as steep canal banks or remote ditches, dry granular formulations may be distributed via hand-cranked cyclone spreaders, offering a practical solution for challenging terrains. The typical application rates average approximately 5,000 pounds/acre, with costs ranging from 2 to 6 cents/yard^2, depending on the chemical and application method. However, the use of soil sterilants requires cautious planning and oversight. These chemicals do not discriminate between weeds and desirable vegetation, and their long-lasting effects can disrupt soil ecology and delay revegetation. Because weed seeds can remain dormant in the soil for years, sterilization should not be considered a one-time, permanent solution. Instead, it must be part of a long-term weed management strategy that includes monitoring and, if needed, follow-up treatments.

Soil type significantly influences the effectiveness of sterilants. Sandy soils, for example, allow for deeper and more uniform chemical movement, whereas clay-rich soils may restrict penetration and reduce efficacy. As such, site-specific soil assessments are recommended before sterilants are selected and applied. Owing to their potent and persistent nature, the use of sterilants, particularly along major irrigation or drainage canals, should be carried out only by qualified personnel. When applied appropriately and under expert supervision, soil sterilization can be a valuable tool for managing aggressive or recurrent weed growth in hard-to-maintain areas.

15.4.7 Grass Seeding

Grass seeding, along the right-of-way adjacent to the conveyance, has proven to be a valuable practice to discourage weed growth. The grasses used for this purpose are typically hardy, wiry species capable of forming dense, competitive sods that smother undesirable vegetation while offering excellent erosion resistance. Specific grass varieties may be selected to match challenging site conditions, including sandy, dry, or saline soils, thereby expanding their applicability across diverse irrigation landscapes.

The primary challenge in implementing grass seeding lies in establishing a sufficiently dense and uniform stand. Many canal banks and adjacent lands are characterized by irregular surfaces, rock outcrops, tree stumps, or steep slopes susceptible to erosion, all of which complicate seedbed preparation and seed–soil contact. Additionally, moisture availability at the time of germination is often limited, further reducing the chances of successful establishment.

Several seeding methods can be employed, depending on terrain conditions and equipment availability, including (1) drilling (which can be performed advantageously only on favorable terrain and where the strips are wide enough to accommodate the machine); (2) broadcasting seeds by (a) harrow into the soil, (b) hand raking into the soil, (c) snow surfaces (as snow melts, the seeds are carried into the top layers of the soil and germinate in the soil moistened by snow melt); and (3) improved seeders and methods (special machines, hand-cranked seeders).

Achieving a robust grass stand often requires repeated efforts, especially in arid regions where rainfall is erratic and surface erosion displaces seeds. However, once established, these vegetative covers provide long-term suppression of weed emergence and reduce the need for chemical or mechanical interventions. In areas where flame throwers, mowers, or herbicide sprayers cannot reach, such as steep embankments or narrow ditch banks, manual grass seeding remains one of the most viable control measures. With thoughtful species selection, site preparation, and timing, grass seeding serves as both a preventive and corrective method in integrated weed management strategies.

15.5 WATER WEEDS AND THEIR CONTROL

Aquatic or water weeds can cause considerably more trouble than those that grow on banks, especially in irrigation canals and drainage systems. These aquatic weeds vary significantly in their growth habits and are commonly classified into four types (Figure 15.2): (1) floating weeds. Plants that float freely on the water surface, e.g., water hyacinth; (2) fully submerged weeds. These weeds can remain entirely under water, e.g., *H. verticillata*; (3) Emergent weeds. These plants have parts above the water surface while the rest of the plant is submerged, e.g., cattails; (4) bank weeds with aquatic roots. These are often shrubs or small trees growing on banks with roots extending into the water, also known as "waterline" weeds, which will be discussed in Section 15.7.

Water weeds are more prevalent in conveyances with low gradients that carry clear water, as these conditions are beneficial to their growth. In contrast, higher

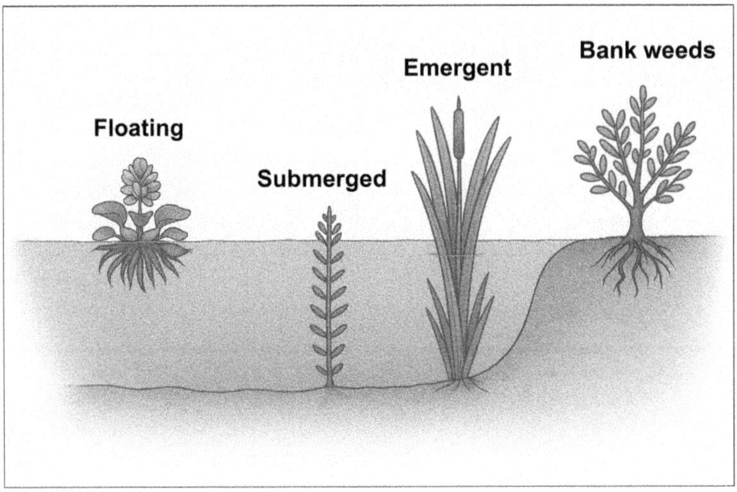

FIGURE 15.2 Classification of aquatic weeds.

water velocities and muddy water tend to discourage their growth. Aquatic weed control is typically more difficult than terrestrial weed control because of environmental constraints and the diversity of weed types.

The presence of various types of aquatic weeds, including channel, floating, submerged, and emergent weeds, can significantly impact water conveyance systems:

1. **Reduction of water carrying capacity**: These weeds can obstruct the flow in channels, reducing the amount of water available for agricultural irrigation.
2. **Increased water consumption**: Emergent weeds consume large quantities of water, thus decreasing the availability for crops.
3. **Silt accumulation**: Vigorous growth can lead to water overflow and increased silt deposition in canals and ditches.
4. **Drainage impedance**: In some cases, weeds can completely block drainage ditches, preventing field drainage and potentially leading to increased soil alkalinity, which can result in crop failure and soil sterility.

Submerged aquatic weeds in irrigation canals generally fall into two categories: (1) rooted weeds and filamentous algae. Rooted weeds are often referred to as pondweeds, and these plants anchor to the bottom soil. Algae grow attached to concrete canal linings and structures within the water system.

Table 15.2 provides a summary of different water weed control methods, which are discussed in detail in this section.

15.5.1 Biological Control

Biological control methods employing natural predators or selected species have achieved significant cost savings, with reductions of up to 80% reported in some

TABLE 15.2
Comparison of Different Water Weed Control Methods

Control Method	Primary Targets	Advantages	Limitations
Biological control	Floating, submerged, and emergent weeds	Eco-friendly, cost-effective, long-term suppression	Species specific, may require time to establish
Shutting of water flow	Submerged and emergent weeds	Simple, low-cost in seasonal/ intermittent systems	Not feasible in continuous flow canals during peak irrigation
Dragging of chains	Submerged weeds (entire channel)	Physically dislodges weeds, works in unlined canals	Labor-intensive, slow, expensive
Weed cutting	Submerged and emergent weeds	Mechanized, efficient in accessible areas	Limited by debris, root regrowth, and submerged obstacles
Cleaning	All types (plus sediment)	Dual-purpose: weed and silt removal	Not ideal for dense root mats or algae
Chemical control	Submerged weeds, algae, emergent weeds	Fast-acting, scalable, proven cost-saver in some cases	Environmental risks, cost varies, possible water contamination

cases in the management of aquatic weeds. Notable examples include water hyacinth (*E. crassipes*), hydrilla (*H. verticillata*), and rigid hornwort (*Ceratophyllum demersum*).

15.5.1.1 Case Studies

Control of water hyacinth: In Mexico and Florida, the introduction of specific insects, such as *Neochetina bruchi* and *Neochetina eichorniae*, has been highly effective in controlling water hyacinth populations. These biological control agents drastically reduce infestations from 80% to 5% (Gettys et al., 2009).

Use of grass carp in Mexico: Grass carp (*Ctenopharyngodon idella*) has been used to manage various aquatic weeds, such as pondweed (*Potamogeton* sp.), *Zosterella dubia*, *Chara* sp., *Najas* sp., and *Hymenachne amplexicaulis*. The introduction of these carps has significantly reduced the need for mechanical control methods and has led to an 80% decrease in the requirement for physical weed extraction (Mexican Institute of Water Technology, 2018).

The adoption of biological control strategies provides an eco-friendly alternative to mechanical methods, offering sustainable and cost-effective solutions for managing aquatic weed problems in various environments.

15.5.2 Shutting of Water Flow

Temporarily halting the flow of water in canals and waterways is a recognized method for controlling aquatic weeds. When the water flow stopped, the weeds deprived of their aquatic environment soon dried out. Upon resuming water flow, much of the dried vegetation either floats to the surface or is washed away, effectively clearing the waterways.

However, the practicality of this method varies, depending on the nature of irrigation system.

1. **Limited applicability in continuous systems**: In many irrigation systems where there is a consistent need for water delivery, especially during peak growing seasons, this method may not be feasible. The demand for water often coincides with a period of vigorous weed growth, making it challenging to interrupt water flow without disrupting agricultural needs.
2. **Effectiveness in intermittent systems**: For canal systems that are used for storage purposes or those that operate on an intermittent basis, shutting off the water flow can be an effective weed control measure. In these scenarios, the lack of continuous water requirements allows for temporary cessation of flow, providing an opportunity to manage weed growth without significant disruption to the system.

15.5.3 Dragging of Chains

An old standby method was to drag a very heavy chain (up to several tons) down the conveyor with two tractors positioned on either bank to serve as the primary driving force pulling the chain. This method has proven to be both time-consuming and expensive. Despite these drawbacks, chain dragging remains in use in certain parts of the world where it is deemed effective. In some instances, human labor is employed as the primary force for moving the chain, especially in regions where mechanization is less accessible.

15.5.4 Weed Cutting

A variety of tools and machinery have been developed for mechanical cutting of aquatic weeds, particularly those anchored to the bottom of ditches and canals. This category includes different types of mowers, such as those with reciprocating knives and rotary blades. The primary challenge in using these mechanical tools is the often harsh and unfavorable conditions under which they operate, which can affect their efficiency and effectiveness. To address these challenges, a range of specialized cutting tools have been engineered. These include implements with V-shaped blades, cutters equipped with rigid and flexible tines featuring various cutting points, and disk-based systems. Despite these innovations, only a few designs have proven successful in effectively managing aquatic weed growth.

15.5.5 Cleaning

Periodic cleaning, especially during intervals between irrigation seasons, is an important practice for controlling aquatic weed growth. This routine maintenance not only addresses weed issues but also involves clearing silt and restoring the proper profile and grading of the ditches.

Modern ditching machines are often equipped to handle both soil removal and weed extraction, making them versatile tools for maintaining open ditches.

These machines simultaneously remove weeds while cleaning ditches containing accumulated silt. However, it's important to note that the effectiveness of standard ditching machines in controlling weed growth can be limited. These machines are designed primarily for cutting and lifting soil, and while they can assist in weed management, their capabilities may not fully extend to comprehensive weed control in more challenging scenarios.

15.5.6 Use of Chemicals

Chemical control has become a widely adopted strategy for managing water weeds in irrigation systems, particularly in regions where infestations significantly affect hydraulic efficiency and maintenance costs. In the western United States, the problem became so acute that extensive research was initiated to develop effective chemical treatments—and these efforts have yielded valuable solutions that remain in use today (USBR, 1963).

Among the most commonly used chemical agents are xylene (aromatic solvent) and acrolein for controlling rooted submerged water weeds and copper sulfate for suppressing algal growth. Some of the most recent research efforts that have shown promise in the aquatic weed control field include the application of antifouling paint to irrigation structures to prevent algal colonization and the development of continuous low-concentration copper sulfate feed systems to control rooted aquatic vegetation (USBR, 1967a). Paints incorporating toxicants such as copper or tributyl tin oxide in vinyl resin bases have also successfully prevented the growth of filamentous green algae (USBR, 1967b).

To understand the extent of the aquatic weed problem, consider this example in 1962 within the North Side Canal Company's 160,000-acre irrigation system in Jerome, Idaho. At that time, 15% of the entire operation and maintenance budget was allocated to controlling submerged aquatic weeds. In that year, 7,335 gallons of aromatic solvent replaced the need for a full chaining crew, typically comprising two caterpillar tractors and seven personnel. This substitution led to the early completion of weed removal and a total savings of 13,709. More importantly, the improved flow conditions enhanced water delivery efficiency, reduced water losses, and ultimately contributed to better crop yields. Beyond direct financial savings, the most significant benefit, although harder to quantify in monetary terms, was the improvement in water delivery efficiency. This led to reduced water losses and, consequently, increased crop yields, highlighting the broader impact of effective aquatic weed management.

Algae control in lined canals and structures: Control of algae in lined canals, around structures, and even in closed conduits (although chlorine is more generally used in the latter) can be accomplished rather simply and inexpensively. Copper sulfate crystals are applied at a rate of 1 lb/cubic foot per second (cfs) of water flow, typically at intervals of 5–6 miles along the canal. A treatment frequency of every 14 days has proven effective. The cost of treatment typically ranges from 5 to 10 cents/cfs.

Antifouling paint for measuring structures and short canal reaches: In measuring devices, short reaches of hard surface linings in canals and structures with antifouling paint have been successful in controlling algae. Surfaces are typically cleaned and painted with two coats of paint. The cost of applying paint is approximately 10–11 cents/ft^2. The oxidation and deterioration of the paint film and the oxidation of copper provide preventive effects. Thus, the paint must be replaced every 1 or 2 years.

Submerged weed control (e.g., sago and leafy pondweeds): Among the submerged water weeds, the most troublesome are most likely sages and leafy pondweeds. Aromatic solvents or xylene plus a good emulsifier introduced into canals or drained at a rate of approximately 10 gallons/cfs of flow usually restore the treated stream to full capacity within 24–36 hours. Booster applications varying from 50% to 100% of the original application are required downstream at 3–4-mile intervals. Costs vary considerably due to the method of purchase and transportation distance for the chemicals.

Although lined canals experience fewer weed-related issues, they are still susceptible to algal growth and colonization by organisms, such as freshwater sponges, bryozoans, and mollusks. As such, even these systems require periodic chemical treatment.

One alternative method, injecting silt into the flow to reduce light penetration and suppress photosynthetic weed growth, has shown limited success and is not widely practiced.

15.6 BIOLOGICAL GROWTH IN CLOSED CONDUIT SYSTEMS

Comparatively few studies have been published on the biology of aqueducts and pipes in irrigation systems. Biological growth in closed conduit systems can be particularly important from a hydraulic viewpoint in the operation of an irrigation system. While both plants and animals develop in open systems, such as flumes and canals, only animals, fungi, and some microbes can survive in the dark, enclosed conditions of closed conduits.

Organisms commonly found in covered systems include sponges, bryozoans, mollusks, crustacea, insect larvae, worms, crustaceans, fungi, and protozoa. In contrast, open systems support chlorophyll-bearing organisms such as algae and higher aquatic plants. The absence of light in closed systems leads to a biological shift. Photosynthetic organisms are replaced by heterotrophic animals and fungi that thrive on organic material and shelter attached growths.

Environmental factors, such as water velocity, temperature, and wetted perimeter of the conduit, significantly influence the type and extent of biological colonization. Once conditions are favorable for high biological productivity, some animal forms are likely to infest the system and sometimes develop in great numbers. To effectively prevent biological growth in closed conduits, it is important to identify and address the primary operational challenges. These include the following:

1. **A reduction in conduit capacity**: The presence of these organisms can reduce flow capacity and block pipes and structures, especially those with small diameters.
2. **Fouling of systems**: Sponges, bryozoans, and other organisms can clog water pipes and conduits, sometimes necessitating extensive cleaning efforts. For example, gelatinous bryozoan colonies can obstruct water intake screens and hydroelectric plant grates.

3. **Blockages by clam shells and insect larvae**: Instances of clam shells obstructing pipes have been noted, and aquatic insect larvae can also impede water flow.

Although biological growth in closed systems has not yet caused major disruptions in most irrigation projects, the increasing trend toward pressurized and enclosed irrigation systems, combined with rising nutrient levels in surface water, increases the risk of future infestations.

To minimize these hydraulic problems, the design of new systems should prioritize high-velocity flows, smooth interior surfaces, and the avoidance of structures and pipe sections prone to clogging. Regular maintenance should include mechanical cleaning via scrapers and brushes, as well as chemical treatments when needed. Among chemical agents, chlorine remains the most commonly used, whereas copper sulfate and xylene-based herbicides have been shown to be effective against organisms such as clams and bryozoans. New pesticide formulations are also emerging as viable control tools.

Different types of weed growth are found in closed conduit irrigation systems, a short discussion of which is in order.

15.6.1 Freshwater Sponges

A sponge represents the simplest form of animal life, comprising an intricate network of cells that form its skeletal structure. It resembles a complex communal entity characterized by a delicate network of colorless channels, columns, and cavities. While the majority of sponges inhabit marine environments, select few thrive in freshwater. The green color of freshwater sponges is due to algae, which are single-celled plants that engage in a symbiotic relationship with the sponge. The algae supply oxygen and nourishment to the sponge, which, in turn, provides the algae with a habitat. This mutually beneficial interaction is known as symbiosis.

Approximately 150 freshwater sponge species are found across the United States, predominantly in the clear, shallow waters of ponds, lakes, and calm stream backwaters. Montana, for example, is home to species such as *Spongilla lacustris*. Freshwater sponges are typically a few inches in length and often exhibit crust-like growth on twigs or other solid surfaces. These vary in size from a few square centimeters in areas up to 40 m^2 under favorable conditions. The thickness, which projects into the flow, was measured up to approximately 3 cm in a tunnel west of Fort Collins, Colorado, United States.

15.6.2 Bryozoa

Bryozoa comprises approximately 5,000 existing species and numerous fossil species. These aquatic organisms typically exist in colonies comprising interconnected individuals, with a single colony hosting anywhere from a few to several million individuals. Bryozoans are versatile in their habitation, with some species encrusting rocks, shells, or algae, whereas others, akin to the fossil bryozoans displayed here, form delicate, fan-like colonies that significantly contribute to limestone formations in certain regions. Although these colonies can range from a few millimeters to several meters in size, individual bryozoans are usually no larger than a millimeter and are often mistaken for hydroids, corals, or seaweeds.

In some contexts, bryozoans are considered problematic. Over 125 species are known to grow on ship hulls, creating drag, thereby reducing vessel efficiency and maneuverability. They also tend to accumulate on pilings, piers, and docks. Sometimes referred to as pipe moss, approximately 14 species have been reported in the fresh waters of the United States. In freshwater environments, certain species can form large, jellylike colonies that are capable of obstructing public or industrial water pipes. They are never found in polluted water. Light is not necessary for their growth. They multiply best in 66°F–73°F (19°C–23°C) range and die off at about 50°F (10°C).

Bryozoans have been found growing in a number of situations related to irrigation projects, such as the pipeline carrying irrigation water through the city of Kennewick, Washington; the louvers of the Tracy Fish Collection Facility near Tracy, California; and the concrete runnel walls of the Colorado–Big Thompson project near Loveland, Colorado, United States. The pipeline in Kennewick was being plugged. Bryozoan growth usually occurs in a thin thread-like mat and may reach a thickness of 2 in. or more.

15.6.3 Mollusks

Bivalve mollusks, commonly known as clams or mussels, are often found in irrigation systems. In the United States, particularly in states such as Arizona, California, Oregon, Washington, and Idaho, the Asiatic clam is a prevalent species. These clams inhabit major rivers, such as the Colorado, Snake, Columbia, and Sacramento. Another species, the zebra mussel, is notorious for clogged pipes in Europe. The zebra mussel is a small freshwater species native to the lakes and river systems of southern Russia and Ukraine. However, it has been inadvertently introduced to numerous regions worldwide, where it has become an invasive species. The spread of zebra mussels began to significantly impact various ecosystems in the United States in the 1980s, with notable invasions occurring in the Great Lakes, Hudson River, and Lake Travis. This expansion has raised environmental concerns because of the impact of mussels on local habitats and ecosystems.

The Asiatic clam seeks areas where it can burrow into the bottom sediment and feed on plankton and other food, as it is carried to them in the water. An open system is therefore a more favorable habitat for their growth and development. Since the larval stage of the clam is readily transported by water, the closed system is infested with immature clams. If these larvae find a suitable environment in a closed system, they will mature. In some cases, the adult clam may be flushed into a closed system. The clam may not survive under these conditions, but their shells may become a problem by clogging certain structures or by forming a deposit of shells if they are not flushed through the system.

A significant instance of Asiatic clam infestation occurred in the cooling water system of the synchronous condenser at the Tracy Switchyard in Tracy, California. In this case, the clams multiplied extensively, leading to substantial blockage in the cooling system pipes.

Zebra mussels have become a major concern in the management of pressurized collective irrigation networks, especially those drawing water from infested reservoirs, rivers, or canals. It is challenging to eradicate these mussels because of the large volume of infested water, high reproductive rates, and adaptability.

The U.S. Geological Survey Initiative: A network was established by the U.S. Geological Survey to monitor the presence of zebra mussels across various water bodies in the United States. If irrigation systems are located downstream of infested waters, they are highly susceptible to infestation.

In Canada, particularly in the Alberta Irrigation Districts, there is growing concern about the potential infestation of their extensive irrigation network, including underground pipelines. Efforts are being made to prevent the introduction of zebra mussels into irrigation water supply reservoirs, with recommendations for control and eradication measures.

Several river basins in Spain, such as the Ebro, Jucar, Segura, and Guadalquivir River basins, have reported zebra mussel colonization in reservoirs supplying water to irrigation districts. Monitoring networks have been set up to detect and measure infestation levels. The Ebro River basin, in particular, has the highest number of colonized reservoirs in Spain (CHE, 2018). In the Riegos del Alto Aragón project, which covers 120,000 ha in the Ebro River basin, zebra mussels infest 66% of the irrigated area.

15.6.4 Insects

During their developmental stages, aquatic insects often inhabit freshwater environments. These insects, particularly their larval forms, are known to attach themselves to various submerged structures, such as rocks, logs, or masonry, within these environments. The presence of these larvae can lead to significant issues in water management systems, particularly in closed structures such as pipelines. The accumulation of numerous insect larvae in pipelines and other closed structures can lead to a measurable decrease in their capacity to transport water. Dobsonflies, caddisflies, and trueflies are groups that have been found to infest pipelines. The Southern California Electric Company reported that larvae of caddis and Dobson flies caused up to an 8% reduction in the capacity of water conduits in one of their plants. A power plant in Nebraska, United States, has experienced significant issues due to caddisfly larva infestations in its water conduits. Caddisfly larvae infested the conduits of a power plant in Nebraska. The pupa stage of the black fly, in particular, has been found to attach to irrigation structures in Idaho and surrounding regions, contributing to operational challenges in these water systems.

15.6.5 Other Aquatic Organisms

A number of other organisms, including fungi, diatoms, worms, crustacea, protozoa, and ratifiers, are often found on the walls of closed systems. These types of organisms, many of which are microscopic in size, are often found growing and living in association with what has a brown to gray color and a matlike appearance. An encrustation of this type ranging from 1/8 to 1/4 in. (2.2 to 6.4 mm) thick was recently collected from a tunnel in the Colorado–Big Thompson Project and examined. It was found to contain organisms belonging to the aforementioned groups. The dried encrustation mixture was analyzed and was found to contain 8.8% ferric oxide and 8.7% manganic oxide. Copper, sodium, calcium, aluminum, magnesium, and silica were also found in the dried sample. The filamentous bacteria observed in the sample were probably responsible for the high content of iron and manganese found in the enriched material. These bacteria use the oxidation of these elements as an energy source, leading to the formation of insoluble compounds that contribute to encrustation.

15.7 WATERLINE WEEDS AND THEIR CONTROL

The term "waterline" weeds is used because plants of this type may not be of the general class of either land or water weeds. These are the weeds that grow on ditch banks at the water surface and are dependent on the water in the canal for continuous growth. They cannot survive when completely submerged, nor can they survive or propagate on dry land or soil that is not saturated with water. This group includes plants, such as sedges, barnyard grass, willows, and cottonwoods. These plants cause the development of silt berms and are very difficult to eradicate.

Fortunately, this class of weeds does not perform well along canals where there is a fluctuating water surface, where the water surface varies either up or down at least 1 ft (30 cm). When waterline weeds can be kept either submerged or left high and dry for a period of 10 days to 2 weeks, they do not grow. However, they become troublesome if the water level is held near constant for a month or more.

These weeds are very troublesome because they hang into the water, thus slowing the velocity and inducing silt deposits. This, in turn, causes a berm to build out into the flow, and a roughly rectangular cross section, rather than the normal trapezoidal section, is rapidly developed. In the Colorado–Big Thompson Project in northern Colorado, a canal was constructed in 1956, but no attempt was made at waterline weed control until after the 1958 irrigation season. During this time, an almost continuous infestation of carex water sedges developed along the water line. There were also several other varieties of waterline weeds, e.g., water grass, phragmites, willows, and cottonwoods. During the 1958 season, these plants caused silt berms to develop to the extent that, in places, the canal's cross-section was reduced by as much as 40%. This not only decreased the carrying capacity of the canal but also resulted in costly maintenance and draglining efforts before the following irrigation season.

The control of waterline weeds is critical for maintaining the efficiency and functionality of irrigation canals. To manage these weeds, the commonly used methods include (1) weed burners, (2) herbicides, and (3) soil sterilants.

15.7.1 Weed Burners

Propane weed burners have proven effective in controlling waterline weeds. The procedure typically involves applying the burner once to kill the plants. Approximately 2 weeks later, a second pass is made to incinerate and remove the dead vegetation. However, this method leaves behind roots and stumps of larger plants, which can slow down water flow, lead to silt deposits, and potentially encourage new weed growth. These remnants often require removal by dragline to ensure complete weed control.

15.7.2 Herbicides

Contact herbicides have also been used to manage waterline weeds and have produced the desired results with some classes of vegetation. A notable challenge with herbicides is their tendency to be washed away by water flow. To mitigate this, the water level may need to be temporarily lowered during and after the application of herbicides for them to be effective.

15.7.3 Soil Sterilants

Soil sterilants have shown high effectiveness in dealing with waterline weeds. For example, in the Colorado–Big Thompson Project, a soil sterilant, as a liquid, was applied to a canal at a rate of approximately 12 pounds (dry weight of chemical) per acre (13.5 kg/ha). A narrow band 2–3 ft (0.6–0.9 m) wide at the high-water line on each side of the canal was treated. Thus, only the waterline weeds were covered, and desirable grasses higher on the banks were not destroyed. The liquid was sprayed on an adapted version of a land weed sprayer. The results indicated that a slightly higher concentration would have been desirable, and the best results may have been obtained if the application was made early in the spring.

The use of soil sterilants significantly reduces the costs associated with dragline removal of weeds. In the mentioned project, total costs decreased from $7,000 to $4,000, indicating a 43% cost reduction. When the costs of dragline and sterilant were combined, there was an overall savings of $2,000 compared with scenarios without sterilant use.

15.8 TRANSIENT WEEDS AND THEIR CONTROL

Transient weeds are a distinct category of vegetation that mature on land near or at a distance from water conveyances. These weeds, such as Russian thistles, seep weeds, and ragweed, are characterized by their dense foliage and durable skeletal structure. When mature, they break or are blown away from the soil, often tumbling into nearby canals and waterways due to their ability to roll or tumble along the ground. They are commonly known as "tumbleweeds" and pose significant challenges in water distribution systems.

These windblown weeds can obstruct water conveyance, leading to problems such as the clogging of trash racks, waste, and lodging in various water control structures, such as turnout gates and bridge piers. This issue is particularly notable in many areas of the western United States and possibly other parts of the world.

Grass seedings along conveyances will aid in the control of these transients. This approach is not entirely effective because weed skeletons may originate at some distance from canals. Most likely, the best means to combat this problem is the automatic weed rake, which carries the weeds out of the canal through the use of chain-drawn drags operating over a weed screen. Additionally, weed screens placed at a flat vertical angle force them out of the water where they may be removed. Screens installed at an angle to the flow may be used to direct weeds toward the sides of the conveyances for collection and removal. Fortunately, this problem does not occur year-round and is most severe in spring.

REFERENCES

CHE (2018). Control larvario de especies exóticas invasoras en las masas de agua superficiales (embalses) de la cuenca del Ebro. Campaña, p. 135.

FAO (1961). *Methods and Machinery for Cutting and Cleaning Irrigation and Drainage Channels*. Land and Water Development Division, Food and Agriculture Organization of the United Nations, Informal Working Bulletin No. 13. Food and Agriculture Organization, Rome.

Gettys, Lynn A., Haller William T., and Bellaud Marc (Eds). 2009. *Biology and Control of Aquatic Weeds. A Best Management Practices Handbook*. Aquatic Ecosystem Restoration Foundation, Marietta, GA; Gaines, FL, 214 pp. Available from https://plants.ifas.ufl.edu/

MIWT (2018). *Comprehensive Control of Aquatic Weeds in Irrigation Districts*. Mexican Institute of Water Technology, Mexico.

NISIC (2025). National Invasive Species Information Center, United States Department of Agriculture. National Agricultural Library. https://www.invasivespeciesinfo.gov/

Texas A&M AgriLife Extension (2025). Aquaplant. https://aquaplant.tamu.edu/

University of Florida (2025). Center for Aquatic and Invasive Plants. University of Florida, IFAS. https://plants.ifas.ufl.edu/

USBR (1963). Chemical Control of Submerged Water weeds in Western Irrigation and Drainage Canals. Joint Report ARS 34–57, October 1963. Agricultural Research Service, Department of Agriculture and Bureau of Reclamation, Department of the Interior, United States.

USBR (1967a). Progress Report on Antifouling Materials for Algae Prevention. Water Conservation Report No. WC-30. U.S. Department of the Interior, Bureau of Reclamation, Division of Research.

USBR (1967b). Progress Report on Evaluation of Copper for Aquatic Weed Control and Herbicide Residues on Irrigation Systems. Water Conservation Report No. WC-32. U.S. Department of the Interior, Bureau of Reclamation, Division of Research.

Virginia Tech (2025). Weed Identification Website and APP. https://agweedsci.spes.vt.edu/extension/weedid.html

16 Rehabilitation and Betterment of Irrigation Systems

NOTATION

IE	irrigation efficiency (%)
MIE	modified irrigation efficiency (%)
ΔS	change in soil water storage in the crop root zone during the irrigation event (m^3)
θ	soil moisture (cm^3/cm^3)
R_d	root zone depth (m)
A_L	irrigated area (m^2)
OPE	overall project efficiency (%)
WUE	water use efficiency (kg/m^3)
IWUE	irrigation water use efficiency (kg/m^3)
Y_i	economic yield for irrigation level i (kg/ha)
Y_d	dryland yield or crop yield without irrigation (kg/ha)
ET_i	evapotranspiration for irrigation level i (mm)
ET_d	evapotranspiration of the dryland crops (or of the ET without irrigation) (mm)
I_i	total amount of irrigation applied for irrigation level i (mm)
AE	application efficiency (%)
DU	distribution uniformity of the low quarter (%)
CU	Christiansen's uniformity coefficient (%)
X_i	depth of water in equally spaced catch cans at the i-th grid (m)
X_m	mean depth of water of the catch in all cans (m)
n	number of catch cans
CU_H	Heermann and Hein coefficient of uniformity (%)
d_i	distance from the pivot point to each catch can (m)
x_m^*	weighted mean depth of water in all catch cans (cm)
E_c	water convenience efficiency (%)

16.1 INTRODUCTION

An irrigation system may be rehabilitated and bettered in many ways. It must be acknowledged that all constructions, especially structures, will wear out and/or pass their period of usefulness sooner or later. They must then be repaired, replaced, or abandoned. Among some of the considerations that management must deliberate and act upon are (1) replacements (immediately or over a period of time), (2) water conservation (various means), (3) modernization and automation, (4) consolidation of

DOI: 10.1201/9781003747444-19

facilities (within the district and/or with adjoining districts), (5) economic aspects, and (6) legal aspects and public relations aspects. It is difficult to separate rehabilitation and betterment. Additionally, major rehabilitation and betterment programs have become design and construction projects rather than operation and management projects. This chapter aims to provide a comprehensive guide for effectively rehabilitating and improving irrigation systems.

16.2 IMPROVEMENT PLANS

Once an irrigation system is in place and is in operation, plans for improvement or betterment must be initiated. Improvement is an ongoing process that must be an integral part of the project. The first step in this direction is to gather as much information and data about the irrigation system and its performance as possible. Although information gathering is an ongoing activity, for purposes of analysis, at least a few years of data should be available. The second step is to analyze the gathered data and develop as complete an understanding of the system as possible. The third step is to delineate areas for improvement. For example, in a particular lateral or minor case, the seepage loss may be unacceptably high, as shown by the data. To reduce seepage, an economically viable solution must be considered. Similarly, a particular distributary is able to maintain the desired irrigation head. Then, an improvement scheme should be conceived. A particular minor is eroding too fast. Then, a scheme must be devised to reduce erosion in the channel. Thus, evaluation, monitoring, and feedback provide vital information to launch an improvement program.

However, each irrigation system is unique. Therefore, improvement solutions are bound to be site specific. Nevertheless, the conceptual framework for developing an improvement program is about the same. There can be more than one improvement solution or alternative. Among these options, the most cost-effective is chosen. Sometimes there is no way to tell for sure as to which solution is the most preferred, and only experience can attest to it. This is another reason why monitoring, evaluation, and interaction/feedback are so important.

Irrigation system improvement implies improvement at all levels, i.e., improvement of not only the physical system but also the management, administration, and public relations and interaction. Sometimes, physical improvements are characterized as "hardware" solutions, and nonphysical improvements are characterized as "software" solutions. In general, in the early years of an irrigation system, most of the solutions will be physically improved. In later years, as greater understanding of the system is gained, the improvements will be a combination of hardware and software solutions.

16.3 EVALUATION OF POTENTIAL IMPROVEMENT

The first step in developing a plan for potential improvements is to formulate short-term and long-term objectives for irrigation system improvements. To that end, two factors must be kept in mind. First, the objectives are dynamic and change in time. Second, improvement is an ongoing process. For example, a long-term objective is to keep the entire irrigation system (physical) in as good a condition as possible.

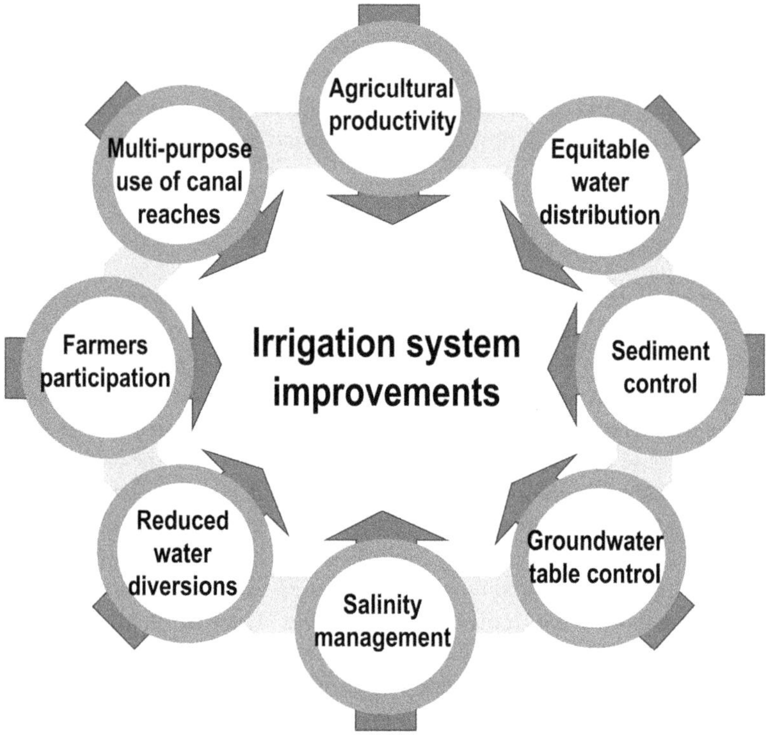

FIGURE 16.1 Objective of irrigation system improvements.

A short-term objective is to distribute water equitably, depending on the weather and crop conditions. The objectives can be defined as follows: (1) agricultural productivity, (2) equitable water distribution, (3) sediment control, (4) groundwater table control, (5) salinity management, (6) reduced water diversions, (7) farmer participation, and (8) conversion of canal reaches to multipurpose resources. These objectives are briefly discussed in order, as shown in Figure 16.1.

16.3.1 Agricultural Productivity

Irrigation is used to increase agricultural productivity. If agricultural productivity can be achieved without irrigation, there will be no need for irrigation. Likewise, any improvement in an irrigation system can be justified only if agricultural productivity is either sustained or increased. The aim is always to at least maintain current levels of productivity. For example, agricultural productivity must not be allowed to decline.

16.3.2 Equitable Water Distribution

The equitable distribution of water is a fundamental aspect of effective irrigation management. Importantly, water is allocated fairly to all parts of an irrigation project, ensuring that every area receives the necessary amount of water for optimal crop growth. This principle is vital for maintaining the sustainability and efficiency of the irrigation system as a whole. The criteria set for water distribution should be consistently applied across the entire irrigation project. This ensures fairness and transparency in how water is allocated.

Given the size and complexity of large irrigation projects, different areas might have varying water needs. The criteria for water distribution must be allowed to change with changing needs due to changing cropping patterns, climatic changes, changing land use, etc.

16.3.3 Sediment Control

Sediment control refers to total sediment management. This can be efficiently and effectively dealt with only if sediment management is undertaken in the entire watershed, and this is quite complicated. The major source of sediment is the watershed, and the source of sediment can be eliminated or diminished by proper soil conservation measures. In irrigation projects, the emphasis is on the management of sediment loads in canal and channel systems. This process entails the removal of sediment from these conveyances. Details of the methods for sediment removal from canals and channels can be found in Chapter 13.

16.3.4 Groundwater Table Control

Irrigation practices can significantly affect groundwater levels, often leading to a rise in the water table. This rise typically results from inefficient water management and can lead to various problems, including drainage issues and negative impacts on crop fields. The best way to control the rise of the water table is to employ improved water management practices. If the water table has already risen, then some kind of drainage must be used to remove the excess water. This, depending on the situation at hand, can be accomplished through pumping from vertically installed wells or subsurface pipe drainage. However, drainage is very expensive and must be employed when absolutely essential, and there is no other economical method. Details of the methods used for groundwater table control can be found in Chapter 9.

16.3.5 Salinity Management

Soil salinity is a significant challenge in irrigated agriculture, with profound impacts on soil health and agricultural productivity. Countries, such as India, have witnessed fertile lands becoming barren due to salinization, a process exacerbated by waterlogging and the accumulation of salts in the root zone. The best way to reduce soil salinization is by employing a combination of drainage and improved water management practices throughout the irrigation system. In some cases, managing salinity requires a broader approach that encompasses the entire river basin. This comprehensive strategy involves coordinating water use and drainage across the basin to address the causes of salinization effectively. Details of the methods for salinity management can be found in Chapter 14.

16.3.6 Reduced Water Diversions

In most developing countries, irrigated agriculture is the largest consumer of water. The demand for scarce water resources is increasing each year. This is due to increasing population, urbanization, rising standards of living, industrialization, etc. Some of the water will have to be diverted for domestic water supplies and for groundwater recharge. This means that irrigated agriculture runs the risk of receiving reduced water and will have to be practiced within this limitation without affecting yield. Thus, water management has become even more important now and will be so in the future.

Enhanced water use management requires (1) implementing advanced irrigation methods, such as drip and sprinkler systems, that minimize water losses; (2) optimizing irrigation scheduling to match crop water requirements; (3) promoting water-saving agronomic practices; (4) selecting crops and designing rotations on the basis of water availability and climate suitability; (5) encouraging the reuse and recycling of irrigation return flows; (6) incorporating innovative technologies such as soil moisture sensors, remote sensing tools, and smart irrigation systems to guide decision-making; (7) investing in farmer education and capacity building to support efficient water use; and (8) establishing supportive policy frameworks and regulatory mechanisms. These strategies collectively help ensure that agriculture can continue to thrive even under reduced water allocations.

16.3.7 Farmers' Participation

The active involvement of farmers in irrigation management is a key factor in achieving efficient and sustainable system operation. When farmers are involved in the planning, operation, and maintenance of irrigation systems, performance tends to improve, and resource use becomes more efficient. Increasingly, there is recognition that involving users, particularly through participatory management or partial privatization, can enhance accountability, responsiveness, and long-term viability. Privatization, in this context, refers to shifting operational responsibilities from centralized government agencies to user associations or other decentralized, market-oriented entities.

Despite its potential benefits, privatization often effectively faces resistance, especially in regions with large, state-managed irrigation systems. For example, in India, each state operates its own irrigation bureaucracy, which is often burdened by inefficiencies and resistance to reform. Political challenges and institutional inertia make full-scale privatization difficult to implement. However, alternative pathways exist for enhancing farmer participation without dismantling existing structures entirely.

Effective strategies for overcoming resistance to privatization may include (1) building consensus through stakeholder engagement, emphasizing the long-term benefits of efficient irrigation management; (2) implementing pilot projects to demonstrate the effectiveness of privatization; and (3) encouraging public–private partnerships as a transitional approach toward full privatization. Empowering farmers also involves investing in education and training in water-saving techniques, supporting their role in decision-making on water distribution and system maintenance, and encouraging the formation of farmer cooperatives or water user associations. These approaches help build local capacity, foster a sense of ownership, and support more resilient and adaptive irrigation systems.

16.3.8 Multipurpose Use

Traditionally, irrigation systems in developing countries, such as India, have been designed and operated exclusively for agricultural water delivery. However, growing interest in resource optimization and rural economic development has prompted a shift toward treating these systems as multipurpose infrastructures. Dams, reservoirs, and canal systems often occupy scenic or strategically located areas, making them well suited for recreational uses such as boating, fishing, swimming, picnicking, and ecotourism. By diversifying the functions of irrigation systems, additional revenue streams can be generated to support system maintenance, improve services, and stimulate local economies.

Transforming irrigation infrastructure into multi-resource systems provides significant opportunities, but it also introduces new challenges. Planners must ensure that the primary objective, i.e., reliable agricultural water supply, is not compromised. Environmental impacts, social equity, and community needs must be carefully considered. For example, increased recreational use may raise concerns about water quality, access rights, or ecological disturbance.

Effective integration of multipurpose functions requires thoughtful policies, stakeholder consultation, and coordinated management strategies. Examples of successful multipurpose water projects, as highlighted in Figure 1.12, demonstrate how irrigation infrastructure can support agriculture, recreation, tourism, and even hydropower simultaneously, while contributing to broader goals, such as climate resilience, food security, and rural development.

16.4 IRRIGATION SYSTEM EVALUATION

Evaluating an irrigation system is a critical step in identifying opportunities for improvement and determining how to maximize performance with available resources. The evaluation process involves a thorough assessment of the system's physical components, operational practices, and overall efficiency. The objective is to identify areas where strategic investments can deliver the greatest benefits, economically, environmentally, and operationally.

The areas for potential improvements have already been discussed. The solutions or ways by which improvements or upgrades to the physical system may be accomplished, while considering the minimum investment needed to produce the greatest impact. Several investments can be made, including (1) canal linings to reduce seepage losses, (2) the installation of new control structures and measuring devices to improve flow regulation and monitoring, and (3) upgrades or substitutions of outdated farm irrigation systems with more efficient technologies. These investments should be prioritized on the basis of their potential to increase water conservation, improve agricultural productivity, enhance management efficiency, and support environmental sustainability.

System evaluation should consider improvements at all levels: (1) farm subsystems, which directly affect crop yield and resource use; (2) tertiary subsystems, which handle water delivery at the field or block level; and (3) main subsystems, which include major conveyance and control infrastructure. Each subsystem presents unique challenges and opportunities, and a short discussion of these categories follows to guide-targeted assessment and intervention planning.

16.4.1 Farm Subsystems

Farm subsystems represent the final stage of the irrigation delivery network, where water is applied directly to crops. Improvements at this level have a direct effect on crop yield, water productivity, and farmer profitability. Therefore, enhancing farm-level irrigation practices is often the most effective way to improve overall system performance. However, achieving meaningful improvements requires a holistic approach that considers agronomic practices, soil characteristics, crop selection, and the design and management of the irrigation system itself.

16.4.1.1 Water Use Efficiency

Data collection and evaluation are essential for identifying the best water management strategies at the farm level. One of the most commonly used performance indicators is irrigation efficiency, which is defined as the ratio of water beneficially used to the total water delivered to the field. Beneficial uses include crop evapotranspiration (ET), salt leaching, and water used for seed germination or nutrient delivery. Nonbeneficial uses include deep percolation beyond the root zone, tailwater runoff, and evaporation losses. Efficient on-farm irrigation depends on water use, energy use, labor and capital investment, and how these factors are related to production and profitability. There are several definitions for irrigation efficiency, and there is not a single definition that covers all aspects of irrigation efficiency. The traditional definition of irrigation efficiency (IE) (from ASCE, 1978) is as follows:

$$\text{IE} = \frac{\text{Volume of water beneficially used}}{\text{Volume of water delivered to the field}} \qquad (16.1)$$

In traditional irrigation efficiency calculations (Equation 16.1), all applied water that is not immediately used by the crop is considered a loss. However, this may not reflect actual system performance in cases where part of the applied water is stored in the soil profile and remains available for future crop use. To address this limitation, Burt et al. (1997) introduced a modified definition of irrigation efficiency that accounts for the change in soil water storage during an irrigation event. This approach is especially useful when evaluating a single irrigation or when comparing systems with different soil–water dynamics. The modified irrigation efficiency (MIE) is calculated as follows:

$$\text{MIE} = \frac{\text{Volume of water beneficially used}}{\text{Volume of irrigation water applied} - \Delta S} \qquad (16.2)$$

where ΔS represents the change in soil water storage in the crop root zone during the irrigation event.

$$\Delta S = (\theta_{\text{after}} - \theta_{\text{before}}) \times R_d \times A_L \tag{16.3}$$

where θ represents the soil moisture (cm^3/cm^3), R_d represents the root zone depth (m), and A_L is the irrigated area (m^2).

Example 16.1

A farmer applies water to a 1-ha (10,000 m^2) wheat field using a surface irrigation method. To assess irrigation efficiency, the following field data were collected:

Volume of irrigation water applied: 1,200 m^3;
Volume of water beneficially used: 800 m^3 (used for ET and salt leaching);
Soil moisture before irrigation: 0.20 cm^3/cm^3;
Soil moisture after irrigation: 0.25 cm^3/cm^3;
Effective root zone depth: 0.3 m.

Compare the irrigation efficiency (Equation 16.1) and modified irrigation efficiency (Equation 16.2).

Solution:

Irrigation efficiency using Equation (16.1):

$$\text{IE} = \frac{\text{Volume of water beneficially used}}{\text{Volume of water delivered to the field}} = \frac{800}{1,200} \approx 0.667 = 66.7\%$$

Based on Equation (16.1), 66.7% of the water was effectively used, and 33.3% was considered to be lost.

Since $R_d = 0.3$ m and $A_L = 10,000$ m^2, the change in soil water storage is as follows:

$$\Delta S = (\theta_{\text{after}} - \theta_{\text{before}}) \times R_d \times A_L = (0.25 - 0.20) \times 0.3 \times 10,000 = 150 \text{ m}^3$$

Therefore, 150 m^3 of water was stored in the soil profile, which is available for future crop use.

Modified irrigation efficiency using Equation (16.2):

$$\text{MIE} = \frac{\text{Volume of water beneficially used}}{\text{Volume of irrigation water applied} - \Delta S} = \frac{800}{1,200 - 150} \approx 0.762 = 76.2\%$$

The traditional IE is 66.7%, which does not account for the water stored in the soil profile. The MIE increases to 76.2%, acknowledging that 150 m^3 of applied water is stored in the root zone and will be available for future crop use. The modified IE reflects a more realistic measure of system performance in this scenario.

Another broadly used term is overall project efficiency (OPE), which assesses the effectiveness of an entire irrigation system—from water diversion at the source to final use by crops. An analysis of the efficiency of each element, including storage in the supply system, conveyance, distribution, and application processes, is beneficial. This approach enables the identification of components that function optimally and those that require improvement. Consequently, the overall performance of the irrigation system can be understood as the cumulative result of the efficiencies of these individual components, which is calculated as the ratio of water beneficially used by crops to the total water diverted into the system:

$$\text{OPE} = \frac{\text{Crop water used}}{\text{Total inflow into supply system}} \tag{16.4}$$

In practice, OPE can also be calculated as the product of efficiencies across each subsystem, including storage efficiency, conservation efficiency, and application efficiency. This breakdown helps determine where water losses occur, such as seepage in unlined canals, overirrigation in fields, or malfunctioning distribution structures. By identifying underperforming components, system managers can prioritize interventions to improve water use efficiency and overall project performance.

Another widely used metric in agricultural water management is water use efficiency (WUE), which relates the economic yield of a crop to the amount of water used during the growing season. WUE is particularly useful for comparing the effectiveness of different irrigation methods, cropping systems, or water management strategies. According to Howell (2001), the WUE is defined as follows:

$$\text{WUE}\left(\text{kg/m}^3\right) = \frac{\text{Economic yield}\left(\text{kg/ha}\right)}{\text{Crop water use}\left(\text{m}^3\text{/ha}\right)} \tag{16.5}$$

Here, crop water use is the total amount of water used to produce crops. To simulate calculations in practical settings, losses, such as deep percolation, groundwater uptake, and surface runoff, are often neglected. Therefore, crop water use can typically be estimated as the sum of effective rainfall, irrigation water applied, and net soil water depletion from the root zone during the season. Experienced practitioners can use the WUE to compare different irrigation methods, crop selections, and scheduling strategies within a region.

To isolate the yield benefit gained specifically from irrigation, irrigation water use efficiency (IWUE), proposed by Bos (1979), compares the additional yield achieved under irrigation relative to rainfed conditions. It also considers the additional water use resulting from irrigation. IWUE is calculated as

$$\text{IWUE}\left(\text{kg/m}^3\right) = \frac{Y_i\left(\dfrac{\text{kg}}{\text{ha}}\right) - Y_d\left(\dfrac{\text{kg}}{\text{ha}}\right)}{\text{ET}_i\ (\text{mm}) - \text{ET}_d\ (\text{mm})} \tag{16.6}$$

where IWUE is the irrigation water use efficiency (kg/m^3), Y_i is the economic yield (kg/ha) for irrigation level i, Y_d is the dryland yield or crop yield without irrigation (kg/ha), ET_i is the evapotranspiration (mm) for irrigation level i, and ET_d is the evapotranspiration of the dryland crops (or of the ET without irrigation). Since 1 mm of water over 1 hectare equals 10 m^3, evapotranspiration values in mm can be easily converted to m^3/ha.

Equation (16.6) can be simplified as

$$\text{IWUE}\left(\text{kg/m}^3\right) = \frac{Y_i\left(\frac{\text{kg}}{\text{ha}}\right) - Y_d\left(\frac{\text{kg}}{\text{ha}}\right)}{I_i\,(\text{mm})} \tag{16.7}$$

where I_i is the total amount of irrigation applied for irrigation level i (mm).

IWUE is particularly useful for evaluating the marginal benefit of irrigation in terms of yield gain per unit of irrigation water consumed. It plays a key role in water-scarce environments and when assessing the return on investment in irrigation infrastructure.

Example 16.2

A field trial was conducted to evaluate the performance of wheat grown under irrigated and rainfed (dryland) conditions (Table 16.1). The irrigated wheat yielded 7,000 kg/ha, whereas the rainfed wheat yielded 3,000 kg/ha. Both treatments received an effective rainfall of 140 mm during the growing season. The irrigated field received an additional 300 mm of irrigation water. The soil moisture content was measured at planting and harvest. At planting, the volumetric soil moisture was 0.25 m^3/m^3 in both treatments. At harvest, it was 0.20 m^3/m^3 in the irrigated plot and 0.11 m^3/m^3 in the rainfed plot. The effective root zone depth was 0.5 m. Use the data to compute the water use efficiency (WUE) for both treatments and the irrigation water use efficiency using Equations (16.6) and (16.7).

Solution:

Step 1: Calculate soil water depletion

Irrigated:

$$\text{Depletion} = \left(\theta_{\text{planting}} - \theta_{\text{harvest}}\right) \times R_d = \left(0.25\text{ m}^3/\text{m}^3 - 0.20\text{ m}^3/\text{m}^3\right) \times 0.5\text{ m}$$

$$= 0.025\text{ m} = 25\text{ mm}$$

Rainfed:

$$\text{Depletion} = \left(\theta_{\text{planting}} - \theta_{\text{harvest}}\right) \times R_d = \left(0.25\text{ m}^3/\text{m}^3 - 0.11\text{ m}^3/\text{m}^3\right) \times 0.5\text{ m} = 0.07\text{ m} = 70\text{ mm}$$

Step 2: Calculate crop water use based on soil water balance:

TABLE 16.1
Field Data Summary for Example 16.2

Parameter	Irrigated	Rainfed
Yield (kg/ha)	7,000	3,000
Effective rainfall (mm)	140	140
Irrigation applied (mm)	300	0
Soil moisture at planting (m^3/m^3)	0.25	0.25
Soil moisture at harvest (m^3/m^3)	0.20	0.11
Effective root zone depth (m)	0.5	0.5

$$\text{Crop Water Use (mm)} = \text{Effective Rainfall} + \text{Irrigation Applied} + \text{Soil Depletion}$$

$$\text{Irrigated: Crop Water Use (mm)} = 140 + 300 + 25 = 465 \text{ mm} = 4{,}650 \text{ m}^3/\text{ha}$$

$$\text{Rainfed: Crop Water Use (mm)} = 140 + 0 + 70 = 210 \text{ mm} = 2{,}100 \text{ m}^3/\text{ha}$$

Step 3: Calculate the water use efficiency (WUE)
WUE for irrigated wheat:

$$\text{WUE}\left(\frac{\text{kg}}{\text{m}^3}\right) = \frac{\text{Economic yield}}{\text{Crop water use}} = \frac{7{,}000}{4{,}650} \approx 1.51 \text{ kg/m}^3$$

WUE for rainfed wheat:

$$\text{WUE}\left(\frac{\text{kg}}{\text{m}^3}\right) = \frac{\text{Economic yield}}{\text{Crop water use}} = \frac{3{,}000}{2{,}100} \approx 1.43 \text{ kg/m}^3$$

Step 4: Irrigation water use efficiency (IWUE)
Method 1: Using Equation (16.6):

$$\text{IWUE}\left(\frac{\text{kg}}{\text{m}^3}\right) = \frac{Y_i\left(\frac{\text{kg}}{\text{ha}}\right) - Y_d\left(\frac{\text{kg}}{\text{ha}}\right)}{\text{ET}_i\,(\text{mm}) - \text{ET}_d\,(\text{mm})} = \frac{7{,}000 - 3{,}000}{4{,}650 - 2{,}100} \approx 1.57 \text{ kg/m}^3$$

Method 2: Equation (16.7):

$$\text{IWUE}\left(\frac{\text{kg}}{\text{m}^3}\right) = \frac{Y_i\left(\frac{\text{kg}}{\text{ha}}\right) - Y_d\left(\frac{\text{kg}}{\text{ha}}\right)}{I_i\,(\text{mm})} = \frac{7{,}000 - 3{,}000}{3{,}000} \approx 1.33 \text{ kg/m}^3$$

Equation (16.7) is simpler and more practical for farm-level decision-making, especially when detailed data on soil moisture or crop ET are unavailable. It relies only on the volume of irrigation water applied and the corresponding yield increase. In contrast, Equation (16.6) incorporates changes in soil water storage and provides a more accurate and realistic estimate of crop water use by irrigation and rainfed lands.

16.4.1.2 Application Efficiency (%)

When the efficiency of irrigation is the focus of a single irrigation event or a single part of the irrigation system, the following term can be used:

$$\text{AE}\,(\%) = \frac{\text{Water applied that is stored in crop root zone}}{\text{Total amount of water delivered to the farm}} \tag{16.8}$$

The term involves the water percolated below the root system, the tail water and evaporation. A comparison of the different application efficiencies for different irrigation methods is presented in Table 16.2. The application efficiency can be affected

TABLE 16.2
Example of Attainable Field Application Efficiencies

Irrigation Method	Attainable Application Efficiency (%)	Range (%)	Average (%)
Surface			
Graded furrow	75	50–80	65
Graded furrow with tail water reuse	85	60–90	75
Level furrow	85	65–95	80
Graded border	80	50–80	65
Level basin	90	80–95	85
Sprinkler			
Periodic move	80	60–85	75
Side roll	80	60–85	75
Moving big gun	75	55–75	65
Center pivot			
Impact head with end gun	85	75–90	80
Impact head without end gun	95	75–95	90
LEPA without end gun	98	80–98	95
Lateral move			
Spray heads with hose feed	95	75–95	90
Spray heads with canal feed	90	70–95	85
Microirrigation			
Trickle	95	70–95	85
Subsurface drip	95	75–95	90
Microspray	95	70–95	85

Source: Adapted from Howell (2003).
Note: LEPA refers to low-energy precision application.

by many factors, such as winds for sprinkler systems and the texture or topography of surface systems.

Example 16.3

A 1 ha cotton farm (10,000 m^2) located in a semi-arid region was used to evaluate the performance of three different irrigation systems: a graded furrow, sprinkler, and center pivot (LEPA). The following data were recorded for a single irrigation event (assuming that no rainfall occurred during the event):

Irrigation Method	Water Delivered to Farm	Water Stored in Root Zone
Grade furrow	40	24.0
Sprinkler	30	21.5
Center pivot (LEPA)	25	22.5

The seasonal crop water requirement (net irrigation depth) is 300 mm. The farmer wants to calculate the application efficiency for each method and determine the total irrigation water (gross depth) required throughout the season.

Solution:

The application efficiency is calculated using Equation (16.8):

$$\text{Graded furrow:}\quad \text{AE}(\%) = \frac{\text{Water applied that is stored in crop root zone}}{\text{Total amount of water delivered to the farm}}$$

$$= \frac{24.0}{40} = 0.600 = 60.0\%$$

$$\text{Sprinkler:}\quad \text{AE}(\%) = \frac{21.5}{30} \approx 0.717 = 71.7\%$$

$$\text{Center pivot (LEPA):}\quad \text{AE}(\%) = \frac{22.5}{25} = 0.900 = 90.0\%$$

To meet a seasonal net irrigation depth of 300 mm, the gross irrigation depth is calculated using

$$\text{Graded furrow:}\quad \text{Gross depth (mm)} = \frac{300}{\text{AE}(\%)} = \frac{300}{60.0\%} = 500\text{ mm}$$

$$\text{Sprinkler:}\quad \text{Gross depth (mm)} = \frac{300}{\text{AE}(\%)} = \frac{300}{71.7\%} \approx 418.4\text{ mm}$$

$$\text{Center pivot (LEPA):}\quad \text{Gross depth (mm)} = \frac{300}{\text{AE}(\%)} = \frac{300}{90\%} \approx 333.3\text{ mm}$$

Since 1 mm over 1 ha = 10 m^3, the total amount of water applied per season is

$$\text{Water volume}\left(\text{m}^3\right) = \text{Gross depth}(\text{mm}) \times 10\ (\text{ha})$$

The gradient furrow uses the most water (5,000 m^3/ha) because of significant losses from runoff and percolation. The sprinkler is moderately efficient (4,184 m^3/ha). The LEPA center pivot system is the most efficient system (3,333 m^3/ha), requiring 33% less water than furrow irrigation to meet the same crop demand.

16.4.1.3 Distribution Uniformity

An important aspect not addressed in the previously mentioned definitions of irrigation efficiency is distribution uniformity, which refers to the evenness of water application in a field. While the total volume of water delivered to a field may meet crop water requirements on average, uneven distributions can result in localized areas receiving too much or too little water. This situation is detrimental to plant health and system performance. Overirrigated areas may lead to surface runoff, nutrient leaching, or waterlogging, whereas underirrigated zones may experience crop stress and yield loss.

Uneven distribution often results from inadequate field levelling, inconsistent pressure or flow within the irrigation system, or variable infiltration rates due to soil heterogeneity. These discrepancies can lead to waterlogging in some areas, causing oxygen deprivation and root stress, whereas other areas may remain too dry, limiting plant water uptake. Additionally, excessive deep percolation in overirrigated zones can contribute to rising water tables or the formation of perched water tables, depending on the soil profile conditions. However, it is important to note that some level of percolation is necessary for leaching salts from the root zone—particularly in arid regions.

Various methods have been developed to quantify distribution uniformity. For surface irrigation systems, uniformity is assessed by measuring the advance time and infiltration depth at multiple points along furrows or borders. Differences in water contact time reveal how evenly water is distributed along the length of the field. For pressurized systems such as sprinklers and pivots, distribution is assessed via catch cans or collectors that measure the depth of applied water across the field.

Among the common metrics, the distribution uniformity of the low quarter (DU) is widely used across all irrigation systems. Although it is best suited for surface irrigation systems, such as furrows and borders, DUs can also be applied to sprinkler and drip irrigation systems. However, it can also be used to evaluate all other irrigation systems. It evaluates how well the driest portion of the field (typically the lowest quarter) was irrigated relative to the average, using the following equation:

$$\text{DU} = \frac{\text{Average lowest quartile depth of water applied to the crop}}{\text{Average depth of water applied to the crop}} \tag{16.9}$$

A DU value approaching 1.0 or 100% indicates a highly uniform water distribution. The DU method is simple, robust, and widely used for both field evaluations and system comparisons. This is especially practical where infiltration patterns vary along the field due to slope or advance time. An example of the DU method is shown in Figure 16.2 in which 20 observations are ranked from the highest depth of information to the lowest and the lowest quarter depth of irrigation.

For sprinkler and drip irrigation systems, Christiansen's uniformity coefficient (CU) is the preferred metric (Christiansen, 1942). It measures the consistency of applied water volumes using a grid of catch cans, which is calculated as

$$\text{CU} = 100 \times \left[1 - \frac{\sum_{i=1}^{n} \left| (x_i - x_m) \right|}{n x_m} \right] \tag{16.10}$$

where CU is the Christiansen uniformity coefficient (%), x_i is the depth of water in equally spaced catch cans at the ith grid, x_m is the mean depth of water of the catch in all cans, and n is the number of catch cans. CU values above 85% indicate acceptable performance, whereas values above 90% are considered excellent. The CU is ideal for periodic-move and solid-set sprinkler systems and is sensitive to operational factors, such as wind or pressure variations. It is particularly useful in systems

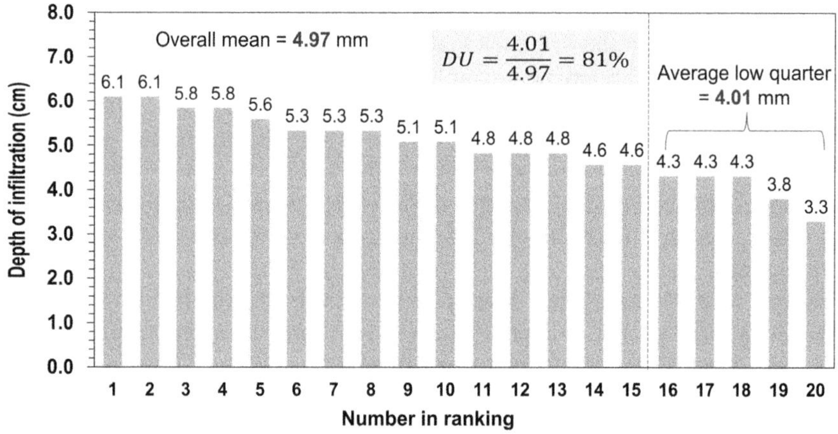

FIGURE 16.2 An illustration of the lowest quarter depth of irrigation and the average depth applied to the entire field.

Note: Values are scaled to 0.1 cm for visualization; the actual irrigation depth was rounded to 0.01 cm

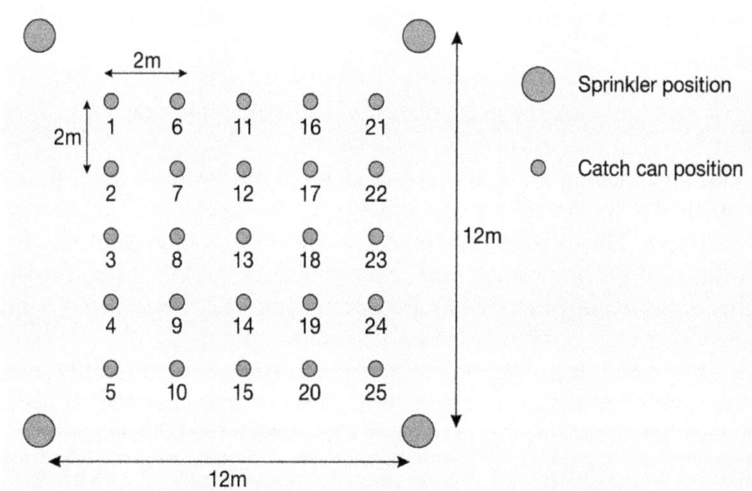

FIGURE 16.3 Catch can setup under a solid-set sprinkler system.

where uniform emitter performance and minimal wind drift are desired. A typical catch-can setup used to evaluate water application uniformity in a solid-set sprinkler system is shown in Figure 16.3.

For center pivot systems that apply water in circular patterns, the Heermann and Hein coefficient of uniformity (CU_H) (Heermann and Hein, 1968) is more appropriate. This method accounts for the spatial variability in catch can location and area represented by each can. It is calculated as

$$\mathrm{CU}_H = 100 \times \left[1 - \frac{\sum_{i=1}^{n} \left| \left(x_i - x_m^* \right) \right| d_i}{\sum_{i=1}^{n} x_i d_i} \right] \tag{16.11}$$

$$x_m^* = \frac{\sum_{i=1}^{n} d_i x_i}{\sum_{i=1}^{n} d_i} \tag{16.12}$$

where d_i is the distance from the pivot point to each catch can, x_i is the *ith* measured depth of water in each can, and x_m^* is the weighted mean depth of water in all catch cans. This method ensures that uniformity is evaluated relative to the actual irrigated area, which is critical for center pivots where the outer edge of the field receives proportionally more water.

Example 16.4

Evaluate the uniformity of a solid-set sprinkler system using the distribution uniformity (DU) and Christiansen uniformity coefficient (CU). The measured depths of water collected from 25 evenly spaced cans are given in Table 16.3.

Solutions:

Arrange the depth values in descending order and calculate the mean depth of the lowest quartile depth (i.e., the lowest six values):

$$\frac{4.55 + 4.56 + 4.56 + 4.93 + 4.94 + 5.01}{6} \approx 4.76 \text{ cm}$$

The mean depth of all 25 observations is 5.39 cm.

The distribution uniformity (DU):

$$\mathrm{DU} = \frac{\text{Average lowest quartile depth of water applied to the crop}}{\text{Average depth of water applied to the crop}} = \frac{4.76}{5.39} \approx 0.883 = 88.3\%$$

TABLE 16.3
Measured Depth of Water in 25 Evenly Spaced Cans (cm)

Can #	Depth (cm)	Can #	Depth (cm)	Can #	Depth (cm)	Can #	Depth (cm)	Can #	Depth (cm)
1	5.36	6	5.09	11	5.88	16	5.69	21	5.99
2	4.93	7	5.85	12	5.01	17	5.77	22	4.56
3	5.97	8	5.35	13	5.30	18	4.94	23	5.96
4	4.55	9	5.88	14	5.44	19	5.65	24	5.70
5	5.71	10	5.17	15	5.01	20	4.56	25	5.43

The Christiansen Uniformity Coefficient (CU):

$$\text{CU} = 100 \times \left[1 - \frac{\sum_{i=1}^{n} |(x_i - x_m)|}{n x_m}\right] = 100 \times \left[1 - \frac{9.68}{25 \times (5.39)}\right] \approx 92.8\%$$

Example 16.5

A center pivot system has a total pivot length of 1,200 ft and is composed of eight spans, each 150 ft in length. To assess uniformity, 50 catch cans were placed at 20-ft intervals, starting at 150 ft from the pivot point and extending to 1,130 ft. In total, the measured depths of the water collected in each can are recorded in Table 16.4. Calculate its distribution uniformity.

TABLE 16.4
Measured Depths of Water in Each Catch Can (cm) along a Pivot

Can #	Distance from the Pivot Point (ft)	Depth (cm)	Can #	Distance from the Pivot Point (ft)	Depth (cm)
1	150	1.21	26	650	1.23
2	170	1.23	27	670	1.05
3	190	1.28	28	690	1.29
4	210	1.07	29	710	1.29
5	230	1.12	30	730	0.88
6	250	1.26	31	750	0.96
7	270	1.18	32	770	0.92
8	290	1.15	33	790	0.89
9	310	1.20	34	810	0.82
10	330	1.02	35	830	0.91
11	350	1.09	36	850	0.96
12	370	1.16	37	870	0.85
13	390	1.26	38	890	0.82
14	410	1.12	39	910	0.82
15	430	1.16	40	930	1.11
16	450	1.20	41	950	1.09
17	470	1.00	42	970	1.12
18	490	1.01	43	990	1.02
19	510	1.14	44	1,010	1.26
20	530	1.25	45	1,030	1.17
21	550	1.27	46	1,050	1.23
22	570	1.09	47	1,070	1.25
23	590	1.04	48	1,090	1.06
24	610	1.17	49	1,110	1.19
25	630	1.19	50	1,130	1.07

Solutions:

To evaluate distribution uniformity, we apply the Heermann and Hein coefficient of uniformity (CU_H), which is suitable for center pivot systems, with the weighted mean depth of water in all catch cans:

$$x_m^* = \frac{\sum_{i=1}^{n} d_i x_i}{\sum_{i=1}^{n} d_i} \approx 1.08$$

The detailed calculations are shown in Table 16.5. The Heermann and Hein coefficient of uniformity (CU_H) is as follows:

$$\mathrm{CU}_H = 100 \times \left[1 - \frac{\sum_{i=1}^{n} \left| \left(x_i - x_m^* \right) \right| d_i}{\sum_{i=1}^{n} x_i d_i} \right] \approx 88.9\%$$

This result indicates relatively good uniformity for a center pivot system, as values above 85% are typically considered acceptable.

TABLE 16.5
Calculation Results of Example 16.5

No. of Cans	x_i (cm)	$d_i^* x_i$	$\lvert x_i - x_m^* \rvert$	$\lvert x_i - x_m^* \rvert d_i$
1	1.21	181.5	0.13	18.90
2	1.23	209.1	0.15	24.82
3	1.28	243.2	0.20	37.24
4	1.07	224.7	0.01	2.94
5	1.12	257.6	0.04	8.28
6	1.26	315.0	0.18	44.00
7	1.18	318.6	0.10	25.92
8	1.15	333.5	0.07	19.14
9	1.20	372.0	0.12	35.96
10	1.02	336.6	0.06	21.11
11	1.09	381.5	0.01	2.11
12	1.16	429.2	0.08	28.13
13	1.26	491.4	0.18	68.65
14	1.12	459.2	0.04	14.77
15	1.16	498.8	0.08	32.69
16	1.20	540.0	0.12	52.21
17	1.00	470.0	0.08	39.47
18	1.01	494.9	0.07	36.25
19	1.14	581.4	0.06	28.57
20	1.25	662.5	0.17	87.99
21	1.27	698.5	0.19	102.31
22	1.09	621.3	0.01	3.43

(*Continued*)

TABLE 16.5 (*Continued*)
Calculation Results of Example 16.5

No. of Cans	x_i (cm)	$d_i^* x_i$	$\lvert x_i - x_m^* \rvert$	$\lvert x_i - x_m^* \rvert d_i$
23	1.04	613.6	0.04	25.95
24	1.17	713.7	0.09	52.47
25	1.19	749.7	0.11	66.79
26	1.23	799.5	0.15	94.91
27	1.05	703.5	0.03	22.77
28	1.29	890.1	0.21	142.15
29	1.29	915.9	0.21	146.27
30	0.88	642.4	0.20	148.91
31	0.96	720.0	0.12	92.99
32	0.92	708.4	0.16	126.27
33	0.89	703.1	0.19	153.25
34	0.82	664.2	0.26	213.83
35	0.91	755.3	0.17	144.41
36	0.96	816.0	0.12	105.39
37	0.85	739.5	0.23	203.57
38	0.82	729.8	0.26	234.95
39	0.82	746.2	0.26	240.23
40	1.11	1,032.3	0.03	24.19
41	1.09	1,035.5	0.01	5.71
42	1.12	1,086.4	0.04	34.94
43	1.02	1,009.8	0.06	63.34
44	1.26	1,272.6	0.18	177.78
45	1.17	1,205.1	0.09	88.60
46	1.23	1,291.5	0.15	153.32
47	1.25	1,337.5	0.17	177.64
48	1.06	1,155.4	0.02	26.14
49	1.19	1,320.9	0.11	117.68
50	1.07	1,209.1	0.01	15.80

16.4.2 Delivery System and Water Conveyance Efficiency

To develop a range of improvements, field studies and consequent data collection will have to be undertaken, e.g., 2–5 years. The data will show whether the solutions will be acceptable to the farmers as well as to the project personnel. It takes a great deal of time and effort to implement any solution throughout the tertiary systems. Thus, it is very important to thoroughly understand the solution and the ensuing consequences. Any improvement to the main subsystem is normally expensive and must therefore be justified. This should be done on the basis of field data, which should be collected on a regular basis.

The best performance indicator for screening the problem and proposing solutions is the water convenience efficiency (E_c). The water conveyance efficiency is defined as the ratio between the water that reaches a farm or field and that diverted from the irrigation water source. It is defined as

$$E_c(\%) = \frac{\text{Volume of water that reaches the farm or field}}{\text{Volume of water diverted from the source}} \times 100 \qquad (16.13)$$

Conveyance losses include seepage, evaporation losses, and canal spills (operational or accidental) that might result from management. Conveyance losses are much lower for closed conduits or pipelines than for unlined or lined canals. Canal systems with low efficiencies may require the maintenance of major rehabilitation work.

Example 16.6

An agricultural field consists of 100 border strips. The irrigation channel supplies water at a rate of 20 L/min to 40 of the strips and at 25 L/min to the remaining 60 strips. The channel itself receives 3,000 L/min from the source. Calculate the conveyance efficiency of the system.

Solution:

Water delivered to the field:

$$20\text{ L/min} \times 40 + 25\text{ L/min} \times 60 = 800 + 1{,}500 = 2{,}300\text{ L/min}$$

Conveyance efficiency:

$$E_c(\%) = \frac{\text{Volume of water that reaches the farm or field}}{\text{Volume of water diverted from the source}} \times 100 = \frac{2{,}300}{3{,}000} \times 100 \approx 76.7\%$$

16.4.3 Seepage Measurement

The ponding method is a widely used technique for measuring seepage in canal systems. This approach involves isolating a segment of the canal by constructing sealed earthen dams at both ends and then filling the segment with water. Prior to the test, detailed surveys are conducted to determine the dimensions of the canal. During the testing phase, the water level within the canal segment is monitored, either manually or automatically, to track the rate at which water infiltrates or seeps away. By observing the daily water loss per meter of canal length, it becomes possible to evaluate the extent of seepage. Measurements are useful for prioritizing which canals need to be lined. Details of measuring seepage losses from canals via the ponding method can be found in Leigh and Fipps (2019). New methods for seepage detection have been introduced in Chapter 13. In addition to traditional methods, modern seepage detection techniques, such as geophysical sensing and drone-based thermal imaging, have been introduced in Chapter 13, offering enhanced accuracy and efficiency in seepage diagnostics.

16.5 POTENTIAL SOLUTIONS FOR IMPROVEMENT

To achieve the objectives established for improvement, there can be more than one way; indeed, there can be a whole range of solutions. Keeping in mind the site-specific conditions, several alternative solutions can be selected. Each proposed solution should be carefully designed and subjected to rigorous field evaluation over a period of time. This evaluation process helps assess the technical feasibility, cost-effectiveness, farmer acceptance, and long-term sustainability of the intervention. Site-specific variables, such as soil type, water availability, topography, cropping patterns, and existing infrastructure, must be factored into both the design and implementation stages.

Only those solutions that demonstrate clear, measurable benefits under real-world conditions should be adopted. A well-documented, evidence-based decision-making process increases the likelihood of long-term success and stakeholder support.

16.6 ECONOMIC ANALYSIS

Financial assistance for irrigation system improvements can often be sourced from both private and government entities. In the United States, the federal government provides loans under specific conditions for such projects. An alternative method is to implement a program gradually, funding it over time through increased operational and management charges. This latter approach is commonly adopted by many districts in the western United States.

16.6.1 Benefit-Cost Analysis

Any improvement has a cost associated with it and must be accompanied by a benefit. Unless the benefit exceeds the cost, it is difficult to justify the improvement. If data are collected appropriately, it may be possible to link improvements to irrigation practices and to crop yield.

16.6.2 Cost-Effective Analysis

There must be a measure to evaluate the efficacy of an improvement plan, in addition to determining the cost of the plan. For example, in a scheme to control the groundwater table, the depth to the water table measured from the ground surface can be used as a measure. Salinity management can be viewed in terms of the total dissolved solids (salts) in the root zone in metric tons per acre. For sediment control, a useful measure is the reduction in sediment yield in metric tons per year. These measures can be translated into costs.

16.6.3 Improvement Report

An irrigation system improvement plan must be prepared. Depending on the availability of resources and needs, improvements must be prioritized for investigation and eventual implementation. The plan must include a discussion of present, historical, and potential project performance; alternative irrigation improvements with their impact on project performance and their costs, such as capital costs and operation and maintenance costs; cost-effectiveness of alternative improvement plans; priority of alternative improvements; and recommendations for implementation of priority improvements.

16.6.4 Learning While Implementing

The implementation of improvement plans should be viewed as an exercise in learning. Fundamental to this learning exercise is data and visual experience through unrecorded observations. The data gathering should commence from the beginning of the system operation. Thus, operation and maintenance should yield valuable data that can serve as a basis for developing potential improvements and ways to implement them. When improvements are implemented, data collection should continue. In other words, a monitoring and evaluation program should be instituted to evaluate the actual effectiveness of improvements and their relationship with benefits and costs. In addition, a feedback and refinement program should be incorporated into the implementation of improvement learning, and the improvement plan should be further improved. This will have to continue in the future. With changing needs and changes in climate and cropping patterns and agricultural inputs, the types of improvement solutions will also change.

16.7 BETTERMENT INCLUDING MODERNIZATION

Many of the items mentioned under rehabilitation are for improving the irrigation system. In fact, no rehabilitation program should be undertaken without the idea of betterment being accomplished. There are numerous improvements that can be made to existing irrigation systems that may pay dividends in terms of lower costs of operation and management, reduced use of water, and better utilization of land. Road improvements, bridge and flume replacements using long-life materials or substitute structures, wastewater reservoirs, the use of pipes and linings in conveyance systems, and the installation of automation and remote controls on storage, diversion, and conveyance facilities are suggested. There are undoubtedly other cost-saving improvements that might be considered.

The improvements to reduce operation and management costs should include (1) automation and telemetering, (2) linings and pipe conveyances, (3) realignments of conveyances, (4) changes in points of delivery of water, (5) building desilting works, (6) changes in measuring devices to those requiring less maintenance, (7) changes for permitting increasing use of sprinkler irrigation, (8) changes necessary to accommodate changing land use, and (9) consolidation of works both within the project and with adjoining developments.

Even comparatively new irrigation systems can be improved to permit better operation. Certainly, in older irrigation systems, there are numerous instances where rehabilitation, modernization, betterment, and improvement can be practiced. Management must be forward-looking to keep the system in top condition to meet present and future needs. As competition for water increases, waste, pollution, and poor management practices cannot be tolerated. By keeping the system up to date and at top efficiency, these challenges can be met. Modernization, improvement, rehabilitation, and betterment cost money but are well worth the expense.

16.8 REPLACEMENT

The replacement of a system and its equipment is a matter of economics and judgment when the annual costs of maintenance approach the annual costs of replacement and when something must be done. Records must be kept so that management can make informed decisions. To that end, the following must be considered: (1) replace the structure or build a different structure, for example, replace a flume with an inverted siphon, replace a series of drops with a chute and stilling pool; (2) use the same materials or another material, e.g., replace aging wooden structures with precast or prestressed concrete for longer service life and reduced maintenance; (3) modify the system so that the structure or reach of canal or other may be abandoned, e.g., a reach of canal may be abandoned in favor of pressurized or piped systems; (4) financing for (a) programs for replacement to cover a period of years on a pay-as-you-go basis; (b) sinking funds; and (c) borrow funds to complete replacement in one phase, with repayment spread over time; and (5) legal and public relations problems. All changes should comply with applicable water rights, environmental regulations, and landowner agreements. Stakeholders should be encouraged early in the process of managing expectations, addressing concerns, and building support.

16.9 WATER CONSERVATION

Enhancing water conservation in irrigation systems encompasses a variety of strategies that address various aspects of water use, distribution, and management. These approaches aim to reduce water loss, increase delivery efficiency, and promote sustainable water use at the system and field levels.

16.9.1 Canal Linings

Canal lining is an important method for reducing seepage and improving the overall conveyance efficiency. Before canals are lined, losses in canals per unit length should be measured, and efficiencies should be estimated. The goal is to identify and implement solutions that minimize seepage, minimize environmental impacts, and align with the operational dynamics of the irrigation district.

Several options can be considered for rehabilitating old, deteriorating canals. Canals can be relined with concrete (Figure 16.4), replaced with pipelines or lined with other synthetic canal lining materials. The seepage rates for different soil types and lining materials are discussed in Chapter 13. Some irrigation districts can experiment with different materials to evaluate the ease of installation and maintenance of the materials, their relative performance, and their life durability. Askar Karimov et al. (2009) evaluated the durability of different synthetic materials under various stresses, including UV, animal traffic, intentional and unintentional vandalism, and normal irrigation district operational and maintenance activities, such as mowing

FIGURE 16.4 Patching an irrigation canal with concrete.

TABLE 16.6
Description of the Composition of Each Lining Material

Material	Description
Polyester with protective barrier	A geocomposite consisting of two layers (top and bottom) of 8 oz/yd^2 nonwoven polyester bonded to an olefinic copolymer geomembrane, 20 mil thick. The protective barrier consists of 2–3 inches of shotcrete
PVC with protective barrier	Non-reinforced Poly Vinyl Chloride (PVC). The protective barrier consists of a wire mesh with 2.5 inches of shotcrete
Polypropylene	A reinforced polyester scrim 16 oz/yd^2 between polypropylene layers, 24 mil thick
PVC Alloy	A polyvinylchloride blend, reinforced with a polyester scrim, 40 mil thick
EPDM Rubber	A non-reinforced EPDM (ethylene propylene diene monomer), 45 mil thick
Polyurethane	Two layers of 3 oz/yd^2, heat-bonded, non-woven geotextile saturated with liquid polyurethane, 40 mil thick

Note: A mil is a unit equivalent to 0.001 inches or 1,000th of an inch.

and cleaning of canals (Table 16.6). Their findings highlighted that the best lining system was a synthetic liner with a protective barrier of shotcrete (Figure 16.5). The synthetic liners (PVCE and polyester) significantly reduced seepage, with the shotcrete providing protection against physical damage and needing little to no maintenance. The study also emphasized that the performance of synthetic liners without a protective barrier varied dramatically according to the location of the project. In high-traffic areas with dense people and animals, liners are more prone to damage than they are in remote areas. The PVC alloy was the most durable material of the evaluated liners without a protective barrier, being more difficult to cut and less

FIGURE 16.5 (a) Laying a geosynthetic membrane covering an irrigation canal and (b) pouring shotcrete as a protective cover. Note that the concrete adheres to the surface and does not slip.

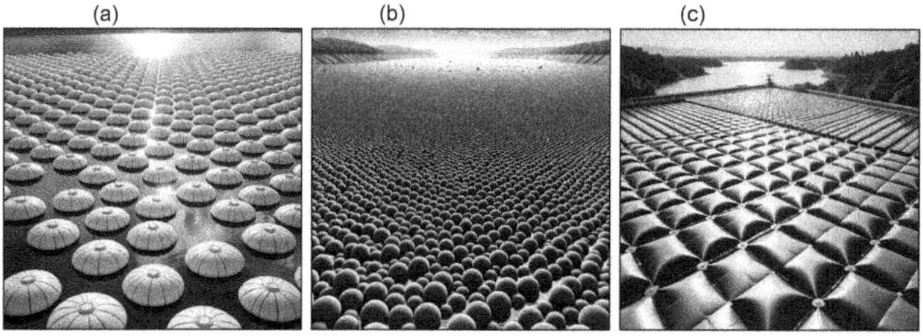

FIGURE 16.6 Evaporation reduction using (a) Aquacaps, (b) shade balls, and (c) shade cloth.

likely to be damaged by unintentional vandalism. They noted the importance of good installation practices, such as properly stretching the fold to avoid pulling it with machinery or by children swimming in the canals.

16.9.2 Evaporation Reduction in Storage

Reducing evaporation from reservoirs and storage ponds is a key conservation strategy, especially in arid regions. The methods include covering water surfaces with floating covers, such as Aquacaps and shade balls (Figure 16.6a and b); suspending covers, such as shade cloth made of polyethylene (Figure 16.6c); and applying monolayer chemical films that reduce evaporation. These techniques can be particularly effective in arid regions where evaporation rates are high. However, these methods must be implemented, while considering the potential impact on water quality and aquatic life.

16.9.3 Non-Economic Vegetation Control

Effective vegetation management in and around canals, reservoirs, and storage structures helps reduce water loss due to transpiration and maintain clear channels for water conveyance. Regular monitoring and the use of eco-friendly herbicides, mechanical removal, or controlled burning help maintain an optimal balance between vegetation and water flow. This approach also aids in maintaining biodiversity and preventing erosion along banks.

16.9.4 Change of Irrigation Methods

Changing from traditional surface irrigation methods, such as furrow or border irrigation, to more efficient systems, such as sprinklers or drip irrigation, can greatly reduce water losses due to evaporation, deep percolation, and surface runoff. Among these, drip irrigation offers, in particular, the application of water directly to the root zone, reducing losses from evaporation and runoff. This transition often requires initial investment but pays off in the long run through increased water efficiency and crop yield. These systems also fit well with closed conduit systems.

16.9.5 Automation of Irrigation System

Implementing automated systems using sensors, timers, and remote-control technology can optimize water usage. For example, automated irrigation gates equipped with water level sensors can be used to maximize water delivery while reducing water loss (Blair, 2010). Smart irrigation scheduling Apps can help optimize irrigation scheduling, based on real-time data, such as soil moisture levels, weather forecasts, and crop water requirements, which significantly improve water use efficiency.

16.9.6 Combined Facilities

Consolidating resources and infrastructure, both within and between irrigation districts, can lead to more effective water usage. Shared facilities, such as storage reservoirs, distribution pipelines, and water treatment plants, allow for optimized resource allocation, reduce redundancy, and increase system resilience during droughts or peak demand periods. Collaborative planning and governance are essential to ensure equitable access and cost sharing among stakeholders.

16.9.7 Salvaging Return Flows

Capturing and reusing return flows, i.e., water that runs off or drains from irrigated fields, is a practical and sustainable strategy for maximizing water availability. These return flows may include tailwater from surface irrigation or subsurface drainage. To effectively utilize this water, collection infrastructure (e.g., tailwater pits, sump pumps, reuse ditches) must be in place, and water quality assessments may be necessary to determine whether treatment is required before reuse. Salvaged water can be redirected back into the system for supplemental irrigation, helping reduce dependency on primary water sources and enhancing overall system efficiency.

16.9.8 Rehabilitating Measuring Devices

Accurate water measurement is essential for efficient distribution, fair allocation, and system accountability. Rehabilitating or upgrading outdated measuring devices helps ensure that water delivery matches crop requirements and irrigation schedules. Key upgrades may include flow meters (e.g., propeller, magnetic, or ultrasonic), water level sensors (manual or automated), and telemetry systems for real-time monitoring and data transmission. Modern measurement technologies support precision irrigation, reduce losses from overapplication, and help managers evaluate system performance more effectively.

16.9.9 Water Pollution Control

Maintaining water quality is fundamental to water conservation. Efforts should focus on reducing agricultural runoff, controlling the use of fertilizers and pesticides, and preventing industrial pollutants from entering water bodies. Clean water is easier to manage, requires less treatment, and is more suitable for reuse.

Protecting water quality is a critical component of water conservation. Polluted water not only limits agricultural use but also harms aquatic ecosystems and human health. Conservation-oriented pollution control efforts should focus on (1) reducing agricultural runoff through buffer zones, contour farming, and improved irrigation techniques; (2) optimizing fertilizer and pesticide application to minimize leaching and surface wash-off; and (3) preventing industrial and urban pollutants from entering irrigation waterways. Maintaining clean water reduces treatment requirements, supports sustainable reuse, and extends the lifespan of irrigation infrastructure by preventing clogging and corrosion.

16.10 LEGAL ASPECTS

Legal and public relations problems will arise and must be addressed. In the end, no improvement program can exceed the water users' willingness and ability to repay costs. The current policy from the U.S. Bureau of Reclamation policy mandates that all new canals and laterals be lined or piped, unless a compelling justification is provided. This policy is based on the demonstrated benefits of reducing seepage, improving delivery efficiency, and minimizing maintenance costs. In most cases, pipeline substitution is considered superior or at least equal to canal lining, offering advantages in durability, space efficiency, and water quality protection. Irrigation managers and planners should be fully aware of these regulatory requirements and incorporate them into system design and funding plans.

16.11 CHANGES IN IRRIGATION PRACTICES

Management must also consider evolving irrigation technologies and user preferences when planning infrastructure upgrades. One of the most significant shifts in recent decades has been the widespread adoption of sprinkler irrigation systems, which require cleaner water and more reliable delivery conditions than traditional surface methods do. To support efficient sprinkler operation, system improvements, such as canal lining or pipeline installation, are essential to minimize silt, algae, weeds, and other contaminants that can clog sprinkler nozzles or damage equipment. These upgrades not only enhance system performance but also reduce labor and maintenance costs.

The shift toward sprinkler irrigation has been notable since the 1960s. By 2015, approximately 34 million acres in the United States were equipped with sprinkler systems, representing 54% of all irrigated land. This trend reflects a growing emphasis on precision agriculture, water conservation, and mechanization. Irrigation district managers should closely monitor these trends and adjust system designs, maintenance programs, and long-term planning strategies accordingly.

REFERENCES

ASCE (1978). Describing irrigation efficiency and uniformity. *Journal of Irrigation and Drainage Engineering, ASCE*, 104(1), 35–41.

Askar, K., Leigh, E., and Fipps, G. (2009). Evaluation of Canal Lining Projects in the Lower Rio Grande Valley of Texas. Texas AgriLife Extension Service. Irrigation Technology Center. https://idea.tamu.edu/

Blair, A. (2010). Low-Cost Automatic Gates for Irrigation Canals, prepared for Harlingen Irrigation District under a Texas Water Development Board Innovative Technologies for Agricultural Water Management and Flow Measurement grant. https://www.twdb.texas.gov/conservation/agriculture/demonstration/doc/Factsheet_AutomatedGates.pdf

Bos, M.G. 1979. Standards for irrigation efficiencies of ICID. *Journal of Irrigation and Drainage Engineering*, 105(IRI), 37–43.

Burt, C.M., Clemmens, A.J., Strelkoff, T.S., Solomon, K.H., Bliesner, R.D., Hardy, L.A., Howell, T.A., and Eisenhauer, D.E. (1997). Irrigation performance measures: Efficiency and uniformity. *Journal of Irrigation & Drainage Engineering*, 123(6), 423–442.

Christiansen, J.E. (1942). *Irrigation by Sprinkling*. California Agricultural Experiment Station Bulletin 670. University of California, Berkeley.

Heermann, D.F. and Hein, P.R. (1968). Performance characteristics of the self-propelled center pivot sprinkler irrigation system. *Transaction of the ASAE*, 11(1), 11–15.

Howell, T.A. (2001). Enhancing water use efficiency in irrigated agriculture. *Agronomy Journal*, 93, 281–289.

Howell, T.A. Ed. (2003). Irrigation efficiently. In: *Encyclopedia of Water Science*. Marcel Dekker Inc., New York, pp. 467–472.

Leigh, E. and Fipps, G. (2019). Measuring seepage losses from canals using the ponding test method. Texas AgriLife Extension, Texas A&M System, B-6218.

17 Finances

NOTATION

AR annual repayment ($)
TCC total construction cost ($)
CRF capital recovery factor
i annual interest rate (%)
n number of years
AP annual payment ($)
FP future required amount ($)

17.1 INTRODUCTION

Financial management is fundamental to the effective operation and sustainability of irrigation systems. Just as water flows through canals and pipelines to reach crops, financial resources must be strategically allocated and accounted for to ensure that the system functions smoothly. From the construction of infrastructure to its ongoing operation, maintenance, and eventual rehabilitation, every stage of an irrigation project requires thoughtful fiscal planning, monitoring, and control.

Historically, irrigation enterprises started as small-scale operations, often privately or mutually managed, diverting water from natural sources to irrigate limited adjacent farmlands. Consequently, capital investments and operational expenses were relatively modest. Today, most simple, low-cost irrigation systems have already been established. Modern projects are significantly more complex and often involve substantial infrastructure, such as storage reservoirs, lined canals, extensive distribution networks, pressurized irrigation systems, and advanced automated control mechanisms. These large-scale systems necessitate considerable capital investment, accompanied by substantial and recurring operational and maintenance expenditures.

The key financial responsibilities of modern irrigation enterprises include (1) billing and collection for water sales; (2) accounting for multiple income streams (e.g., leasing water rights, recreational fees, hydropower revenues); (3) annual budgeting for routine operations and establishing emergency reserves; (4) managing the repayment of construction loans, grants, or other financial obligations; and (5) ensuring legal compliance through regular audits and accurate financial recordkeeping. Due to increased infrastructure complexity and rising operational costs, the cost per acre to deliver irrigation water, whether to new or existing farmland, is notably higher today. Consequently, achieving a sustainable financial balance between income and expenditure is more challenging and critical than ever before.

A robust financial management system must be (1) tailored to match the specific size, scope, and nature of the organization; (2) transparent, facilitating clear internal communication and external audits; (3) flexible, allowing for adjustments due to

DOI: 10.1201/9781003747444-20

unforeseen needs or system expansions; and (4) comprehensive, enabling detailed cost tracking for informed economic and engineering analyses.

Good fiscal management also involves diligent record maintenance. Detailed records should be kept, and costs should be separated so that engineering economic studies can be made intelligently on separate parts of the system, on structures, on major items of equipment, etc. Records should be maintained, and those which do not serve a useful purpose should be destroyed, or those which become a burden. Audits may be required by law and, even if not, should be carried out for good business practice. Management must pay close attention to fiscal matters and keep abreast of sources of funds to meet emergencies or other needs. An irrigation enterprise can be a large business, and professionals may need to keep it running smoothly. This chapter focuses on the financial aspects of managing irrigation projects, covering various topics that are important for understanding the different costs associated with irrigation projects.

17.2 FISCAL ISSUES

Effective fiscal management in irrigation systems requires careful categorization of costs, long-term financial planning, and strategies for sustainability. Most irrigation enterprises encounter three broad categories of financial obligations (Table 17.1): (1) repayment of construction and development costs, (2) operating and maintenance costs (to include some reserve or emergency funds), and (3) rehabilitation and betterment costs. It is not difficult to differentiate between the costs under categories 1 and 2. However, some rehabilitation and betterment costs may be included in the operation and maintenance costs each year, so there may not be a clear line of demarcation between funds designated under category 2 and those under category 3.

There are many who believe that reserves for depreciation should be established in every irrigation project and that the charges to depreciation expense, collected each year together with operation and maintenance expenses, should be sufficient to provide for all replacement of structures and other properties. The fact remains that in many older projects, this has not been done, and the need for major rehabilitation of the system grows to a point where the cost cannot be met on a "pay-as-you-go" basis and where a sizeable amount of financing must be arranged. In other instances,

TABLE 17.1
Categories of Financial Obligations in Irrigation Enterprises

Category	Description	Frequency	Example
1. Construction and development	Initial infrastructure investment	One-time (long-term)	Canal networks, dams, reservoirs
2. Operating and maintenance	Daily, seasonal, and annual operating costs	Recurring	Labor, energy, weed control, repairs
3. Rehabilitation and betterment	Major upgrades, modernization, or deferred maintenance	Periodic (10–30 years)	Canal lining, SCADA installation, pump replacement

betterment or modernization of the project is considered. Repayment of the cost of any of these programs will then constitute the third category of fiscal considerations.

17.3 BUDGET PREPARATION AND PLANNING

Budgeting is a vital financial tool for irrigation system management. This ensures that all anticipated expenses, such as routine expenses and non-routine expenses, are planned in advance and matched with available or projected revenues. A well-prepared budget promotes fiscal discipline, supports informed decision-making, and provides a financial roadmap for sustainable operations.

An excellent procedure is to project the estimates of annual budgets (especially estimated expenditures) for periods of, say, 5–10 years. The purpose of these projections over a relatively long period is to aid management in the development of an orderly schedule of replacements, structures, major pieces of equipment, or other property. It should be obvious that expenditures on such replacements must not be allowed to reach a high point, or points, in any one year or successive years. If this is allowed to happen, the burden on water users might exceed their ability to meet their obligations.

Budgets are drawn up by the manager and staff, reviewed by the governing board, adjusted and revised, and finally approved. This is a fair procedure because the governing board generally comprises representatives elected or selected by water users, tasked with safeguarding their interests. If these representatives fail to adequately represent their constituents, mechanisms exist to replace them, ensuring continuous accountability and alignment with water users' needs.

17.3.1 Accounting for Costs

There are two types of costs associated with irrigated land under an irrigation system: (1) construction costs and (2) operation, maintenance, and management costs. Construction costs are the initial investments to establish an irrigation system, including infrastructure such as canals, dams, reservoirs, gates, and other critical structures. Such costs are typically financed through loans, bonds, or governmental grants, and repayment terms may vary significantly. Operation, maintenance, and management costs are ongoing expenditures that ensure effective daily operations of the irrigation system, covering labor, energy, equipment maintenance, weed control, and minor upgrades. Budgeting these costs requires careful forecasting, inflation, and market fluctuations.

The distribution of these costs and the terms under which they are recovered from users may differ from country to country and from system to system. However, a common practice is to base the annual budget on projected operational expenditures adjusted for cost escalation, approved improvements, and the establishment of emergency or reserve funds. Figure 17.1 illustrates a detailed cost breakdown for a 105-ha drag-hose sprinkler irrigation system, which is based on the Mutange irrigation scheme in Zimbabwe (Savva and Frenken, 2002). Costs are categorized into construction and operation, maintenance, and management, including replacement, maintenance, energy use, and technical support.

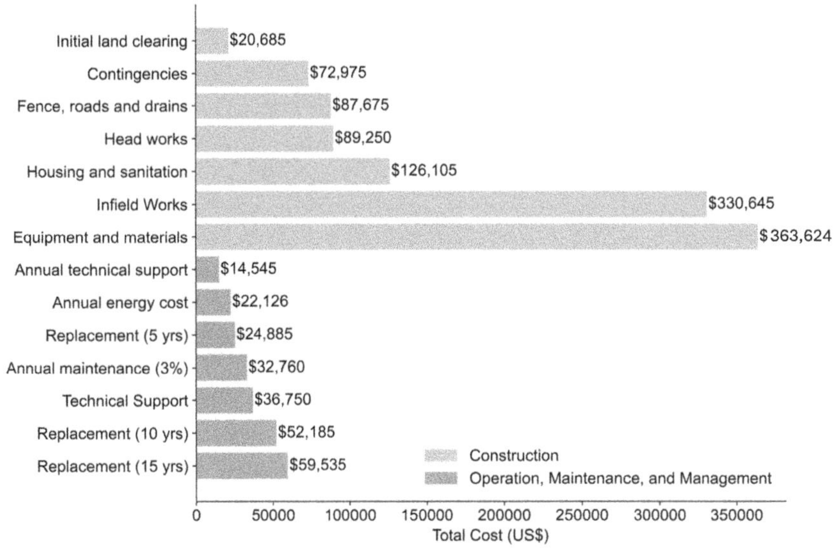

FIGURE 17.1 Total cost breakdown for the Mutange irrigation scheme. (The data used in this figure are from Savva and Frenken, 2002.)

Accurate and detailed cost records play a crucial role in budgetary control and are essential for determining equitable water charges. These records also support transparent financial assessments and allow for informed decision-making. For example, the Mutange Irrigation Scheme implemented a 21-year financial plan to track both costs and benefits over time. In Figure 17.2, annual costs are visualized using stacked bars, whereas projected benefits—with and without the irrigation intervention—are represented by line plots. This comparative analysis highlights the importance of long-term financial planning in irrigation management. It also emphasizes how strategic investments in infrastructure and ongoing operational support contribute to economic viability, sustainability, and return on investment in irrigation development.

The operation and maintenance of the drainage system are normally included in the operating budget (Hill, 1952). Additionally, a minor amount may be allowed each year for the construction of needed drainage systems. When sinking funds are established for major equipment purchases, system rehabilitation, or other long-term capital improvements, these contributions must also be included in the yearly budget. Allocating funds in advance for such future expenses promotes financial resilience, reduces the risk of deferred maintenance, and supports the sustainable management of irrigation systems.

17.3.2 Accounting for Income

Notably, assessments of water users may not be the only source of income for the project. Other miscellaneous sources of income might include recreational activities, the sale or leasing of excess water, and land rental for grazing or cropping. Income might

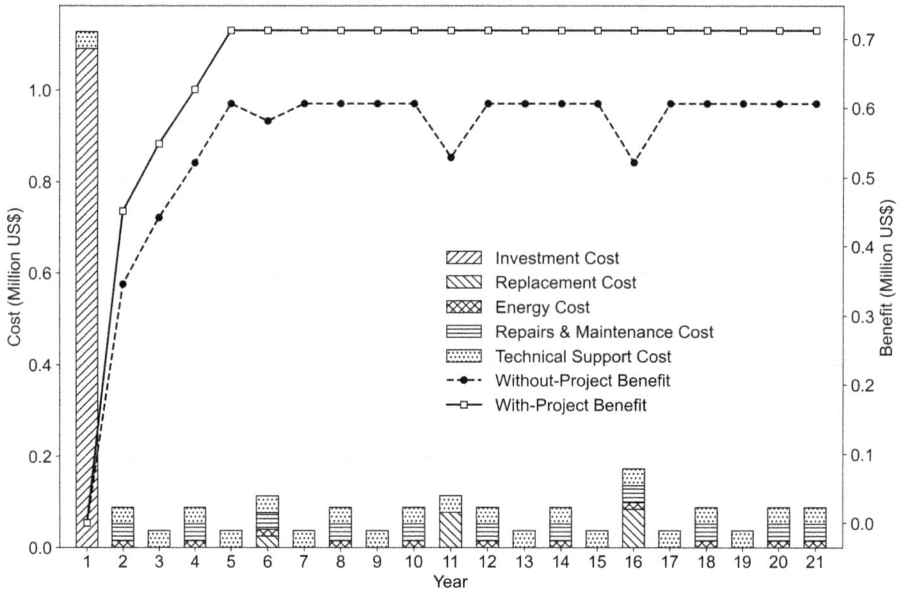

FIGURE 17.2 Annual cost and benefit analysis of the Mutange Irrigation Scheme (105 ha). (The data used in this figure are from Savva and Frenken, 2002.)

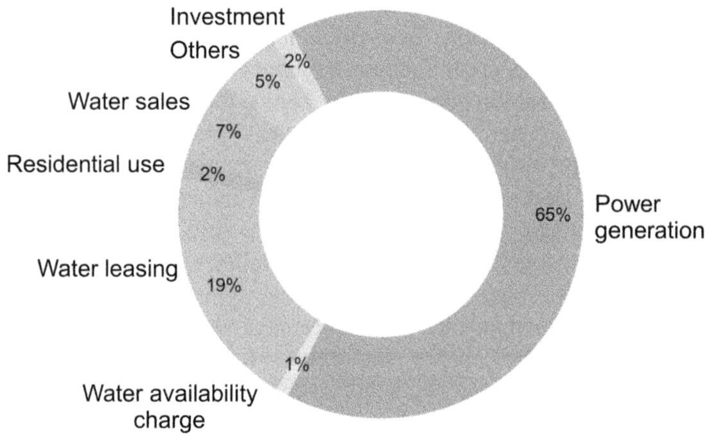

FIGURE 17.3 Income composition of the Imperial Irrigation District, California (2020).

also come from fees related to fish and wildlife enhancement and power generation. In general, incomes from these sources are managed in separate accounts and used to support various activities involved. In some cases, these funds can be used to reduce assessments against water users. It is worth noting that the generation of hydropower within a district often results in profit. Most districts use this profit to reduce assessments, although some have used it for rehabilitation and betterment of the irrigation system. Figure 17.3 illustrates the revenue structure of the Imperial Irrigation District

(IID) in California for the year 2020 (IID, 2020). The income streams include water sales, water transfers to other districts, residential use, water availability charges, power generation, investment income, and other sources. The distribution reflects IID's strategic diversification of revenue sources, where power generation emerges as the dominant contributor. This highlights the district's ability to support irrigation operations through a balanced mix of agricultural, utility, and programmatic income sources.

17.3.3 Accounts Keeping

Accurate and organized financial recordkeeping is essential for the sound management of any irrigation system. These records are critical not only for day-to-day operation and budget preparation but also for meeting legal and regulatory requirements. Irrigation organizations, especially large districts, often manage financial operations on a scale comparable to that of municipal governments. As such, the development and maintenance of a professional accounting system is a necessity—not a luxury.

A system of accounts and financial records is of prime importance to management regardless of the size of the organization. The details and complexity needed vary between organizations depending on the nature and extent of irrigation work and the number of water users served. A very simple system is adequate for an organization serving a small area where management has an opportunity to observe all facets of the day-to-day operations. An organization serving a large area will generally require a more elaborate system to guide management in the performance of its responsibilities. In any event, the system should be designed to permit an orderly expansion as the need for additional information arises.

Record keeping can become a wearisome job, unless it produces information, which can be used later. Records must have a purpose and be understandable to those supplying the data and those maintaining them if they are to be of value. Those that do not yield the needed information must be eliminated, and new information must be added to produce the desired information.

Many water user organizations on irrigation projects tend to maintain only basic records of income and expenses. On the other hand, privately owned domestic water companies and companies developing hydropower must report their costs in considerable detail to their state or federal regulatory bodies. Many irrigation districts keep their records of account in sufficient detail to permit effective management and control of costs.

The system developed for the project must be designed to fit the needs. The system used and the details included vary considerably with the size of the project, the use of water for purposes other than irrigation, any activities other than irrigation, and the type of organization formed for managing the district. Certainly, it is not economically feasible for small organizations to adopt accounting systems used by very large enterprises or the federal government.

17.3.4 Audits

Water user organizations are generally public or quasi-public corporations organized under state law. The provisions of law governing these organizations vary according to the type of organization and between the states. All states have provisions for periodic examination or audit of the organizations' records by either a representative of the state or a public accounting firm. The results of the examination are reported to those with the authority to take any necessary action to correct irregularities, if such action is found. Such audits ensure that the securing of funds for financial operations is in accordance with the law and is utilized for their intended purposes. The efficiency of operations or management is usually not investigated, although many public accounting firms are quite willing to volunteer for information on procedures that may benefit the organization.

17.4 CLASSIFICATION OF COSTS

The cost of operation of irrigation systems can be divided into four general classes: (1) fixed costs, (2) operational costs, (3) maintenance costs, and (4) general expenses. Each of these costs is an actual part of the cost of service. If the construction cost is already paid, the annual charge will be the annual cost of operation and maintenance. These can be based on any appropriate division of the cost between acreage and the use of water.

Fixed costs are constant and do not vary with the output of the irrigation system. They include charges or expenses, such as interest in the system's value, which can be considered a fixed charge, especially if the construction cost has already been paid off. Operational costs vary on the basis of both the acreage and the actual usage of water. They include the costs associated with the day-to-day functioning of the irrigation system. The maintenance costs are also related to the acreage, and these costs cover the maintenance of the system to ensure its continued operation and efficiency. Unlike operational costs, maintenance expenses are not directly linked to the amount of water used. General expenses are also dependent on the acreage and include administrative and other overhead costs associated with managing the irrigation system.

For billing and cost allocation purposes, it is essential to understand the demand and service portions of these costs. Fixed costs, maintenance costs, general expenses, and a portion of operational costs can be combined as a demand charge. The remaining operational costs, related to water acquisition, distribution, and measurement, can be considered direct delivery or service costs. The exact proportions and amounts of these cost categories can vary significantly from one system to another.

17.4.1 Construction Cost

The construction or building cost includes the cost of construction of the irrigation system, i.e., the cost of construction of the canal system from start to finish, including the cost of gates, outlet works, bridges, etc. This cost is paid only once, and the payment usually extends over a period of years. The method of payment varies according to the ownership structure of the system. If the irrigation system is owned by the government, as is the case in India, Pakistan, Nepal, etc., the entire cost or a major portion thereof is borne by the government and not by landowners or farmers. If the system is owned by landowners, the construction cost is equal to the actual cost of building the system, the cost is shared by the owners, and the payment is distributed over a number of years, depending on the interest rate. When the system is owned by a private company, its motive is to make a profit. Thus, the cost of construction is less than the price actually being charged to the landowner.

17.4.2 Apportionment of Construction Cost

Two methods are employed for the distribution of the construction cost to irrigable land: (1) a uniform rate per acre of irrigable land and (2) a variable rate per acre or a rate based on the benefits received. The first method applies a consistent charge to each acre of irrigable land, regardless of its location within the system. Despite the fact that the actual construction costs may vary across different parts of the system, this method simplifies cost recovery by charging all lands equally. Due to its simplicity, it is more commonly used. The variable rate method is based on the benefit received by the landowner. Land closer to the canal typically has a greater guarantee of water access than land situated further away. However, this proximity also has a higher risk of damage from canal failure or breaches. Additionally, lands near the upper parts of the system often incur lower servicing costs than those further away. This makes it challenging to calculate charges based on the precise construction costs for each area, as the benefits and risks vary significantly depending on location.

17.4.3 Effect of Form of Organization

If the owner of an irrigation system is a cooperative organization, such as a mutual or stock company, the construction costs are assessed against the stock rather than directly against the land. However, the stock owners are the landowners, with the stock being held in proportion to the irrigable land owned. Thus, the construction cost is assessed directly or indirectly against land.

17.4.4 Comparison of Flat Rate and Benefit Charges

In small irrigation projects, where variations in topography and soils are small, the flat rate method of charging construction cost is more common. Even for large systems, this is the usual method, except for projects divided into separate units. Within these large projects, variations in soils and topography are quite large. There are cases of poorer or rougher lands on which the benefit or returns per acre of cultivation may be lower and the cost of construction greater than those for lands with more uniform topography. In such cases, the benefit-based method of charge, rather than the average cost of construction, may be more appropriate or even necessary. The complication, however, is due to the variable nature of benefits from year to year.

17.4.5 Classification of Irrigation Area

The rules for classifying irrigable land and assessing and charging construction costs vary from system to system. In some cases, all land with the irrigation project is included for cost sharing. In some cases, only the net irrigation land is included, excluding the areas occupied by roads or nonfarmable lands. In other cases, land that is rocky, alkaline, seeped, high, or filled with burrow pits or creek channels should be considered. This land may be deducted from irrigable land. Similarly, land with a slope of 10% or 20% may be excluded. Likewise, land with a soil depth of, say, 1 ft (30 cm) may be deducted. All rights-of-way may also be deducted. It may be noted that the irrigable land area on any farm is not a fixed quantity. This decrease may be due to road construction or water logging. It may also increase because of the grading of the knolls, levelling of the burrow pits, etc.

17.4.6 Terms of Payment

The terms of payment by landowners vary with the form of organization. In the case of a cooperative company, where the systems are constructed by individual farmers through the company, the collective credit of these owners may be used to finance construction, and the time of payment may be extended over relatively long periods of time by means of bonds. Interest must be paid in such funds. With cooperative companies, borrowed funds are more usually secured on notes for shorter periods of time. In some cases, the construction cost is charged to the stock, and the individual stockholders have to finance their own portion of the cost.

17.4.7 Operation and Maintenance Costs

There are two methods for charging operation and maintenance costs to the land: (1) a flat rate per acre and (2) a quantity rate. In the flat rate per acre method, the charges are computed on the basis of the area irrigated irrespective of the amount of water used, which is limited by the maximum quantity specified by the water right. This is more common with cooperative companies, where all costs are assessed uniformly against the stock of the company. In the quantity-meter method, the amount of water actually delivered to a tract of land determines the charge. The actual charge, however, is composed of a minimum charge per acre plus a rate per acre-foot of water used in excess of a certain amount. For systems supplied from storage or by pumping, the cost of operation may be determined more by the amount of water delivered than by the area covered.

17.4.8 Limits of Beneficial Use

The legal right to use water and the requirements of beneficial use are interconnected. The use of water includes the rate of use, the length of time of use, or the total quantity taken per acre per year. The requirements of beneficial use vary from one system to another. However, some systems restrict the maximum rate of water use and limit the total amount per acre that can be diverted per season or determine the extent of beneficial use in terms of the quantity used per acre rather than the maximum rate of use.

17.4.9 Factors Affecting Use of Water

The amount of water used on a farm depends on the soil, topography, climate, water table, crops, land preparation, and method of applying water. The factors may differ across different parts of an irrigation system or different farms. Some of the factors are fixed, some are variable, and some are controlled by the irrigator. If the variations in use result from factors under the control of the landowner, then the quantity rate would be preferable because this would encourage beneficial and wise use of water.

17.4.10 Effect of Rates on Water Measurement

The measurement of water is essential when the quantity method is used. In many systems, accurate measurements are difficult. This explains the popularity of the flat rate per acre of charging. If the cost to the user is determined by the recorded amount of water delivery to the user, a more careful and accurate measurement will be needed. This will lead to improved measurement techniques.

17.4.11 Flat Rates

One way to compute flat rates is by dividing the total charges by the acreage. Usually, the irrigable area defines the acreage, but the area actually irrigated, sometimes, is also used. In some irrigation systems, flat rates may differ for different parts because of mixed ownership or other reasons.

17.4.12 Quantity Rates

The quantity-based rate comprises two charges: (1) demand charge and (2) service or delivery charge. The demand or steady charge is a fixed amount per acre of irrigation or irrigable land. Upon payment, the owner is entitled to a certain amount of water per acre. For water in excess of the minimum amount, a certain rate per acre-foot is charged.

With respect to the service charge, various methods are used for determining rates. In the first method, the rates for different classes are proportional to the cost of offering such services. In the second method, the rates are adjusted among different classes of service in such a manner as to fix the rate for each class based on the value of the service rather than its cost. The quantity rate is not suitable for localities where irrigation is only supplemental to precipitation and where the use of irrigation water varies widely from year to year.

17.4.13 Terms of Payment

Operational costs may be paid monthly or once for the whole season. The monthly mode of payment is used as a quantity rate method. The flat rate per acre is more suitable for advance payments. For delinquent payments, penalties must be assessed and charged.

17.5 REPAYMENT OF CONSTRUCTION AND DEVELOPMENT COSTS

17.5.1 Planning for Costs

A plan for repayment of construction and development costs of the project must be included, at least in skeleton form, in the definite plan for the project. This ensures that stakeholders understand how capital costs will be recovered, regardless of whether funds come from federal or state programs, private investors, or the organization's capital reserves. The repayment contract builds upon this initial plan and formalizes the financial obligations.

In multipurpose water development projects, repayment planning is often integrated with broader basin development strategies. For example, in the Columbia River Basin, hydropower revenues are used to subsidize irrigation development costs. The Bonneville Power Administration (BPA), for example, redistributes excess power revenues to help cover irrigation-related construction expenses, thus easing the financial burden on farmers (Comptroller General of the United States, 1957).

However, financial relief through cross-subsidization should not be a substitute for sound project planning. Construction must not begin until funding is secured and repayment strategies are in place. For federally funded projects, a legally binding repayment contract must be executed before awarding construction contracts. For example, under the Water Conservation and Utilization Act, no federal funds are released unless the water users' organization signs a repayment contract detailing the schedule, interest (if any), and method of cost recovery.

17.5.2 Repayment

Repayment obligations typically begin after construction is completed and the project is turned over to the operating organization. By this stage, significant financial decisions have been finalized. Therefore, the terms of repayment, whether it includes interest or not, and the method of repayment are determined by the contract agreement. These contracts are tailored to the financial capabilities of the water users and the source of capital. They may stipulate full or partial cost recovery, interest-free or interest-bearing terms, and differing repayment durations—commonly 25, 40, or 50 years. In federally funded projects, such as those managed by the U.S. Bureau of Reclamation, repayment terms are defined under provisions of acts such as the Reclamation Project Act of 1939 or the Water Conservation and Utilization Act.

The operation and management personnel must prioritize these repayment obligations in their long-term financial planning. Annual repayment amounts must be budgeted alongside operation, maintenance, and reserve fund allocations to ensure financial stability. An example of such a contract can be found in the documentation of the U.S. Bureau of Reclamation (1964). This is an agreement between a water user's organization and the U.S. Government. It's important to note that in the United States, construction funds can be sourced from various entities, including the federal government and state governments (with states such as Wyoming and Utah providing loan funds for water development), as well as other sources. These diverse funding sources contribute to the variation in repayment contract terms.

In contracts that require interest-bearing repayments, annual payments are often calculated using the capital recovery factor (CRF) method. This approach allows the borrower to spread the repayment evenly over the entire period, covering both the original investment and the interest to the remaining unpaid balance. The CRF method results in a fixed annual repayment amount and is particularly useful for budgeting large infrastructure investments. The method is represented by the following equations:

$$\text{Annual repayment }(\text{AR}) = \text{Total construction cost}(\text{TCC}) \times \text{CRF} \tag{17.1}$$

$$\text{CRF} = \frac{i}{1-(1+i)^{-n}} \tag{17.2}$$

where AR is the annual repayment amount (\$), TCC is the total construction cost (\$), i is the annual interest rate (%), and n is the repayment period in year.

Figure 17.4 provides a visual reference of capital recovery factors over a range of interest rates and repayment durations, which help estimate annualized capital costs for on-farm infrastructure and equipment.

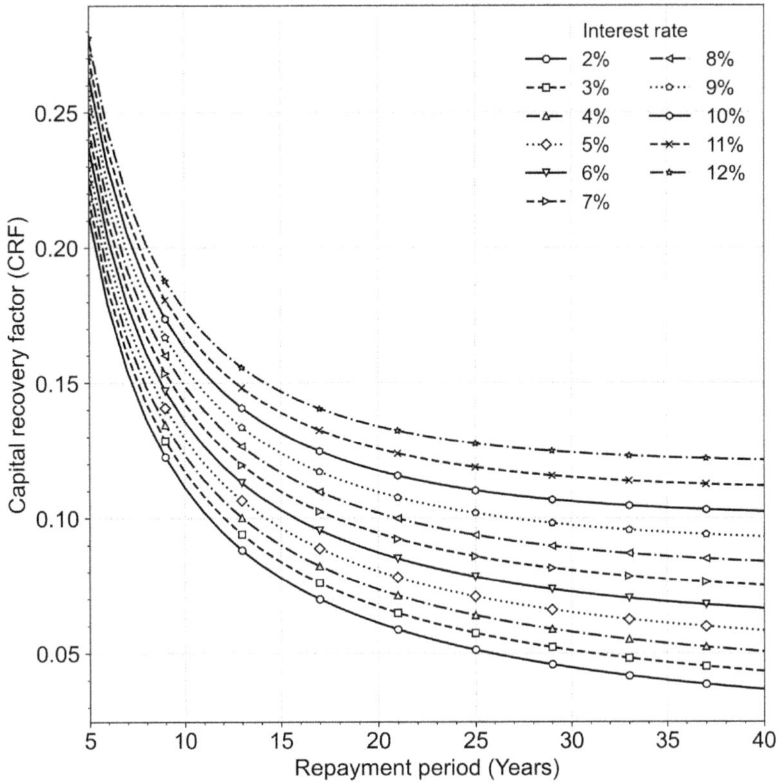

FIGURE 17.4 Capital recovery factor by interest rate and year.

Example 17.1

An irrigation district has received a loan of $5,000,000 for the construction of canal infrastructure. The repayment term is 30 years at an annual interest rate of 3%. Calculate the annual repayment amount the district must pay using the CRF method.

Solution:

Given $i = 3\% = 0.03$ and $n = 30$, the capital recovery factor is

$$\text{CRF} = \frac{i}{1-(1+i)^{-n}} = \frac{0.03}{1-(1+0.03)^{-30}} \approx 0.0510$$

Given TCC = $5,000,000, the annual repayment is

$$\text{AR} = \text{TCC} \times \text{CRF} = \$5,000,000 \times 0.0510 = \$255,000$$

This means that the organization must allocate approximately $255,000/year for 30 years to fully repay the construction costs under the agreed-upon terms. Such calculations are essential for developing financially sustainable repayment plans, particularly when the project involves interest-bearing loans.

17.5.3 Allocation of Repayment of Construction Costs

The usual practice in the United States is to allocate repayment of construction costs on the basis of acres of land actually receiving irrigation benefits. This approach follows the principle that only landowners who directly benefit from the irrigation system should bear financial responsibility. Land within project boundaries that is deemed non-irrigable or currently unserved is typically excluded from repayment obligations. However, if the status of such land changes over time—for example, if it is later developed or converted into irrigated farmland—it becomes necessary to adjust the repayment structure so that these newly benefited landowners contribute their fair share of the costs.

There is one major exception to this general rule. Conservancy districts have legal power and valorem taxes for all property in the district, regardless of whether they directly receive irrigation water. This broader taxing authority is justified by the fact that conservancy districts are created to serve regional, multipurpose objectives, such as flood control, municipal water supply, recreation, and ecosystem protection, in addition to agricultural irrigation. As a result, their financial responsibilities are spread across a more diverse group of stakeholders, including urban residents, utilities, and industrial users.

17.5.4 Sale of Stocks

In the case of commercial-type irrigation companies, a common approach to finance construction costs is selling stocks to water users. Each share of stock entitles the holder a prescribed amount of water. The revenue generated from these stock sales is then allocated to retire the construction costs. The charge for the water is on the basis that it includes construction, operating and management costs, and all other costs incurred by the district. There are other methods for recovering construction costs. Regardless of the chosen method of recovery, it is the obligation of management of the enterprise to collect costs and disburse them in the manner set forth in the contract.

17.5.5 Contract Renegotiation under Exigencies

The real test for the management of an irrigation enterprise comes when, due to an economic depression, a drought, or other reasons, the ability of the project to meet the payments specified in the contract is exceeded. Adjustments must somehow be negotiated, and management personnel must know what the district is capable of doing and, at the same time, be appreciative of the position of those who have provided capital. This is not an easy task. Because of escalating construction costs, underestimation of costs by the designer or builder, or for other reasons, costs envisioned at the time of signing the original contract have been appreciably exceeded. Consequently, one of the primary responsibilities of the management team may involve negotiating a revised contract. This new agreement must meet the satisfaction and capabilities of all parties concerned. This process is crucial for ensuring the continued viability and success of irrigation projects under challenging conditions.

17.6 OPERATION AND MAINTENANCE CHARGES

Most irrigation enterprises are mutual of some sort. That is, they are not running to make a profit. Additionally, any charges must be based on some equitable distribution of costs among the participants, and each pays his fair share. Assessing the cost of operation and maintenance of facilities for water users is no exception. These charges vary considerably from place to place. Additionally, the method of assessment varies considerably throughout the world. However, there are, in general, four categories into which the majority will fall: (1) a charge based on acres or area irrigated, (2) a charge based on water delivered, (3) a charge based on the acreage of particular crops matured, and (4) a charge based on each irrigation for a given area (Table 17.2).

TABLE 17.2
Comparison of Operation and Maintenance Assessment Methods

Method	Basis	Common Regions/ Use Cases	Pros	Cons
Area irrigated	Acres of land served	Traditional districts (U.S. West)	Simple to administer	Ignores actual water use
Water delivered	Volume of water used	Water-scarce regions (CA, TX)	Promotes water conservation	Requires accurate metering
Acres of crops matured	Type and area of crop	Asia (India, Pakistan, Bangladesh)	Reflects actual crop demand	Requires surveys and variable rates
Charge per irrigation event	Per-event per area basis	Semi-arid regions with seasonal use	Suitable for irregular irrigation needs	Less precision in water use tracking

The subsequent discussion shows that the method of assessing the operation and maintenance costs of the irrigation system against water users will vary. The ultimate goal is to allocate these costs in the most equitable way possible so that each user will pay his fair share in proportion to the benefits he receives. This is not an easy task and always results in much discussion between users and management. It is very probable that the objective is never precisely reached, but the closer the approach is, the better the relationship between the users and management tends to be.

17.6.1 Area Irrigated

In the United States, the oldest method, which is still used by certain types of organizations, is the method that assesses changes based on acres or irrigated areas. This method is particularly common in older stock-based irrigation companies. In these companies, the stock was held in proportion to the land irrigated by each owner. It was natural then to assess the operation and management costs on the basis of stock held, which indirectly meant the acreage owned. These older companies usually had rights to an adequate amount of water, and the actual acre feet of water used were not critical. Because construction costs are usually based on acreage, bookkeeping is rather simple when this manner of assessment is used.

17.6.2 Water Delivered

Assessing charges based on the volume of water delivered is probably the most extensively used method in the United States. In general, this method may be administered in two ways. A charge may be made for all the water delivered to the irrigator, within the limits of the water rights. This means good measurements of all deliveries. It is more common to assess a demand charge based on a fixed amount per acre of land irrigated, or irrigable land, held by the irrigator. This amount varies depending on the amount of water available for the season, water rights, climates, soils, and crops. What is generally calculated is the optimum amount of water to produce the crop, with other considerations taken into account. To this amount is added a delivery charge, which is the charge for the amount of water delivered in excess of the amount if he knows that he will be charged for it. In areas of acute water deficiency, delivery charge rates may increase as the amount of water exceeds the base amount. This further encourages the careful use of water.

17.6.3 Acres of Crops Matured

Assessing charges, based on the acreage of crops matured, is used in Asia but has never been used extensively in the United States. Rates vary in accordance with the crops grown. Items such as the amount of water necessary to produce a certain crop, the season when most water is necessary for the crop (winter or summer, or during peak demand and periods of less demand), the value of the crop, and perhaps other factors, all enter the formula from which the charge is derived. This method is quite equitable and does not require precise measurements of flow; however, it does require crop surveys and involves additional bookkeeping tasks.

17.6.4 Individual Irrigation for an Area

The method of assessing a charge, which is based on each irrigation for a given area, is used in climates where irrigation is not necessary at all times for the successful growth of crops or in locations where the same crop is grown in large areas year after year and where a fixed charge is equitable. This method is used in some areas in the Lower Rio Grande Valley of Texas, where water application to the land is primarily conducted through sprinkler systems. This method suits the specific agricultural and climatic conditions of the area, balancing the need for irrigation with the practicalities of water usage and cost allocation.

17.6.5 Collection of Dues

The efficient collection of assessments is crucial, as they constitute a significant portion of the income of an irrigation enterprise. Ensuring uniform collection and minimizing payment delinquencies are vital for ensuring the financial stability of operations. One method to discourage late payments is by imposing penalties on overdue assessments. Typically, these assessments are levied and due annually. However, some districts offer flexibility, allowing payments to be made in two installments, e.g., one in the fall and another in the spring.

The system established for collections and disbursements needs to align with the structure of the levies. Specifically, the collection process for repaying construction costs is often managed separately, especially when third-party lenders or government agencies are involved. The irrigation enterprise typically collects construction repayments and transfers them to the appropriate financing institution under separate contractual terms. This system ensures that each aspect of the financial operation, from regular operational dues to construction cost repayments, is handled efficiently and in accordance with the set contractual obligations and financial structures.

17.7 REHABILITATION AND BETTERMENT

Over time, all irrigation systems experience wear, deterioration, or functional obsolescence due to aging infrastructure; deferred maintenance; evolving agricultural practices; or environmental stressors such as drought. Just as a transportation system requires continual upgrades to remain efficient, an irrigation system must be periodically rehabilitated and modernized to ensure reliable service.

Rehabilitation refers to restoring worn or damaged components (e.g., canal linings, pumps, gates), while betterment involves upgrades that improve performance, such as automation, water conservation, or energy efficiency. Engineering economic studies are central to evaluating whether to maintain current operations, undertake major rehabilitation, or pursue replacement. These studies rely on operation and maintenance records and cost–benefit analyses to guide investment decisions that balance functionality with economic sustainability.

17.7.1 Assessment of Costs

After the facts have been determined and the willingness and ability of the water users to pay the costs have been ascertained, the matter of capital to cover these costs must be considered. If the district has established a pay-as-you-go policy and each year adjusts the operating and maintenance costs upward in an amount necessary to support a continuing program, no large amount of capital will be required at any one time. This is an excellent procedure and should be followed if practical.

If due to repayment of construction costs set by the contract at a high rate compared with income, a lack of proper maintenance in the early years of the project, economic recession, drought, or other causes, rehabilitation and betterment of the system becomes a major undertaking and requires a large capital outlay. Those who manage the enterprise face the problem of financing the required work. Changes in land use, methods of irrigation, and types of crops raised may also render the system obsolete and necessitate modernization.

17.7.2 Method of Financing

17.7.2.1 Sinking Fund

A sinking fund can be established if the need for major expenditures is identified early through economic studies or other means. This fund accumulates the necessary funds over time, making them available at a predetermined future date for the required improvements. This method distributes the costs over time and can potentially reduce the total costs, as the funds in the sinking fund can be invested to generate additional income. Annual payments (AP) are deposited into an interest-bearing account until the full amount is available in the target year, which can be calculated as

$$\text{AP} = \frac{\text{FP} \times i}{(1+i)^n - 1} \tag{17.3}$$

where AP is the annual payments (\$), FP is the future required amount (\$), i is the annual interest rate (%), and n is the number of years of unit expenditure.

Example 17.2

An irrigation district anticipates the need for $2,000,000 in 10 years to replace its main canal lining. If the district can earn 4% annual interest, how much must it set aside each year?

Solution:

Given FP = $2,000,000, i = 4% = 0.04, and n = 10, the annual payment is

$$\text{AP} = \frac{2,000,000 \times 0.04}{(1+0.04)^{10} - 1} \approx \$166,582$$

The district must deposit $166,582 annually for 10 years at 4% interest to accumulate $2 million.

17.7.2.2 Bonds

Bonds may be sold to finance the work (if the legal framework permits), and a sinking fund may be established to retire the bonds at the time they become due. Interest or other income accumulated on the sinking fund will somewhat offset the interest charges on the bonds. If the economic conditions in the project permit, the bonds may be retired in specified amounts annually without establishing a sinking fund. If this method is followed, annual assessments must be increased and sustained at a higher rate for a period of years. Bonds issued for improvements may also be dated to fall after construction costs have been repaid. If money can be obtained for such longer periods of time and at a reasonable interest rate, this method of financing tends to sustain the annual expense to water users at a more or less even rate.

17.7.2.3 Federal Funding

Under the authority of the Reclamation Project Act of 1939 (Public Law 76–260, 53 Stat. 1187), irrigation projects originally constructed with funds advanced by the U.S. Government through the U.S. Bureau of Reclamation may receive additional funding for rehabilitation and betterment. These supplemental funds are typically allocated for critical infrastructure replacements that can no longer be deferred or for project improvements that enhance water conservation, delivery efficiency, or system resilience.

Such funds may be added to the original construction costs and the original repayment of these funds. Usually, the original repayment period is extended, thus keeping repayment charges at a more or less even rate. Contracts for repayment may also be written so that the annual rate of repayment is increased. All such repayment contracts are negotiated to fit the economic conditions of the project. Therefore, considerable variations are found in terms of such contracts. For example, in some cases, rehabilitation funds have been integrated into existing repayment structures with interest-free terms, provided that the improvements primarily benefit irrigated agriculture and do not serve excess land or nonagricultural users. However, when funds support municipal, industrial, or commercial power benefits, interest-bearing provisions are applied in accordance with federal law. An example of the details included in these types of contracts may be found in the U.S. Bureau of Reclamation (1964).

When a project, for some reason, reaches a state where rehabilitation and betterment can no longer be deferred and reasonable operating efficiency can be maintained, management has a number of ways to turn to. It is then their responsibility to exercise wise judgment in selecting a plan that can be initiated and carried out to reach a successful conclusion. Letting the project die because of a lack of judicious action would be a catastrophe for the entire area.

17.7.3 Small Reclamation Projects Act

Another means of financing the rehabilitation and betterment of small irrigation projects, particularly those built with funds other than those advanced by the U.S. Government, is through the Small Reclamation Projects Act of August 1956 (70 Stat.488), as amended June 5, 1957. This act is designed to encourage state and local participation in developing new irrigation works or improving existing ones. Eligible undertakings may include the construction of an entire irrigation system, a specific portion of a district system, or a rehabilitation and betterment program for an existing project. Federal assistance is limited to $5 million per loan, although multiple loans may be approved if they fund different parts of the same system.

Loan repayment is determined based on the ability of the contracting entity to pay and may not exceed a 50-year repayment period with no development period provided. If any portion of the loan is used for non-agricultural purposes or to irrigate lands exceeding 160 acres in single ownership, that proportion must bear interest at a rate established for each contract under the Act. Loans and grants may be made to irrigation districts, conservancy districts, other public agencies, and states, provided that the total project cost, including any local contributions, does not exceed $10 million, and the applicant finances a portion of the construction costs. No interest is charged on reimbursable costs allocated to irrigating lands of 160 acres or less in single ownership, whereas interest applies to portions allocated for excess lands or for municipal, industrial, or commercial water uses. Further details on the Act and example repayment contracts can be found in U.S. Bureau of Reclamation (1964).

If a loan is used partially for non-agricultural purposes and/or irrigation of excess land, the proportion of the loan so used bears an interest charge, the rate of which for each contract is established in accordance with the act. Loans and grants may be made to irrigation districts, conservancy districts, other public agencies, and states for the above-mentioned purposes. Grants and loans are limited to projects that will not have an overall cost of more than $10 million, inclusive of any local contribution. The law requires that the organization seeking such funds must finance a portion of the construction costs. There is no interest in those portions of the reimbursable cost properly charged to the irrigation of lands not in excess of 160 acres in a single ownership. However, interests should be paid for larger landholdings, e.g., lands in excess of 160 acres in a single ownership, or non-agricultural purposes, e.g., municipal, industrial, or commercial. Additional information on this Act and a loan repayment contract may be found in the U.S. Bureau of Reclamation (1964).

Example 17.3

An irrigation district receives a $5 million federal loan under the Small Reclamation Projects Act to upgrade its canal system. Of this total, $3 million is allocated to irrigating lands not exceeding 160 acres in single ownership (interest-free), and $2 million is allocated for municipal water supply (interest-bearing). The repayment period is 40 years, and the interest rate on the interest-bearing portion is 3% annually. Calculate the annual repayment of the irrigation district.

Solution:

Given $i = 3\% = 0.03$ and $n = 40$, the capital recovery factor is

$$\mathrm{CRF} = \frac{i}{1-(1+i)^{-n}} = \frac{0.03}{1-(1+0.03)^{-40}} \approx 0.0433$$

The annual repayment for interest-bearing portion:

$$AR = TCC \times CRF = \$2,000,000 \times 0.0433 = \$86,600$$

The annual repayment for interest-free portion:

$$AR = \frac{\$\ 3,000,000}{40} = \$75,000$$

The total annual repayment:

$$\$75,000 + \$86,600 = \$162,000$$

The irrigation district must repay $162,000 annually over 40 years, of which $75,000 is for the interest-free agricultural portion and $86,600 covers the municipal interest-bearing portion.

17.8 CASE STUDIES

The financial management of irrigation districts varies greatly depending on their scale, scope of services, and strategic priorities. This section examines three representative examples across the United States: IID (California), a large, diversified district providing both water and electric power services; El Paso County Water Improvement District No. 1 (Texas), a mid-sized, water-only district serving both agricultural and municipal users in an arid region; and Riverside Irrigation District (Colorado), a small agriculture-focused district with a conservative financial approach. These case studies show how variations in operational scope influence revenue composition, expenditure patterns, capital financing approaches, and reserve policies. Together, they provide a comparative perspective on the fiscal strategies that sustain irrigation service delivery under different economic and environmental conditions.

17.8.1 Imperial Irrigation District (California)

The IID, located in the Imperial Valley of southern California, delivers water to approximately 500,000 acres of farmland and provides water to local municipalities within its 1,427 mile^2 service area. In 1936, recognizing the hydroelectric generation potential of the All-American Canal, the District entered the power business. Today, it is the third-largest public power utility in California, supplying generation, transmission, and distribution services to more than 165,000 customers, including residential, commercial, and industrial users. This dual role as a water and energy provider highlights a diversified revenue base.

As shown in Figure 17.3 (Section 17.3.2), IID's operating revenues are derived primarily from water sales and energy services, supplemented by other income streams such as water transfers to other districts, residential water use, water availability charges, investment income, and miscellaneous sources. This diversified structure allows the district to sustain irrigation operations through a balanced mix of agricultural, utility, and programmatic revenues. In 2024, total operating revenues were $972.7 million, a 9% increase over 2023 ($890.1 million). Water operating revenues increased by 25% to $341.3 million, driven by higher transfer volumes and rate adjustments, whereas power operating revenues increased modestly by 2.3% to $631.4 million (Table 17.3).

Operating expenses in 2024 included purchased energy, fuel costs, other power-related expenses, canal operations, operations and maintenance, general and administrative expenses, and depreciation of infrastructure. The largest year-over-year increases were in operations and maintenance (+18.9%) and other power expenses (+13.7%), reflecting expanded system upkeep and possible regulatory or modernization costs. Depreciation and amortization increased by 7.3%, indicating ongoing capital investment.

Non-operating revenue and expenses included investment income, interest expenses, and other miscellaneous items. In 2024, IID recorded $26.3 million in net investment and other revenues, supported by a 21.5% increase in investment income.

TABLE 17.3
Revenues, Expenses, and Net Position of the Imperial Irrigation District (2023–2024)

	Amount (Million $)	
Category	2024	2023
Operating revenues		
• Water operating	341.3	273.1
• Power operating	631.4	616.9
Total operating revenues	**972.7**	**890.1**
Operating expenses		
• Purchased energy	261.8	264.3
• Cost of fuel	80.4	72.3
• Other power expense	184.9	162.6
• Canal operation	11.3	10.9
• Operations and maintenance	245.1	206.2
• Administrative	28.8	26.6
• Depreciation and amortization	131.2	122.3
Total operating expenses	**943.5**	**865.3**
Investment and other revenues (expenses)		
• Investment income (loss)	40.7	33.5
• Interest expense	−22.5	−22.3
• Other income	8.1	16.6
Total investment and other revenues	**26.3**	**27.8**
Capital contributions	**74.9**	**26.8**
Change in net position	**130.4**	**79.4**
Total net position, beginning of the year	**2,015.8**	**1,936.4**
Total net position, end of the year	**2,146.2**	**2,015.8**

Source: Data are from IID (2025).

The district invests its idle cash daily. Investments consist primarily of federal agency securities, bank certificates of deposits, medium-term corporate notes, and obligations of the U.S. government. Investment earnings, including the change in fair value, are allocated to the district's operating department on a monthly basis according to their relative investment balances.

The district maintains a capital improvement program to guide multi-year investments in canals, laterals, pumping stations, and modernization projects. In 2024, IID invested in both the rehabilitation of existing assets and new infrastructure to enhance delivery efficiency. These capital projects were financed through a mix of current revenues, reserves, and long-term debt instruments, primarily revenue bonds. Debt service is scheduled to align with projected cash flows and repayment capacity, while restricted funds are maintained for debt service reserves, capital replacement, and environmental mitigation. At the end of 2024, IID's net position was $2.54 billion, including $1.51 billion in net investment in capital assets, $19.5 million in restricted funds, and $619.7 million in unrestricted resources.

IID illustrates how a large-scale, multi-utility irrigation district can sustain an aggressive capital investment program while maintaining fiscal resilience. Its revenue diversification mitigates sector-specific downturns, but high fixed costs require vigilant expense management to preserve operating margins.

17.8.2 El Paso County Water Improvement District No. 1 (Texas)

The El Paso County Water Improvement District No. 1 (EPCWID#1), founded in 1911, manages water deliveries from the Rio Grande to farmland and municipalities in El Paso County, Texas. It serves more than 69,000 acres of irrigated farmland and provides municipal and industrial water deliveries along the Rio Grande. The district operates in a climate where drought conditions are frequent, and water supply reliability is a constant concern. This hydrologic conditions affects both its operational priorities and financial strategies, as variability in river flows can directly affect delivery volumes, revenue stability, and the timing of infrastructure investments.

Without power generation, EPCWID#1 relies entirely on water-related charges, service contracts, licensing fees, and investment income for operations. In 2024, EPCWID#1 generated $13.88 million in operating revenues. Contract revenue was the largest source at $6.65 million (48%), followed by combined land and water use assessment at $3.06 million (22%), and licensing fees at $2.52 million (18%) (Table 17.4). Smaller sources included special projects, administrative revenue, and reimbursement income. Compared with that in 2023, total operating revenues increased by $2.07 million (by 18%), driven primarily by a $2.69 million (by 68%) increase in contract revenue and a $0.67 million (by 84%) increase in water use assessments.

Operating expenses were $14.63 million, exceeding revenues and producing an operating loss of $0.74 million. Payroll costs were the largest expense at $6.34 million (43%), followed by repair/maintenance expenditures and capital projects at $3.94 million (27%). Depreciation accounted for $1.51 million (10%), with administrative, contractual, insurance, and other costs remaining. Compared with those

TABLE 17.4
Revenues, Expenses, and Net Position of the El Paso County Water Improvement District No. 1 (2023–2024)

Category	Amount (Million $)	
	2024	**2023**
Operating revenues		
• Land assessments	1.59	1.59
• Water use assessments	1.47	0.80
• Other assessments	1.35	1.39
• Contact revenue	6.65	3.96
• Licensing fees	2.52	3.50
• Special projects	0.18	0.27
• Administrative revenue	0.08	0.05
• Reimbursement revenue	0.04	0.24
• Other revenues	0.00	0.00
Total operating revenues	**13.88**	**11.81**
Operating expenses		
• Administrative	0.44	0.39
• Contract, consultant, and legal services	0.62	0.59
• Employee allowances	1.17	1.26
• Governmental fees	0.15	0.24
• Insurance fees	0.44	0.37
• Licensing/special projects	0.02	0.09
• Payroll expense	6.34	5.61
• Repair/maintenance and capital projects	3.94	2.18
• Reimbursable expenses	0.00	0.03
• Depreciation	1.51	1.50
Total operating expenses	**14.63**	**12.25**
Operating income (loss)	**–0.74**	**–0.44**
Non-operating revenues		
• Investment interest	1.13	0.16
• Gain (loss) on sale of assets	0.10	0.00
Total non-operating revenues	**1.24**	**0.16**
Change in net position	**0.49**	**–0.28**
Net position, beginning of the year	**27.54**	**27.83**
Net position, end of the year	**28.04**	**27.54**

Source: Data are from El Paso County Water Improvement District No. 1 (2025).

in 2023, operating expenses increased by $2.38 million (by 19%), with the largest increases in repairs and capital projects (by 81%) and payroll (by 13%).

Non-operating revenues offset the operating loss. Investment income contributed $1.13 million, reflecting higher interest rates on the district's cash holdings, and gains from asset sales added an additional $0.10 million. Together, these factors produced a net positive change in position of $0.49 million, increasing the district's ending net position to $28.04 million, compared with a negative change of $0.28 million in 2023.

Operating in a drought-prone environment demands a strong focus on cost efficiency, infrastructure reliability, and diversification of revenue sources. EPCWID#1's dependence on contract and assessment revenues makes it vulnerable to fluctuations in water availability, but investment income serves as a stabilizing factor in dry years. The district's experience highlights how financial resilience in arid regions often hinges on balancing short-term operational needs with long-term supply uncertainty.

17.8.3 Riverside Irrigation District (Colorado)

The Riverside Irrigation District, established in 1902, is a relatively small but financially stable district serving the South Platte River Basin in Colorado. Its operations focus on the Riverside Reservoir and a distribution network of canals and pipelines, providing irrigation water primarily to agricultural users. The district also leases water to municipalities and maintains additional income streams from land, pumps, and pasture rentals.

Riverside's financial management emphasizes low operating costs, steady reserve accumulation, and no reliance on long-term debt. This approach allows the district to finance capital improvements from cash reserves, eliminating interest expenses and ensuring rapid response to infrastructure needs. In 2023, the district's total revenues were $4.03 million, with water leases contributing 63%, royalties contributing 15%, and investment income contributing 8% (Table 17.5). Smaller but consistent sources included running fees, pump fees, pasture rentals, and miscellaneous income. The total operating expenses were $1.20 million, with the largest category being the conservation of natural resources (72% of total expenses), reflecting the district's investments in water resource management and environmental stewardship. General and administrative costs represented 20%, whereas capital outlays were limited to 8% of expenditures, which is consistent with the district's strategy of periodic, targeted infrastructure investment.

The fund balance increased by $2.84 million in 2023, reaching $27.77 million at year-end, equivalent to nearly seven times the district's annual revenues. Of this, $27.73 million was unassigned and $0.04 was classified as an emergency reserve (in compliance with Colorado's Taxpayer's Bill of Right (TABOR) requirements). This means that nearly the entire fund balance is available for general use or future projects, giving the district the flexibility to address major infrastructure needs without debt financing.

Riverside's conservative model prioritizes financial security over rapid expansion. With reserves equal to nearly 7 years of revenues, it has unmatched flexibility to address emergencies or infrastructure needs without borrowing. The trade-off is a slower pace of modernization compared with debt-financed districts.

17.8.4 Comparative Analysis

These three districts illustrate distinct financial philosophies shaped by their size, service mix, and environmental conditions. IID operates on a massive scale with both water and power utilities, enabling highly diversified revenue streams. Its model

TABLE 17.5
Revenues, Expenses, and Changes in the Fund Balance of the Riverside Irrigation District (2023)

Category	Amount (Million USD)
Revenues	
• Acreage/stock assessments	0.04
• Water leases	2.53
• Running fees	0.04
• Interest	0.34
• Farm	0.02
• Water, easements, land and pump rentals	0.01
• Pasture rent	0.03
• Water well pump fees	0.27
• Royalties	0.60
• Miscellaneous	0.15
Total Revenues	**4.03**
Expenses	
• General and administrative	0.24
• Conservation of natural resources	0.86
• Farm operations	0.01
• Capital outlay	0.09
Total Operating Expenses	**1.20**
Other financing uses	
• Debt service	0.00
Total other financing uses	0.00
Change in fund balance	**2.83**
Fund balance, beginning of the year	**24.93**
Fund balance, end of the year	**27.76**

Source: Data are from Riverside Irrigation District (2024).

emphasizes continuous modernization, financed through a combination of current revenues, reserves, and strategic debt issuance. Diversification mitigates the district against sector-specific downturns, whereas its large unrestricted reserves allow flexibility in responding to unexpected needs. However, the complexity of a dual-utility operation results in high ongoing expenses and requires careful cost control to maintain financial stability.

EPCWID#1 represents a mid-sized, water-only operating in a highly arid, drought-prone region where Rio Grande flows can vary significantly from year to year. This situation highlights the importance of financial reserves, cost control, and investment income to stabilize operations during dry years when delivery volumes exist; thus, revenues may fall.

Riverside Irrigation District exemplifies a small, conservative financial model. By keeping expenses low and building substantial cash reserves, nearly seven times annual revenues, it avoids debt entirely and funds all capital works directly from savings. This approach virtually eliminates interest costs and ensures operational security during low-revenue periods. The trade-off is that large-scale modernization may proceed more slowly without the leverage that debt can provide.

These examples show that there is no single "best" approach to irrigation district financing. Effective management depends on aligning the fiscal strategy with mission, scale, and risk tolerance. Large, diversified utilities such as IID can sustain major infrastructure programs through a mix of revenues and borrowing, whereas mid-sized districts such as EPCWID#1 rely on targeted revenue growth and non-operating income to maintain balance. Small districts such as Riverside demonstrate the power of reserve-driven funding and minimal obligations, prioritizing stability over rapid expansion.

REFERENCES

Bureau of Reclamation (1964). Reclamation repayment contracts: A compilation together with explanatory notes on basic features of several types of contracts most frequently entered into. Prepared by the Bureau of Reclamation, Presented by Mr. Hayden. Senate Document No. 92, 88th Congress, Second Session, U.S. Government Printing Office, Washington DC.

Comptroller General of the United States (1957). Audit report to the congress of the United States. Columbia River Basin Water Resources Development Program Corps of Engineers (Civil functions), Department of the Army Bureau of Reclamation and Bonneville Power Administration Department of the Interior.

El Paso County Water Improvement District No. 1 (2025). Financial report with supplementary information for the years ended October 31, 2024 and 2023. El Paso County Water Improvement District No. 1. https://www.epcwid1.org/static/docs/misc/EPCWID1_Audit_2023-2024.pdf

Hill, R.A. (1952). Operation and maintenance of irrigation systems (with discussions by Howard, Flook, Ballaster, Gaels and Hill) Paper No. 2480. *Transactions of ASCE*, 117, 72–88.

IID (2020). 2020 Annual Report Imperial Irrigation District. https://www.iid.com/home/showpublisheddocument/19700/637725750417070000

IID (2025). Imperial Irrigation District Financial Statements and Supplementary Information. https://www.iid.com/about-iid/news-resources/reports

Riverside Irrigation District (2024). The Riverside Irrigation District Audit Financial Statements. December 31, 2023. Littjohann, Kauffman, and Pederson Certified Public Accountants.

Savva, A.P. and Frenken, K. (2002). Financial and Economic Appraisal of Irrigation Projects. Water Resources Development and Management Officers FAO Sub-Regional Office for East and Southern Africa.

18 Rules and Regulations

18.1 INTRODUCTION

There must be a set of rules and regulations for the operation of irrigation systems, which must be brought home to water users. These must be clear, concise, and to the point. These must cover more essential points, keeping in mind that too many minor regulations not only inconvenience the organization and users but also reduce the relative importance of the more essential rules. Rules and regulations must serve water users and management, not vice versa. These must be applied uniformly, regardless of who the individual is, and must ensure that each person (i.e., land owner) receives the service he or she is entitled to.

The rules and regulations can be divided into two general classes: (1) those governing the relationships between the users and the irrigation management or owner and (2) those specifying the duties of the ditch riders. Since farmers and ditch riders are in frequent contact with each other, the latter class of rules must be disseminated to the land owner. The rules can be posted on the irrigation website so that everyone knows about them and can consult them when needed.

Typical regulatory frameworks in irrigation districts may include (1) water conservation policies, which outline goals, programs, and user responsibilities for efficient water use; (2) drought contingency and water allocation policies, which detail procedures for fair water distribution during shortages; and (3) rules and regulations for water delivery, which specify scheduling, measurement, and service protocols. A well-structured rule system ensures equitable water distribution, promotes resource conservation, and strengthens trust between the irrigation authority and its users.

18.2 WATER CONSERVATION POLICY

A water conservation policy is not simply a list of rules but rather a comprehensive plan that reflects the district's operational philosophy, legal obligations, and commitment to sustainability. The purpose of such a policy is to ensure that water, an increasingly scarce and valuable resource, is used in the most efficient and equitable way possible, while maintaining the productivity and economic viability of the agricultural lands it serves. Effective policies go beyond technical guidelines; they foster a culture of stewardship among both district staff and water users. They also provide a framework for decision-making during both normal operations and periods of shortage, ensuring that conservation is an ongoing priority rather than a temporary response to drought.

Water conservation policy should clearly outline the purpose of the plan and the district of authority, i.e., its establishment date and relevant legislation. It should include (1) the purpose of the plan, (2) general information about the irrigation district, (3) service account types, (4) system inventories of the district, (5) water

 DOI: 10.1201/9781003747444-21

conservation goals, (6) program monitoring and records, (7) leak detection and repair policies, and (8) on-farm assistance programs.

18.2.1 Purpose of the Plan

The purpose of the plan is to provide a clear framework for staff and users to manage water resources effectively. This includes strategies for water diversion, delivery, and application, ensuring that these practices are consistent with regional, state, and federal plans and compliant with water codes. The plan should also clarify the legal and operational authority of the district, linking its operations to the foundational laws and regulations. The plan should also define who within the organization has decision-making authority for implementing conservation measures, whether that responsibility lies with the general manager, the operations supervisor, or the elected board. Linking these responsibilities to enable legislation and district bylaws ensures that conservation objectives have both operational and legal standing.

In the water conservation policy of Harlingen Irrigation District Cameron County #1 (2003), the purpose of the plan is clearly stated in the first section:

> The Board of Directors of the Harlingen Irrigation District Cameron County #1 (District) deems it to be in the best interest of the District to adopt a water conservation policy to be the operational policy of the District. This policy is intended to guide the staff and users in all aspects of the diversion, delivery and application of the waters of the District consistent with the approved regional water plan developed for the Lower Rio Grande Region (Region M). This policy constitutes the District's water conservation plan required under Section 11.1271, Texas Water Code, and associated administrative rules of the Texas Natural Resource Conservation Commission (Title 30, Texas Administrative Code, Chapter 288).

18.2.2 General Information about the Irrigation District

A well-prepared conservation policy situates its strategies within the broader context of the district's geography, hydrology, and operational capacity. General information should offer a comprehensive overview of the district, including the total cropland area covered and the extent of lands with water rights. It should specify the district's water sources, whether they are diversions from a river, deliveries from a federal project, extractions from groundwater wells, or a combination of these, and provide historical data on annual diversions, seasonal delivery patterns, and variability in supply. It should also detail how water distribution is managed, especially during times of shortage, and provide information about water supplied for other uses, such as municipal, rural, and domestic needs.

In the water conservation policy of Harlingen Irrigation District Cameron County #1 (2003), the general information of the district is listed in the third section:

> The District serves 38,025 acres of irrigated cropland and has an authorized water right for 39,574 acres, 1,750 acres of which are subject to proration in periods of shortage. This water right yields 98,232.5 acre feet maximum allocation of water usable per annum if storage lakes were at conservation level. The District is the diverter for 22,541 acre feet of domestic/municipal water that is designated for the municipalities served by the

> District. The District has an additional domestic allocation of 4,692 acre feet which is used for supplementing the needs of these municipalities and for the domestic needs of the rural areas within the District. The boundaries of the District consist of 88.3 square miles and include the cities of Harlingen, Palm Valley, Rangerville and part of Primera, Combs and Los Indios.

Maps are an invaluable addition here, showing the district's boundaries, canal network, pumping stations, reservoirs, and service areas. This information helps users understand how water is physically moved and managed, and it provides an essential context for why certain conservation measures are necessary. For example, in districts that rely heavily on open earthen canals, seepage and evaporation losses may be the primary conservation concerns, whereas in pipeline-based systems, operational efficiency and leak detection may be the focus. In irrigation districts in the Lower Rio Grande Valley, GIS is used to map infrastructure, measure seepage losses along canals, and conduct cost–benefit analyses for proposed rehabilitation or modernization projects (Piccinni et al., 2006).

18.2.3 Type of Service Accounts

The irrigation service accounts represent uses or groups of users who have requested water for various purposes. This categorization is important for understanding the demand and distribution dynamics within the irrigation system. Different types of service accounts need to be detailed here. These accounts could be categorized by the nature of the crop (seasonal, perennial), types of areas (greenhouses, golf courses, parks, schools), and purpose of irrigation (recreational, environmental). Further categorization based on irrigation methods (e.g., furrow, flood, drip, sprinkler) is also crucial for addressing the diverse needs of users.

By tracking these accounts, the irrigation district can efficiently manage water allocation, ensuring that the water supply meets the diverse needs of all users. These data are also crucial for planning, managing resources, and implementing conservation measures. Understanding the number and types of accounts helps design services and infrastructure to suit the specific requirements of different user groups within the irrigation system.

In the Harlingen Irrigation District Cameron County #1, account classification reveals a system heavily oriented toward agricultural production, supplemented by smaller but significant nonagricultural uses (Table 18.1). The district manages 3,309 accounts, consisting of 2,779 individual water orders accounting for crop production on an as-needed basis; 30 orchard and commercial nursery accounts using drip or microspray emitter systems; and 500 accounts for nonagricultural uses, such as lawn watering, golf courses, parks, schoolyards, and ponds.

The district delivers water to an average of 34,750 acres annually, supporting a diverse crop mix that creates varying seasonal water demands. The large-acreage crops include sorghum (10,850 acres), cotton (10,000 acres), and sugarcane (7,000 acres), which together dominate the district's irrigated landscape. Smaller but high-value plantings include citrus (600 acres), onion (600 acres), cabbage (400 acres), and nursery plants (400 acres), whereas specialty crops such as peppers, aloe vera, cucumbers, squash, and green onions collectively account for several hundred acres.

TABLE 18.1
Irrigation Service Accounts and Irrigation Methods in Harlingen Irrigation District Cameron County #1 (2003)

Category	Description	Number of Accounts/Extent
	Service Accounts (Total = 3,309)	
Individual water orders	Accounts receiving water based on individual orders, typically on an as-needed basis	2,779
Orchards and commercial nurseries	Annual permits for orchards and nurseries using drip irrigation or microspray emitter systems	30
Lawn watering, golf courses, parks, schoolyards, and ponds	Annual permits for nonagricultural uses such as turf and recreational areas	500
	Irrigation Methods	
Furrow irrigation	Predominant method for row crops	~95% of total areas
Flood irrigation	Used primarily for orchards and pasture areas	Limited
Drip or microspray systems	Installed in a small number of orchards and commercial nurseries	Few installations
Overhead Low Energy Precision Application (LEPA) systems	Applied in specific crop situations where precision application is beneficial	Several fields

This diversity of crops requires flexible delivery scheduling and infrastructure capable of meeting both peak seasonal demands and the more continuous needs of perennial and specialty crops.

Irrigation methods in the district further satisfy service needs and conservation opportunities. Furrow irrigation is the dominant method, accounting for approximately 95% of irrigated acreage, with flood irrigation used primarily for orchards and pasture areas. Drip or microspray systems are found in a small number of orchards and nurseries, whereas low-energy precision application (LEPA) systems are applied in selected crop situations where uniform application and efficiency are priorities. This mix of methods, coupled with the district's diverse cropping patterns, provides clear opportunities for targeted conservation strategies, such as improving furrow irrigation efficiency, expanding drip systems in high-value crops, and modernizing delivery infrastructure to support precision application technologies.

The crop types shown in Figure 18.1 illustrate how water demand in the Harlingen Irrigation District is shaped by both the scale and type of agricultural production. Large-acreage field crops, such as grain sorghum, cotton, and sugarcane, create significant peak-season irrigation demands during the summer months, placing pressure on the district's delivery system and storage resources. In contrast, perennial crops, such as citrus and nursery plants, as well as specialty vegetables, such as onion, cabbage, and pepper, often require more frequent but smaller water applications, particularly during the establishment and harvest periods. This diversity in crop water needs requires flexible delivery schedules, careful canal flow management, and targeted conservation strategies. For example, efficiency gains in furrow irrigation for

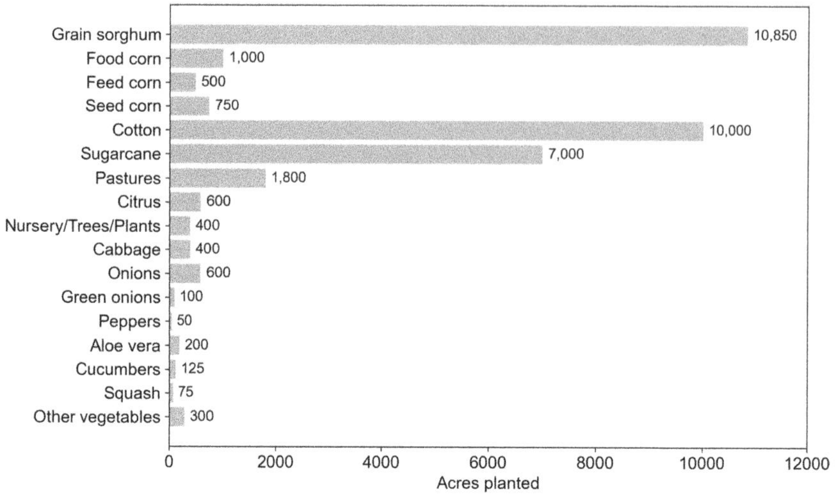

FIGURE 18.1 Crop types and acreage in the Harlingen Irrigation District.

large-acreage crops could significantly reduce total water use, whereas expanding drip or microspray systems for high-value crops could increase water productivity and improve yield quality. Understanding this crop–water relationship is essential for aligning conservation policies with operational realities.

18.2.4 System Inventory of the Irrigation District

The system inventory is a comprehensive record of the physical infrastructure of the irrigation district. This detailed inventory is essential for effective management, maintenance, and planning for future developments or upgrades. The key components to be documented include the following: (1) Pumps: documenting the types, capacities, and locations of all pumps used in the irrigation system. (2) Canals: recording the lengths and sizes of all canals within the district. Information about the lining of these canals, whether concrete or other materials, should also be included, as it affects maintenance needs and water loss rates. (3) Pipelines: detailing the lengths and diameters of pipelines. These data are vital for understanding water flow capacities and for identifying areas that may need upgrades to meet increasing water demands. (4) Reservoirs: documenting all reservoirs, including their locations, capacities, and current usage levels. This information aids in water resources management and planning for periods of high demand or drought. (5) Drainage system: outlining the extent and capacity of the drainage system, including the main drains and subsidiary channels. Understanding the drainage system is key for effective water management and preventing waterlogging or salinity issues.

Table 18.2 shows the system inventory from the Harlingen Irrigation District Cameron County #1 (2003). It is a highly integrated network with diversion, conveyance, storage, and drainage facilities designed to deliver water from the Rio Grande to farms, orchards, nurseries, and municipal landscapes across its service area. The system begins at the river, where six large centrifugal pumps divert up to 470 ft^3/s

TABLE 18.2
System Inventory of the Harlingen Irrigation District Cameron County #1

System Component	Description	Capacity	Dimensions	Power/ Operation	Purpose/ Notes
River pumps	Six centrifugal pumps divert water from the river through a 1,600 ft pilot channel	470 cfs total	Inlet: 41.5–59.5 inch Outlet: 39.5–59.5 inch	300–500 HP Total: 2,200 HP Actual total: 1,768 HP	PROP/ DOPPLER meters
Main canal system	40 miles of earthen canals, mainly constructed in clay soils between 1905 and 1915	N/A	20–45 ft wide, 6–8 ft deep	Gravity flow with check gates	Conveys water to laterals or pipelines
Concrete canal system	20 miles of concrete lined canals, generally elevated above ground level by an average of 6 ft	N/A	10–15 ft wide, 4–6 ft deep	Gravity or relift pumps	Reduce seepage losses
Pipeline system	155 miles of lines extending out from the main canal systems and the concrete canal systems	N/A	12–36 inch in diameters; 14 inch alfalfa type valves with a 3–20 ft head pressure	Relift pumps	Concrete rubber gasket or PVC construction
Relift pumps	55 relift pumps, and most are vertical lift pumps	N/A	8–30 inch pumps powered by electric motors	7.5–75 HP	Gravity feed possible
Reservoir system	Three reservoirs: Rangerville (500 AF), Boggus (160 AF), and McLeod-Hood (280 AF)	940 AF total	Rangerville: 61 ft elevations Boggus: 32 ft elevations	Gravity delivery	Storage, operational flexibility
Drainage system	140 miles of drain ditches to the Arroyo Colorado; 50 miles under Cameron County Drainage District #5	N/A	Earthen ditches	Gravity drainage	Flood/ stormwater removal

(cfs) through a pilot channel into the main canal network. From there, water travels through a combination of earthen canals, concrete-lined laterals, and extensive pipeline systems, with strategically placed relift pumps ensuring reliable pressure and coverage across the district. Three reservoirs provide both storage and operational flexibility, whereas a comprehensive drainage system manages stormwater and excess flows, protecting both infrastructure and farmland.

Each component, from high-capacity river pumps to precision bypasses in the pipeline network, serves an essential function in delivering water efficiently, reducing conveyance losses, and sustaining reliable service during periods of peak demand or drought. Table 18.2 describes these systems in detail, highlighting their physical characteristics, capacities, and operational functions.

18.2.5 Water Conservation Goals

Water conservation goals should be set for the current efficiency of the system, which is important for any irrigation district's long-term resource management strategy. Such goals should be specific, measurable, and aligned with both operational realities and broader legal or environmental obligations. They typically address:

1. **System efficiency targets**: Current and target conveyance efficiency, expressed as a percentage of water delivered compared with the water diverted.
2. **Water duty per acre**: The volume of water diverted or delivered per acre of irrigated land, which can be adjusted to encourage efficient use.
3. **Delivery capacity**: The rate at which water can be supplied per field outlet or turnout.
4. **Infrastructure improvements**: Plans to increase outlet numbers, upgrade conveyance systems, or reduce losses from seepage, leaks, and operational waste.
5. **Measurement and monitoring**: Use of flow meters, telemetry, geographic information systems (GIS), and drainage monitoring programs to identify inefficiencies and prioritize upgrades.
6. **Customer engagement**: Assisting farmers in improving on-farm irrigation practices through training, technology adoption, and data sharing.

Two irrigation districts, Imperial Irrigation District (IID) in California and Harlingen Irrigation District Cameron County #1 (HIDCC1) in Texas, are used to illustrate how conservation goals can differ on the basis of scale, operational context, and legal requirements (Table 18.3). The IID operates at a massive scale, serving approximately 500,000 acres with an annual entitlement of 3.1 million acre-feet (AF) from the Colorado River. Its conservation focus is on maintaining large-scale conveyance reliability, fulfilling binding agreements, such as the Quantification Settlement Agreement (QSA), and implementing major infrastructure and management innovations (Imperial Irrigation District, 2018). In contrast, HID serves a smaller 34,750-acre area with an already high system efficiency of 85%, and its goals are more targeted, such as seeking incremental gains to reach 90% efficiency, reducing water duty from 0.64 to 0.50 AF/acre, and expanding precise metering and drainage monitoring. While IID's approach emphasizes balancing high-volume delivery with legal compliance and broad-scale conservation, HID's strategy centers on localized efficiency improvements, customer engagement, and detailed measurements to optimize every acre-foot delivered.

18.2.6 Monitoring and Records of the Program

Monitoring and record keeping in an irrigation district should use advanced monitoring tools, such as GIS, which can provide detailed mapping and analysis of irrigated

TABLE 18.3
Comparison of Water Conservation Goals in Imperial Irrigation District (California) and Harlingen Irrigation District Cameron County #1 (Texas)

Aspect	Imperial Irrigation District (IID)	Harlingen Irrigation District Cameron County #1 (HIDCC1)
Service scale	~500,000 acres	38,025 acres
Annual water volume	3.1 million AF entitlement from the Colorado River (minus transfer obligations)	~0.64 AF/acre irrigated
Current efficiency	High-volume delivery with extensive automation; efficiency tracked but not expressed as a single %	~85% efficiency
Target efficiency	Maintain high delivery reliability while achieving QSA conservation targets	≥90% efficiency
Primary objectives	Deliver entitlement water economically; meet QSA/transfer obligations; preserve agriculture and the ecosystem; implement major conservation and automation projects	Reduce water duty to 0.50 AF/acre; increase outlet delivery rate to 1 acre/h; targeted upgrades to reduce losses
Measurement practices	Nearly 100% of deliveries metered; SCADA-controlled canal headings; Doppler/propeller meters; CIMIS weather stations; WIS database for quality control	River diversion metering (Doppler and impeller); system- and outlet-level metering; drainage monitoring system
Conservation projects	System Conservation Program: canal seepage recovery, intertie returns, lateral automation, and SCADA upgrades (~103,000 AF/year savings); On-Farm Efficiency Conservation Program: voluntary farmer projects (~200,000 AF/year)	Physical upgrades to pipelines, canals, and outlets; metering expansion; drainage monitoring; training for on-farm efficiency
On-farm support	Incentive payments for water conserved; Tailwater education program; salinity management ("Salty Dawg" assessments)	Technical assistance for efficient irrigation scheduling and system upgrades
Legal/policy drivers	QSA, water transfer contracts, and state/federal water use reporting	District-set operational goals; state reporting requirements
Technology use	SCADA automation, GIS mapping, real-time telemetry, and CIMIS weather integration	Meters, GIS-based mapping, and drainage loss tracking
Conservation focus	Large-scale water delivery optimization with legal compliance and system modernization	Incremental efficiency improvements, precision measurement, and localized loss reduction

Note: WIS refers to an oracle-based system used to collect and process flow data to support water management and conservation; SCADA is a process control software system that collects real-time data from remote field locations, controls equipment and operational conditions, records and log events, and triggers alarms when conditions exceed set safety or operational thresholds (see more details in Chapter 6); CIMIS is the California Irrigation Management Information System.

areas with operational data and enhance the management of water conveyance and drainage systems. High-quality monitoring provides accurate irrigation area measurements, informs fair water fee assessments, and helps identify inefficiencies, water losses, and infrastructure needs. Regular performance evaluations ensure that both operational adjustments and capital investments are data driven, sustainable, and aligned with conservation objectives.

The IID operates one of the most sophisticated monitoring systems in the United States. It integrates SCADA (Supervisory Control and Data Acquisition) for real-time operational control, WIS (an Oracle-based Water Information System) for collecting and processing flow data, and CIMIS (California Irrigation Management Information System) for weather-based irrigation scheduling support. Nearly all deliveries are met with Doppler or propeller meters, and flow data are continuously analyzed to support operations, verify conservation savings, and meet QSA compliance requirements. The GIS layers link canal alignments, delivery points, and on-farm infrastructure, allowing district managers to locate inefficiencies, plan automation upgrades, and optimize water ordering.

The Harlingen Irrigation District Cameron County #1 (HIDCC1) has developed a GIS mapping system integrated with water sales, crop data, user accounts, land ownership, and assessments. This system consolidates operational practices and monitoring tools into a unified platform, enabling the district to assess efficiency in specific system segments, evaluate crop mixes, estimate the number of irrigations per crop, and identify both high-performing and underperforming areas. The GIS platform supports daily decision-making, helps managers prioritize maintenance, address problem areas quickly, and advise farmers on potential risks or opportunities before they make costly crop planning decisions. The HID's metering program, which includes Doppler and impeller meters at all river diversion pumps, combined with the planned expansion of outlet-level metering and a drainage monitoring network, further enhances its ability to fine-tune efficiency and support customers in improving irrigation practices.

IID's monitoring program emphasizes large-scale, automation-driven control, integrating advanced technologies to meet complex regulatory requirements and ambitious conservation targets. In contrast, HID's approach is more localized, prioritizing field-level performance tracking, site-specific efficiency assessments, and direct farmer engagement. Despite these differences in scale and emphasis, both districts share the common goal of maximizing efficiency through integrated data systems, targeted operational interventions, and ongoing performance evaluation.

18.2.7 Leak Detection and Repair Policy

A comprehensive leak detection and repair policy should establish clear procedures for identifying, reporting, and resolving leaks throughout the irrigation system. The key elements should include (1) regular inspections of canals, pipelines, pump sta tions, and structures; (2) use of modern detection technology, such as flow meters, pressure sensors, SCADA alarms, or acoustic leak detection devices, for rapid identification; (3) reporting protocols that allow both district staff and water users to flag suspected leaks promptly; (4) defining repair timelines to minimize water losses and prevent infrastructure damage; (5) record keeping for all leak incidents, including cause, location, repair method, and water loss estimates, to support trend analysis and preventive maintenance; (6) staff training in leak identification, repair techniques, and safety procedures; and (7) resource readiness, ensuring that materials, tools, and personnel are available for rapid response.

Here, the IID and the Harlingen Irrigation District Cameron County #1 (HIDCC1) are used to illustrate how leak detection and repair policies can be implemented in practice. IID integrates leak detection into its larger automation and monitoring framework. SCADA systems continuously track flow rates and pressures across the distribution network, triggering alarms when deviations suggest a leak or structural failure. Data from the Water Information System (WIS) and field inspections are cross-referenced to verify suspected leaks. IID's maintenance crews are equipped for both emergency repairs and scheduled replacements, and leak events are logged for inclusion in annual conservation performance reports required under the QSA.

In contrast, HID operates a low-head system, generally less than 15 ft, where leak detection is the responsibility of canal riders and maintenance personnel. Drains are monitored for unusual flows, and the system is designed with shutoff gates to isolate problem sections, while allowing the rest of the network to remain in operation. District policy calls for leaks to be repaired within 24 hours of detection, and in most cases, repairs are completed within 8 hours of reporting. All river diversions are metered, and a district-wide metering program is being developed to identify distribution losses. HID's GIS platform records leak locations and repair histories, enabling managers to identify recurring problem areas and prioritize targeted infrastructure upgrades (Rio Grande Regional Water Planning Group, 2025).

18.2.8 On-Farm Assistance Program

This section of the plan outlines the potential for an on-farm assistance program aimed at supporting customers in their efforts to conserve water and reduce pollution. The program details the types of services and technologies that the district can offer to its customers. For example, some districts might install specialized outlets that enable consumers to use efficient irrigation systems, such as flexible plastic pipes or gated pipes. These systems are known for their effectiveness in water conservation and can significantly contribute to more sustainable irrigation practices.

Additionally, the district could play a proactive role in educating and informing its customers about the latest advancements in irrigation technology and sustainable farming practices. This could be achieved through the sponsorship of workshops or seminars. These educational initiatives would introduce customers to new technologies and guide them in effectively implementing these solutions in their farming operations. The overall goal of the on-farm assistance program is to foster a collaborative environment where the district and its customers work together toward shared water conservation and pollution reduction objectives, ultimately enhancing the long-term sustainability and resilience of agricultural practices in the region.

18.2.8.1 Example—Imperial Irrigation District (IID)

The IID's On-Farm Efficiency Conservation Program (OFECP) is a large-scale, incentive-based initiative designed to help a district meet its QSA delivery targets, while encouraging growers to adopt proven water-saving practices. The program has evolved through multiple pilot phases, starting in 2008 with irrigation scheduling and event management trials and expanding in 2013 into a fully incentivized program. Participants are paid $285/AF of conserved water, which is verified against historical deliveries for each field and crop. Early solicitations targeted 16,000 AF for conservation in 2013, increasing to 40,000 AF in subsequent years, with caps of up to 4 AF/acre. Eligible practices include precision land levelling, installation of drip or microspray irrigation, tailwater return systems, and conversion from open canals to pipelines. Conservation volumes are independently verified before payments are issued, ensuring accountability and compliance with QSA commitments.

18.2.8.2 Example—Harlingen Irrigation District Cameron County #1 (HIDCC1)

The HID works closely with customers to implement the best available technologies for water conservation and pollution control. In cooperation with the Texas State Board of Soil and Water Conservation Districts and the USDA Natural Resources Conservation Service (NRCS), the HID provides technical assistance and program support for both water conservation and nonpoint source pollution abatement. It actively encourages participation by offering substantial installation support, such as providing and installing pipelines, catch basins, and custom-designed outlet devices at no cost to the customer. These outlet devices are engineered to promote the use of polypipe and gated pipes where applicable, increasing on-farm irrigation efficiency. HID also sponsors workshops and seminars to expose customers to emerging technologies and encourage their adoption in appropriate applications.

HID's program is personalized and infrastructure driven, directly installing water-saving devices and offering hands-on training within a relatively small service area. IID's program is data-driven and incentive-based, operating at a much larger scale and leveraging financial rewards to drive the adoption of efficiency measures. Both approaches demonstrate that on-farm assistance can be tailored to local needs while achieving the shared goals of water conservation and sustainable agricultural production.

18.3 DROUGHT CONTINGENCY AND WATER ALLOCATION POLICY

Drought contingency and water allocation policy are important in the equitable distribution of water resources during times of shortage. The policy provides a structured framework for decision-making, balancing agricultural, municipal, industrial, and environmental needs while preserving the operational integrity of the irrigation system. An effective policy not only establishes rules for allocating water but also promotes transparency, fairness, and user cooperation through clear communication and consistent enforcement. The key components of a comprehensive drought contingency and allocation policy include the following:

1. **User education**: Outreach programs to inform water users about drought conditions, allocation rules, and their roles in water conservation.
2. **Water allocation procedures**: Criteria for initiating and terminating allocations, along with methods for determining allotments among different user categories.
3. **Transfer of allotments**: Rules for transferring water allocations between users or parcels, including limitations, documentation requirements, and accounting for conveyance losses.
4. **Penalties for violations**: Clearly defined consequences for unauthorized actions, tampering with delivery infrastructure, or exceeding allocations.

18.3.1 User Education

User education is the foundation of any effective drought contingency and water allocation policy. By ensuring that stakeholders understand allocation triggers, rules, and operational changes, irrigation districts can cooperate and reduce conflict during periods of shortage. Education programs should be proactive, accessible, and tailored to the district's scale and resources, using a combination of printed materials, digital platforms, in-person meetings, and direct notifications. In practice, education efforts should communicate: (1) the conditions that trigger drought plan activation (e.g., reservoir storage thresholds, as shown in Figure 18.2); (2) allocation procedures,

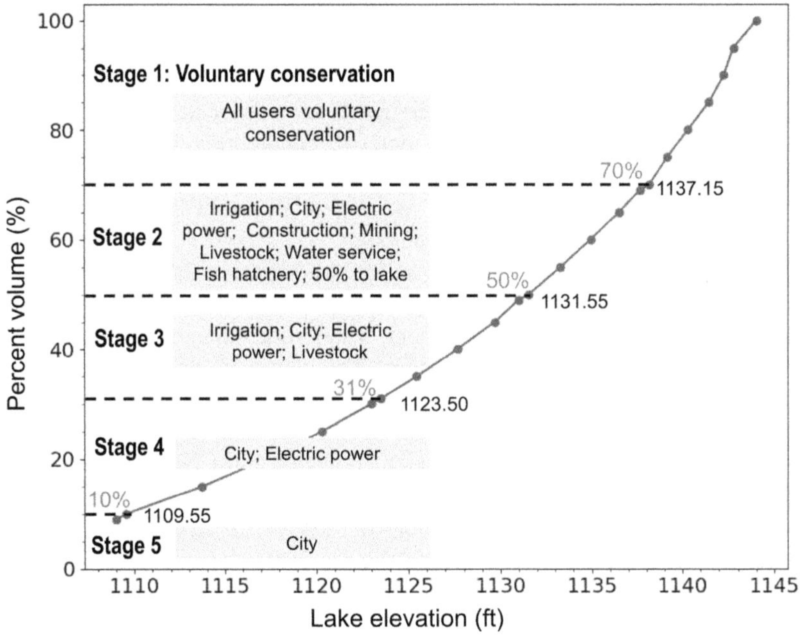

FIGURE 18.2 Drought contingency chart of Lake Kemp in the Wichita County Irrigation district. Note that the progressive stage trigger (55% and 40%) is the required volume where water can be allocated to users when returning to a previous drought contingency state. (Data are from Wichita County Water Improvement District No. 2 (2023).)

including scheduling, priority users, and delivery methods; (3) termination criteria, indicating when normal operations will end; (4) user responsibilities, such as conservation measures or reporting requirements; and (5) penalties for noncompliance.

18.3.1.1 Example—Harlingen Irrigation District Cameron County #1 (HIDCC1)

HID takes a localized, direct approach to user education. The district keeps copies of its drought contingency plan available year-round at the district office and posts it publicly on a bulletin board. When the Board determines that drought conditions warrant implementation, all landowners and water users with current addresses are mailed a notice. These notices summarize the triggering conditions, allocation start or end dates, and relevant rules. By combining the consistent availability of information with targeted outreach during drought events, HID ensures that users can prepare irrigation schedules and cropping plans accordingly.

18.3.2 Water Allocations and Initiation and Termination of Water Allocations

In some irrigation districts, a flat rate per land area is adopted. This approach simplifies the billing process and helps in evenly distributing the fixed costs of maintaining the water distribution systems. However, it also implies that the volume of water allocated should correspond proportionally to the surface areas for which users are paying. This system ensures that all users, regardless of their individual water requirements, contribute equally to the system's maintenance. When determining water allotments, several factors should be considered: (1) storage water availability, (2) distribution efficiency, and (3) total irrigated area. Allocation can be performed on a volumetric basis, if the district has metering capabilities. This approach allows for a more precise and equitable distribution of water on the basis of actual usage. In the absence of metering, allocations are typically made on the basis of the area to be irrigated and the number of irrigation events.

A notable aspect of water allocation policy is addressing the issue of delinquent accounts. Farmers with water rights may sometimes fall behind on their payments. To encourage fairness and prompt payment, water allocation policies may prioritize users who are up to date on their accounts. This approach not only ensures a fair distribution of water but also incentivizes timely payment, which is crucial for the financial stability of the irrigation district.

The initiation and termination of water allocation must be carefully managed. The irrigation manager should oversee the system and recommend when to initiate or terminate water allocations. The factors influencing these decisions include the following: (1) Storage levels: Monitoring the reservoir and storage levels helps in determining when to start and stop water allocation. (2) Water usage per crop and per unit area: Different crops have varying water needs. Understanding these requirements helps optimize water use and distribution. (3) Total irrigated area: Changes in the total area that needs irrigation can influence the timing and volume of water allocation. By carefully monitoring these indicators, the irrigation manager can make informed decisions on when to initiate or terminate water allocations, thereby ensuring optimal use of the available water resources.

18.3.2.1 Example—Harlingen Irrigation District Cameron County #1 (HIDCC1)

The HID's drought allocation policy authorizes the general manager to recommend the initiation of allocations when storage in the district's irrigation water rights account falls to a "one irrigation per acre" level, and the Board determines that supplies are insufficient to complete the crop year. Once approved, the allocation applies to all water users, including individuals, corporations, partnerships, and other entities. Allocations are made pro-rate based on flat-rate acreage, with each acre receiving an allotment equivalent to 0.7 AF, measured at the Rio Grande diversion point. Users with outstanding debt or account balances exceeding one irrigation per acre are ineligible to receive allocations.

To encourage conservation, HID allows metered accounts as an alternative to flat-rate allocation. Users who install and maintain approved meters that are accessible to district personnel can receive credits for documented water savings (which are calculated as metered use plus a 15% delivery loss factor). While metering at every turnout is a long-term goal, the program currently provides a transitional pathway toward volumetric accounting. Violations, such as exceeding allotments or tampering with meters, are subject to district penalties and appealable to the Board.

18.3.3 Transfer of Allotments

A comprehensive water allocation policy should clearly outline the conditions under which water allotments may be transferred between users, as well as any restrictions designed to protect the integrity and efficiency of the distribution system. Key considerations typically include who is authorized to initiate a transfer, whether transfers are permitted only within the district's boundaries, how and when transfers can take place, and how conveyance losses will be accounted for. The Harlingen Irrigation District provides an example of such a policy in action. In the HID, all or part of a water allotment may be transferred from one irrigation account to another within the district but only after a water order has been placed to ensure that the transferred water is tied to an actual irrigation event. Transfers can be initiated only by the landowner or their authorized agent, and all such movements are recorded as irrigation use under district policy. While water transfers to lands outside the district are generally prohibited, the HID allows exceptions in cases where those lands have a historical pattern of receiving out-of-district water through the regular diversion system. Additionally, water originating from outside the district may be brought in for use within HID's boundaries; however, a 15% conveyance loss is deducted from the transferred volume to account for delivery inefficiencies. This framework balances operational flexibility with the need for fairness, transparency, and resource conservation.

18.3.4 Penalties

The plan must specify the penalties for unauthorized actions that threaten the equitable and efficient distribution of water. Such actions may include tampering with, opening, or closing headgates without authorization; diverting water in excess of an allocation; or otherwise interfering with the proper operation of the irrigation system. The clear articulation of infractions and corresponding penalties reinforces the importance of adhering to established water management protocols, deters potential violations, and upholds system reliability, particularly during times of shortage when operational discipline is essential. In the Harlingen Irrigation District, any person who willfully opens, closes, changes, interferes with a headgate, or uses water in violation of the Texas Water Code Section 11.083 may be assessed with an administrative penalty of up to $5,000/day under Section 11.0842. If the violation also involves taking, diverting, or appropriating state water, a civil penalty of up to $5,000/day may be imposed by a court. These penalties, enforceable through legal action in Cameron County courts, may be supplemented by civil remedies pursued by the district, including claims for damage or injunctions to prevent further violations. By combining statutory authority with district-level enforcement, the HID's policy sends a clear message that violations will not be tolerated, ensuring compliance and protecting limited water resources during critical periods.

18.4 RULES AND REGULATIONS FOR WATER DELIVERIES

The rules governing water delivery are the most important ones and must be stated clearly and concisely. These are crucial for facilitating interactions between system management and farmers. The characteristics of rules are influenced by the method of delivery of water and, therefore, vary widely.

The general rules and regulations for water delivery include several key aspects. First, there is a strict implementation of water ticket policies, ensuring that water is allocated exclusively to those with valid water tickets. Second, comprehensive ticket regulations should include detailed provisions about the scheduling of deliveries, resolution of delivery conflicts, procedures for water ticket refunds, and the advance notice required for water delivery. Third, the responsibilities related to water delivery should be clearly defined, with specific officials identified for overseeing the delivery process and issuing water tickets, guaranteeing a systematic, transparent, and well-organized distribution framework.

18.4.1 Control of Canal System

There is a general rule according to which all canals, structures, gates, or other system properties must be under the exclusive control of irrigation system management and its officials, and no one else is permitted to interfere with or change canals or gates. Specifically, only the designated ditch rider is authorized to make changes to gates. These rules are critical for preventing misuse of the canal system. Additionally, the policy clearly outlines what constitutes obstructions to canals and specifies penalties for any violations of these rules, ensuring the integrity and proper functioning of the canal system.

18.4.2 Access to Land

Most irrigation systems should maintain the right to access lands that are being irrigated. This access is essential for various purposes, including inspecting crops, assessing water usage, evaluating the condition of farm ditches, and utilizing adjoining land for emergency repairs to the canal system. Such access rights are crucial for effective system management and maintenance, ensuring the optimal use of water and the overall health of irrigated land.

18.4.3 Rights-of-Way

There must be rules regarding the use of canal rights of way by adjacent landowners. These rules should cover aspects such as the planting of trees, the installation of fences or other obstructions, the protocol for crossing canals, and guidelines for the construction of bridges or similar structures. Proper regulation in these areas is vital to ensure the integrity and safety of the canal system and to maintain clear access for irrigation management and operations.

18.4.4 Maintenance of Laterals

There must be a rule requiring laterals to be in good condition before water can be turned into them. This includes outlining the responsibilities of canal riders, who should supervise the condition of field laterals, including inspecting for weeds or breakages. Such vigilant oversight is crucial to prevent wasteful water delivery. These rules not only ensure the efficiency of the irrigation system but also play a significant role in conserving water resources.

18.4.5 Minimum Period of Delivery

For delivery on demand, there must be a rule governing the limits to both the minimum time of delivery and the minimum quantity to be delivered, particularly for systems operating on a demand basis. This rule varies depending on the type of usage, ensuring that the distribution of water is managed efficiently and aligns with the specific needs of different users.

18.4.6 Measurement of Water

The rules governing the measurement of water must be clearly defined. These rules should specify the location of measurement, the type of measurement, the frequency of measurement, and the type of instrument to be used for measurement. Additionally, they must include guidelines for record keeping and the provision of water delivery information to farmers. This clarity ensures accuracy and transparency in water usage tracking, aiding in efficient water management and equitable distribution.

18.4.7 Interruptions in Service

There must be a rule governing the delivery of water during interruptions, such as canal breaks or other unforeseen issues. This rule should encompass procedures for handling disruptions in laterals, ditches, and main canals, ensuring that there is a clear plan for managing these scenarios and minimizing the impact on water users.

18.4.8 Waste

The rules should clearly outline circumstances under which water service may be refused, such as inadequate preparation for irrigation by farmers or the assessment by ditch riders whose ditches or borders are defective. An important aspect of these rules is the minimization of water waste. This includes provisions to shut off water if it is being wasted and to refuse service to farmers who are found to be wasting water, requiring them to take corrective measures. These guidelines are vital for promoting responsible water usage and ensuring the sustainable operation of irrigation systems.

18.4.9 Complaints

The rules should incorporate a well-defined procedure for addressing complaints. This includes specifying the timeframe within which a complaint must be filed, the duration allowed for resolving the complaint, and the designated office or location where complaints should be lodged. This procedure ensures that grievances are managed efficiently and effectively, maintaining a fair and responsive system for all users.

18.4.10 Liability of Irrigation Management

There must be rules governing the extent of liability or the lack of liability, such as the failure to deliver water due to breaks or a deficient water supply, or damage caused by water after its delivery from the canal.

18.4.11 Liability of Irrigator

The irrigator must be held responsible for any acts that may result in injury to the irrigation system. This includes defining the scope of liability for such incidents. Ensuring that irrigators are aware of and accountable for their impact on the system is essential for its preservation and efficient functioning.

18.4.12 Penalty for Breaking Rules

The enforcement of rules is assured, provided that there are strict penalties for the violation of rules. These penalties should be clearly defined and may include measures, such as imposing additional charges or restricting water access. This approach ensures adherence to the established guidelines and maintains the integrity of the irrigation system.

18.4.13 Miscellaneous Rules for Water Deliveries

The rules should also include provisions to address local conditions and specific scenarios. These may encompass guidelines for managing local emergencies, preventing the entry of polluted water into the canal system, and other relevant situations. The inclusion of such rules ensures the system's adaptability to unique regional requirements and challenges.

18.4.14 Case Study of Harlingen Irrigation District Cameron County #1 (HIDCC1)

HID's delivery rules are built on a strict water ticket system: only bonafide ticket holders may receive water, and tickets must specify a delivery date or be marked "Will Call". Canal riders, the district staff responsible for deliveries, coordinate schedules, resolve conflicts, and may require night irrigation to meet demand. If a user is unprepared or wastes water through poor field preparation, service may be withheld until corrective action is taken. HID also issues credit for unused water, prohibits unauthorized operation of gates, and requires that the field laterals be in good repair before delivery. The district allows several types of permits (annual, temporary, or domestic use) but stipulates that all deliveries occur under canal rider supervision. By integrating order flexibility with strict oversight, HID maintains fairness and efficiency across its diverse service area.

18.5 TERMS OF PAYMENT

In an irrigation project, the terms of payment define how and when users are expected to pay for the water services provided. These terms must be clear, equitable, and enforceable, ensuring that the district can sustain its operations while remaining fair to water users of varying sizes and capacities. A well-structured payment policy should include the following:

1. **Billing schedule**: Clearly defined dates for issuing bills and deadlines for payment, e.g., monthly, quarterly, or annual billing cycles.
2. **Payment methods**: Accepted payment channels, such as direct bank transfers, checks, credit card payments, or in-person cash transactions. Increasingly, districts are adopting online payment portals for convenience.
3. **Advance payments or deposits**: Provisions requiring advance payments for seasonal water allocations or deposits for special orders, which help secure financial stability for the district.
4. **Late payment penalties**: Clearly, specified penalties for overdue accounts, such as interest charges, reconnection fees, or suspension of water delivery until the account is settled.
5. **Incentives for early payment**: Discounts or credits for customers who pay before the due date, encouraging timely settlement.
6. **Special provisions**: Flexible payment plans for small-scale farmers or those facing verified financial hardship, helping maintain equitable access to water resources while preventing service interruptions.

18.6 MISCELLANEOUS RULES

In addition to the core rules and regulations, an irrigation project may require a set of miscellaneous rules to address specific situations, operational contingencies, and unique local requirements that do not fit neatly into the main operational rules.

Although these rules are less frequently applied, they are essential for ensuring that the district can respond effectively to unusual or emergency circumstances. Typical areas covered include the following:

1. **Emergency procedures**: Rules for handling canal breaks, pump failures, drought-related supply reductions, or other operational emergencies. These provisions often authorize district staff to suspend deliveries, reallocate water, or close specific system components without prior notice to protect infrastructure and maintain service to priority users.
2. **Environmental protection**: Measures to prevent pollution from entering the canal system include prohibiting the discharge of chemicals, debris, or contaminated drainage water into district facilities.
3. **Shared resource use**: Rules for shared facilities, e.g., joint-use reservoirs or canals, to ensure fair cost-sharing and coordinated operations with neighboring districts.
4. **Temporary special orders**: Provisions for one-time or nonroutine water delivery, such as construction water, frost protection water, or supplemental deliveries outside normal irrigation schedules.
5. **Amendment procedures**: Guidelines for how and when the rules can be revised, typically requiring approval from the board of directors after public notice.
6. **Public safety provisions**: Restrictions on swimming, boating, or trespassing in canals and reservoirs, as well as liability disclaimers for accidents within district property.

While these miscellaneous rules may seem secondary to core delivery and allocation policies, they play a critical role in ensuring that the irrigation district can operate smoothly under unexpected circumstances and adapt to evolving needs. By addressing emergencies, environmental protection, shared resource management, and public safety, these provisions provide the flexibility needed to safeguard both water resources and infrastructure. Together with the primary water conservation, allocation, and delivery rules discussed in this chapter, they form a unified regulatory framework that supports reliable operations, fair distribution, and the long-term sustainability of irrigation district systems.

REFERENCES

Harlingen Irrigation District Cameron County 1 (2003). Documents for Water Diversions and Deliveries. Water Conservation Policy. September 1, 1999. Amended May 19, 2003. https://share.google/i7JKQXP3qwpmsqhvb

Imperial Irrigation District (2018). 2016 Water Conservation Plan. Imperial Irrigation District Water Resources Section.

Piccinni, G., Supercinski, D., Leskovar, D., Harris, B., and Jones, C.A. (2006). Rio Grande Basin water conservation project. In: G. Lorenzini, I and C.A. Brebbia, Eds., *Sustainable Irrigation Management, Technologies and Policies*. WIT Transactions on Ecology and the Environment, WIT Press, Southampton, pp. 321–330.

Rio Grande Regional Water Planning Group (2025). 2026 Rio Grande Regional Water Plan Initially Prepared Plan. Rio Grande Regional Water Planning Group with administration by Lower Rio Grande Valley Development Council. https://share.google/sokTFJXzS2plIc3Ba

Wichita County Water Improvement District No. 2 (2023). Drought Contingency Chart. https://wichitacountyirrigation.com/drought-contingency-chart/

19 Laws and Legal Issues

19.1 INTRODUCTION

The management personnel of irrigation enterprises should be well acquainted with the laws of the political subdivision within which the operation of the organization is carried out. These laws include those governing the formation and operation of a particular organization, which applies to the enterprise, water laws, drainage laws, and others that affect the operations of the organization. Legal counsel is always advisable and, in many instances, necessary. It is not enough to have knowledge of the current laws because laws are not static and are subject to change. Noteworthy examples include (1) the development of groundwater laws by statute and court rulings, (2) laws and regulatory requirements related to the application of pesticides and herbicides, and (3) laws and regulations to control water and air pollution.

In addition, measures to conserve all natural resources are gaining momentum (conservation measures). Some have reached legal status, and more laws are probably needed in this regard. Another concern that has already affected the development of some water resources, including irrigation, is the increased attention given to aesthetics. With increasing thought being given toward making improvements, particularly in construction, more pleasing to the eye and more tolerable to human desires, it is not beyond the realm of probability that laws and regulations in this regard will vitally affect the operation of irrigation enterprises in the future.

The type of organization used for the operation and maintenance of irrigation varies considerably. The organization chosen to carry out these functions is often determined by state laws. These laws usually permit different types of organizations, such as irrigation districts, conservancy districts, and water districts. The main point is to be familiar with the laws of the state in this regard and to select the permissible type that appears to permit doing the job anticipated. The organization must fit the operation. Once the type of organization has been selected and duly constituted, then the legal framework for this particular organization must be understood and strictly followed.

The organization may be (1) an irrigation district; (2) a water user association; (3) a water, canal, ditch, or reservoir company; (4) a reclamation district; or (5) a conservancy district, or another, as permitted by the laws of the state. All are organized under the laws of the State in which they operate. These laws set forth (specifically or by guidelines): (1) How the organization may be formed (petitions, elections, etc.); (2) within limits what may and may not be done (sphere of activity); (3) who can hold office (applies to directors or the governing body); (4) how finances are to be handled (collections, disbursements, reserves); (5) water rights (how held, total rights, supplemental water rights); and (6) a host of others.

The main point to be made is that management must be thoroughly familiar with the law or laws under which the organization was constituted. Legal counsel is normally advised. This then gives the word on what can and cannot be done under the

DOI: 10.1201/9781003747444-22

law. Bylaws for the organization should be drawn up, approved, and followed. These should be revised periodically. Legal counsel is advisable for these actions.

If the functions of the organization cannot be carried out efficiently and satisfactorily under the type of organization existing, then reorganization may be advisable. This would embrace going through the entire legal procedure of forming a new organization and disbanding the old. There are no general guidelines for doing this, except that it will depend on state laws, and they must be understood. Again, legal counsel is usually necessary.

19.2 AUTHORITY OVER WATER RESOURCES

The management of water resources in the United States involves a complex interplay of federal and state laws, policies, and jurisdictions. The federal government, as outlined in key legislative acts, has primacy in several areas concerning water resources. These areas include navigation, international treaty negotiations, federal water development projects, and water uses associated with federal lands and property. Furthermore, the federal government plays a crucial role in regulating pollution and protecting natural resources (Christian-Smith et al., 2012).

Water rights in the United States are grounded primarily in state constitutions and statutes. These laws are adjudicated by state courts and administered by state agencies (Goldfarb, 1988). This structure reflects the recognition that water resources, and the needs they fulfill, can vary greatly from state to state. States tend to have considerable authority in determining how surface water and groundwater are allocated and managed. This power extends to the creation and enforcement of legal frameworks that govern water rights and usage. These frameworks must balance various factors, such as environmental conservation, public needs, agricultural demands, and industrial usage.

The legal frameworks for water allocation in the United States are diverse, reflecting the unique histories, cultures, economies, and physical characteristics of different regions. For example, the water law in arid West China is markedly different from that in East China. This diversity is a response to the varying climatic and geographical conditions across the country, necessitating distinct approaches to water management. States have developed specific structures for both surface water and groundwater resources. These structures may differ in terms of the principles, the regulatory mechanisms they employ, and the degree of oversight exerted by state agencies. The following sections introduce the typical frameworks used by states for allocating both surface water and groundwater resources.

19.3 SURFACE WATER LAWS

In areas where water is plentiful, there is not much need for laws and regulations to determine who has the right to use the abundant supply, which is adequate for most normal uses. However, in regions where the water supply is inadequate to satisfy the needs of all potential user, water becomes a commodity of considerable value to the livelihood of individuals and the economy of the area. Under these conditions, a system of laws must be developed to determine who has the right to use water when there are shortages. Such systems have been developed in the United States. and, to some extent, throughout the world. These laws are, for the most part, state laws, and they are not static but are changed to fit new conditions as they develop.

In general, water rights (rights to use water) must be firmly established during the irrigation planning stage. Otherwise, it is possible to end with a large amount of very costly construction and no legal right to use the water necessary for the success of the project. In project planning, the vested rights wing should not be taken away without the due process of law being considered. Changes are always possible, but the probability of change, which affects the firm's water supply for the project, must be remote to ensure that the project has the water needed for successful operation.

The source of water for an irrigation enterprise may be surface water, groundwater, or both. Throughout the world, laws pertaining to surface waters have developed to a greater degree than those for groundwater. The United States is no exception.

A brief review of laws in general may be helpful for better understanding water laws. The function of law is to regulate the relations between people or groups of people: (1) The law serves a dual purpose: (a) It provides a mechanism for the solution of conflicts after they have arisen, and (b) it furnishes a guide for keeping future conduct in order. (2) The law at any particular time and place represents the will of the majority for encouraging action deemed desirable and for discouraging or forbidding action considered to be in conflict with some broad public interest.

The law is very complex; the more we know about it, the less apt we are to be in trouble. We know the law requires (1) that we pay our bills when due and (2) that we drive our car within legal limits. These become habits because we know the laws governing these actions. We need to know water law (1) to avoid risks of legal liability, (2) to avoid risks of lawsuits, and (3) to ensure that the project will be within the law and is not lost.

Water law, as treated here, is concerned with the rights of individuals to use surface or underground water. Other water problems, such as floods or unwanted water, which might be included in the broad definition of water law, will not be covered. The water law can be considered a branch of the property law. Water law governs individual rights to use and enjoy property. However, water, unlike land, does not remain stationary. It is in motion; it ignores land property lines.

A water right is a special type of property. The actual water flowing in a stream cannot be owned. However, the right to use water is subject to private ownership. This right to use water is nonetheless a property right. In some states, it is theoretically possible to own underground water while it is in place. This doctrine of absolute ownership of in situ underground water merely makes it easier to recognize a water right as a property right.

Water laws are, for the most part, state laws. Therefore, the law varies from state to state, but principles are often similar or may be grouped into categories on the basis of similar characteristics. The two doctrines generally governing water rights are (1) the riparian doctrine and (2) the appropriation doctrine.

19.3.1 Riparian Doctrine

A riparian water right is based on the ownership of land next to water, is independent of the use or nonuse of the water, and is correlative with other riparian users from the same source if the use is consumptive, regardless of which user first used the water. In all 50 states of the United States, water law is based on either the riparian principle or the appropriation principle in terms of private water rights. In general, the appropriation doctrine seems to be gradually taking over, either through legislation or through court action.

The riparian principle bases the right to use water, which is a real property right for most purposes, on the ownership of land next to, or contiguous to, surface water. Under the common law of riparian doctrine, the right to use water is inherently tied to the land itself. The right is not gained through use, nor is it lost or diminished through nonuse.

In humid eastern states, water rights are governed by the riparian doctrine. Under the common law riparian doctrine, an owner of land contiguous to or abutting a stream has the right to use water. The riparian right is attached to the riparian land and is considered a part and parcel of the land. The riparian owner has the right to require upstream riparian users to allow the stream to flow by or through their land in its natural state. The riparian owner has the duty to allow the stream to leave his land in its natural state. In other words, in its pure form, the riparian doctrine does not permit the consumptive use of water, except for minor amounts, to satisfy domestic and stock watering needs. The riparian owner is not supposed to disturb the stream, even though he may use the water (or water motion), while the water is flowing by or through his land. For instance, the first judicial articulation of the riparian rights doctrine in the United States occurred in the 1795 case *Merritt v. Parker*, in which the New Jersey Supreme Court stated that "In general, it may be observed, when a man purchases a piece of land, through which a natural water-course flows, he has a right to make use of it in its natural state, but not to stop or divert it to the prejudice of another...It cannot legally be diverted from its course without the consent of all who have an interest in it..." (Dellapenna, 2011). The riparian doctrine is suitable for humid areas where irrigation, which consumes a large amount of water, is not necessary.

As soon as irrigation was necessary in a riparian doctrine state, the courts modified the strict riparian doctrine to allow the owner to irrigate his riparian land as long as such irrigation is reasonable in relation to other riparian owners on the same stream. Riparian land is land contiguous to or abutting a watercourse. Typically, only lands within the watershed of the stream can be irrigated under the reasonable use doctrine. The reasonable use doctrine gives a corrective right—a share in the wealth system. In the case of *Scott v. Slaughter* in 1963, the Arkansas Supreme Court articulated the rationale for rejecting rigid doctrines, such as the natural flow rule. It embraces reasonable use theory to "recognize that there is no sound reason for maintaining our lakes and streams at a normal level when the water can be beneficially used without causing unreasonable damage to other riparian owners" (Federal Judicial Center, 2025). Regulated riparianism emerged later to address increasingly competitive water needs among formerly compatible users due to temporal water shortages. Regulated riparianism involves managing and adjusting water use through a collective decision-making process administered by a state agency that issues time-limited licenses on the basis of the reasonableness of the proposed use. This framework enables the state to evaluate both the potential societal benefits and the compatibility of the new use with existing uses. A model law for these statutes is *the Regulated Riparian Model Water Code of the American Society of Civil Engineers* (Dellapenna, 2011).

19.3.2 Appropriation Doctrine

Under the appropriation doctrine, a water right is based upon the beneficial use of water. The first to appropriate and use water has a right to the rights of later appropriators. An appropriation right is independent of the location of the land with respect to surface water, is based exclusively on the beneficial use of water, and recognizes priorities among users according to the first-in-time, first-in-right principle. In the 1882 *Coffin v. Left Hand Ditch Co.* case, the Colorado Supreme Court rejected riparianism and clearly affirmed that water rights in Colorado are governed by the principle of prior appropriation, where the first user to divert water for a beneficial use gains priority over later users, regardless of land adjacency (Casebriefs, 2025). This shift recognized the practical necessity of prioritizing water allocation in the arid West of the United States and was later reinforced by Colorado's constitutional provisions (Aspen Public Radio, 2024).

The appropriation doctrine is well-suited to areas that require the consumptive use of water. Consumptive use refers to withdrawals in which a significantly larger amount of water is taken from the stream system than is returned after use. Many water users do not consume water to any extent. For example, for centuries, the major use of flowing stream water in humid areas was to furnish power for mills. Such a use consumes almost no water. On the other hand, any agricultural use consumes a substantial amount of water. In humid areas, irrigation is not necessary. However, in arid and semi-arid regions, irrigation is a way of life. Irrigation is necessary only where there is a shortage of rainfall, which usually results in a shortage of surface water. The appropriation doctrine is a highly practical way of appropriating waters in water shortage areas, since it is founded on the first-in-time, first-in-right principle. Under the appropriation doctrine, a water right consists of an amount of water and a priority date. The owner of an appropriation right is entitled to use a stated amount of water even if such a use means that other users with later priority dates are deprived of water.

Today, most western states operate under a permit-based appropriation system, where water rights are allocated according to the "first-in-time, first-in-right" rule (Hendrick, 2013). This system requires users to obtain permits specifying the amount, location, and type of use, ensuring that new appropriations do not infringe on senior rights.

19.3.3 Legal Framework in Different States

The majority of western states currently follow the prior appropriation doctrine for water allocation, with some states still using riparian rights from before their transition to this system (Figure 19.1). In the western region, California, Nebraska, and Oklahoma permit riparian landowners to establish new water uses that may supersede appropriative rights under certain conditions. In eastern states, riparian rights

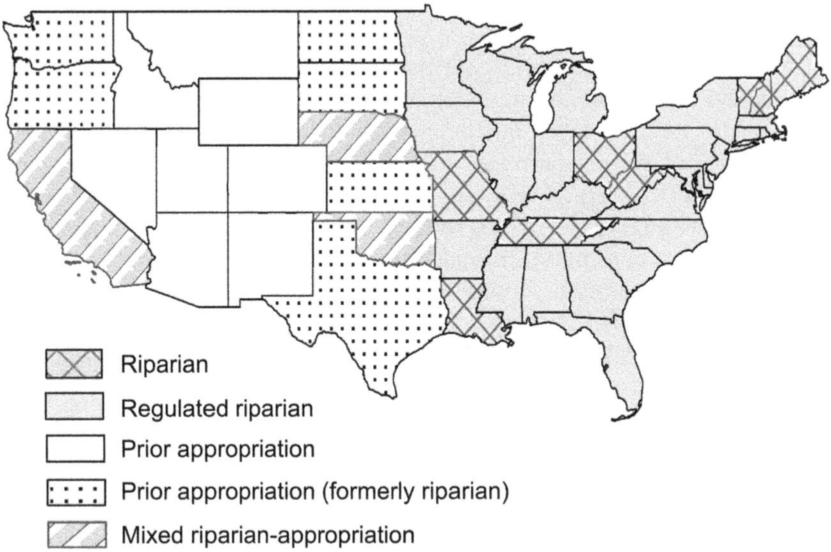

FIGURE 19.1 Water rights doctrines in different states.

prevail (Figure 19.1). However, approximately half of these states have implemented water-use permitting systems, known as regulated riparianism. This approach allows those who do not own riparian land to obtain water rights for reasonable use (Beck and Kelley, 2009). These systems follow the essence of riparian principles but allow for water usage on nonriparian land, issue permits for limited durations, and include conditions to safeguard public values (Beck and Kelley, 2009).

19.3.4 Diffused Water

Water that does not follow a given water course but flows across the surface of land in a variant and unpatterned way is termed diffused water. This water includes rain or snow runoff. Diffused water is the property of the landowner until it enters a natural water course, at which point it becomes a property of the state, as illustrated in Texas water law (TGPC, 2024). Importantly, the "stock tank exception" does not apply to diffuse surface water. This means that a landowner may capture rainwater on their property—whether by harvesting it into the soil or collecting it in storage facilities—before it reaches a natural watercourse.

19.3.5 Rights Based on Need

Private water rights are based on the riparian principle of land location or on the appropriation principle of prior beneficial use. A third principle bases water rights on need. Indian water rights, federal reserved rights, and, to some extent, municipal water rights are based on need. The various aspects of the need principles are beyond the scope of this discussion.

19.4 GROUNDWATER LAWS

Similar to those for surface water, two basic doctrines or legal theories apply to the ownership and use of groundwater. The first of these two doctrines is that of riparian rights, also referred to as the common law doctrine of ownership. The other doctrine is that of prior appropriation. When legal rights and regulations related to groundwater or "underground water" are considered, a distinction is first made between the waters of "underground streams" and those classified as "percolating". Although the waters constituting an underground stream may vary slightly from state to state, either by statutory definition or court interpretation, it usually includes all those waters in a defined channel underlying a flowing stream directly related to and contributing to the surface flow. Because of this direct relationship with surface water, it is understandable that the legal principles and regulations governing underground streams are usually the same as those applicable to surface streams within a particular state.

Percolating waters, by definition, are those waters that pass through the ground under the surface of the earth without any definite channel and are not part of any surface or underground stream. Although artesian waters fall within this definition, some states have separate principles and laws governing their ownership and use. Groundwater fills the pore spaces in rock and soil and supplies wells and springs, with groundwater in underground streams generally regulated similarly to that in surface streams.

The doctrine of absolute ownership is still prevalent in states such as Indiana, Maine, and Texas, which grants landowners the unrestricted right to extract groundwater. This means that landowners in these states can pump groundwater without legal obligations to avoid harming existing uses. In Texas, for example, groundwater is legally subject to the rule of capture. This rule allows landowners to extract all the water beneath their property without owning the water itself and irrespective of the impact on neighboring wells. Such a foundational case in Texas water law is the case of *Houston & Texas Central R.R. Co. v. East, which was* decided by the Texas Supreme Court in 1904 (Texas Supreme Court, 1904). However, there are exceptions to this rule, including special legislation, such as the Edwards aquifer, underground water conservation districts, subsidence exceptions, and restrictions on "malicious pumping" (Kaiser, 1999). For example, owing to severe conflicts over water use, the Edwards Aquifer Authority Act (1993) created a special regulatory body for the Edwards Aquifer, one of Texas's most important groundwater sources. The act established a permitting system to balance pumping with aquifer sustainability, a departure from the pure rule of capture. Overpumping in regions, such as the Ogallala Aquifer (Texas High Plains), has led to severe water table declines. Tension persists between landowners' rights under the rule of capture and the need for long-term aquifer conservation. To address these challenges, the state created groundwater conservation districts (GCDs) to regulate pumping and well spacing, and the Edwards Aquifer Authority Act (1993) introduced a permit system to protect one of Texas's most stressed aquifers. This case illustrates the tension between private rights and sustainable water management.

In arid regions, court decisions regarding groundwater have been shifting toward a doctrine of reasonable use. Under this doctrine, the overlying landowner retains his or her rights to water under his or her property, but they are not permitted to use more than they truly need or to export water to points distant from the source. The California doctrine of correlative rights goes even further in stating not only that the use of water must be reasonable but also that the priorities of all landowners are equal, and that if the supply is not sufficient for all demands, each owner is entitled to no more than an equitable portion of the available water. For example, in the case of the *City of Pasadena v. City of Alhambra* in 1949, the California Supreme Court applied correlative rights principles in the context of overdrafted basins, limiting extractions to safe yields and proportionally allocating water on the basis of historical use (California Supreme Court, 1949). The reasonable use doctrine, which is more common in Eastern states, is akin to absolute ownership but prohibits wasteful practices and requires water to be used on overlying land, unless it can be transported without harming other landowners. Despite its name, this doctrine imposes minimal restrictions on groundwater use on overlying land. Neither absolute ownership nor reasonable use doctrines address the overall demand for aquifers or offer solutions for harm caused by excessive groundwater extraction. In California's Central Valley, decades of largely unregulated groundwater pumping under these doctrines have contributed to severe land subsidence, with up to 80% of serious subsidence incidents in the United States linked to excessive groundwater withdrawal (USGS, 2000).

The water law with respect to groundwater is notably less advanced than that for surface water. This was primarily due to limited understanding of groundwater flow processes, insufficient information about the physical characteristics of groundwater basins, and the relatively modest use of groundwater during the nineteenth century. Many states have enacted new groundwater laws, and others are actively revising them.

19.5 DRAINAGE LAWS

Drainage issues are governed by two primary legal principles, with variations across states due to local practices or interpretations. In important cases, seeking legal advice is always recommended. Some states follow Roman civil law, which specifies that the owners of high land (dominant owners) are entitled to the advantage that this elevation gives them and may discharge their drainage water into lower land through natural depressions and channels without obstruction by the lower (or severe) owner. In the case of Pendergrast v. Aiken in 1977, the North Carolina Supreme Court affirmed that the upper landowner (Pendergrast) holds an easement/servitude over the lower land (*Aiken*), allowing natural surface water to pass without obstruction or interruption; *Aiken* must accept this natural drainage unimpeded (North Carolina Supreme Court, 1977). A dominant owner may accelerate the flow of surface water by constructing ditches or by improving natural channels on his or her property and may install tile drains. He or she may not carry water across a drainage divide and discharge it on land, which would not have received the water naturally, nor may he or she locate the outlet of the area. A service owner can do nothing to prevent natural drainage from entering his or her property from above. Statute law in many states permits the dominant owner to construct drains on the land of a servient owner after a simple eminent-domain proceeding and payment of all costs and damage. Many states also modify the rule of law for cities by relieving the service owner of many restrictions. Otherwise, many lawsuits might develop as a result of grading and developing city lots.

Common English law employs the common-enemy rule. The basic principle here is that water is a common enemy of all, and any landowners may protect themselves from water flowing onto their land from a relatively high elevation. Under this rule, the dominant landowner cannot construct drainage works that results in damage to the property of a service owner without first securing ease. The service owners are allowed to construct dikes or other works to prevent the flow of surface material onto their property. For example, in the 1893 case *Edwards v. Charlotte, Columbia & Augusta R.R.*, the South Carolina Supreme Court held that surface water is a "common enemy", granting each landowner the right to take any measure necessary to protect their property—even if water is pushed onto a neighbor's land (South Carolina Supreme Court, 1893).

Both doctrines of drainage law place the responsibility for damage to any person or organization, altering the natural stream pattern of an area or creating an obstacle that blocks the flow of a natural stream. Common law confers no rights to the control of navigable streams under state jurisdiction except by the construction of levees to keep the stream from overflowing one's land.

Case Study: The San Luis Drainage Problem and Resolution in California's Central Valley

The San Luis Unit of California's Central Valley Project was designed to supply irrigation water but lacked a complete drainage system. By the 1980s, excess irrigation water carrying salts and selenium had been collected in the Kesterson Reservoir. The increase in toxic drainage caused widespread deformities and deaths among migratory birds, sparking lawsuits, closure of the reservoir, and stricter enforcement of the Clean Water Act and California's Porter–Cologne Water Quality Control Act. This crisis underscored the environmental risks of inadequate drainage management.

For decades, the problem has remained unresolved, with farmers, water districts, and federal agencies locked in conflict over responsibility. In 2017, the San Luis Unit Drainage Resolution Act (2017) was introduced by Representative Jim Costa. The act authorized a settlement between the federal government and the Westlands Water District, shifting drainage obligations from federal to local control. Under this agreement, the Westlands assumed responsibility for drainage obligations, committed to more efficient irrigation practices and groundwater reuse, and became subject to strict federal and state environmental laws. Importantly, the act relieved taxpayers of an estimated $3.8 billion federal liability, while ensuring that San Joaquin Valley farmland remained productive.

REFERENCES

Aspen Public Radio (2024). A Water War over Who Gets What and When Is Now Managed with a Smartphone. https://www.aspenpublicradio.org/2024-07-29/a-water-war-over-who-gets-what-and-when-is-now-managed-with-a-smartphone

Beck, R. and Kelley, A., Eds. (2009). *Waters and Water Rights*. Lexis-Nexis, New Providence, NJ.

California Supreme Court (1949). *City of Pasadena v. City of Alhambra*, 33 Cal.2d 908, 207 P.2d 17. https://cite.case.law/cal-2d/33/908/

Casebriefs (2025). Case *Coffin v. Left Hand Ditch Co*. https://www.casebriefs.com/blog/law/property/property-law-keyed-to-cribbet/interests-in-land-of-another-and-in-natural-resources-affecting-anothers-land/coffin-v-left-hand-ditch-co/

Christian-Smith, J., Gleick, P.H., Cooley, H., Allen, L., Vanderwarker, A., and Berry, K.A. (2012). *A Twenty-First Century U.S. Water Policy*. Oxford University Press, Oxford.

Dellapenna, J.W. (2011). The evolution of riparianism in the United States. *95 Marq. L. Rev.*, 53, 57; Merritt v. Parker, 1 N.J.L. 526 (1795).

Edwards Aquifer Authority Act (1993). Edwards Aquifer Authority. https://www.edwardsaquifer.org/wp-content/uploads/2024/07/EAA-Act.pdf

Federal Judicial Center (2025). Water and Law Sidebar: Books – Reasonable Use and Beneficial Use. https://www.fjc.gov/content/376829/water-and-law-sidebar-books-reasonable-use-and-beneficial-use

Goldfarb, W. (1988). *Water Law*. Lewis Publishers, Chelsea, MI.

Hendrick, S. (2013). State Water Withdrawal Regulations. National Conference of State Legislatures. Archived from the original on 7 Mar 2016.

Kaiser, R. (1999). A Primer on Texas Water Law. Presented at the Water Conflicts in an Era of Regional Planning. *Water Issues Management. Identifying and Facilitating Water Disputes. Presidential Conference Center*, April 14–16, 1999, Texas Agricultural Extension Service and The Bush School of Government of Public Service's Center for Public Leadership Studies Program in Conflict and Dispute Resolution, College Station, TX, 22 p.

San Luis Unit Drainage Resolution Act (2017). San Luis Unit Drainage Resolution Act. H.R. 1769, 115th Congress. H. Rept. pp. 115–349. https://www.congress.gov/bill/115th-congress/house-bill/1769/titles

South Carolina Supreme Court (1893). *Edwards v.* Charlotte, Columbia & Augusta R.R., 39 S.C. 472, 18 S.E. 58 (S.C. 1893). https://www.masc.sc/sites/default/files/uploads/affiliated-associations/for-web-common-enemy.pdf

Texas Supreme Court (1904). Houston & Texas Central R.R. Co. v. East, 98 Tex. 146, 81 S.W. 279. https://cite.case.law/tex/98/146/

TGPC (Texas Groundwater Protection Committee) (2024). Texas Law–FAQs. https://tgpc.texas.gov/POE/FAQs/WaterOwnership_FAQ.pdf?utm_source

USGS (2000). Land Subsidence in the United States. USGS Fact Sheet-165-00. https://water.usgs.gov/ogw/pubs/fs00165/

20 Safety

20.1 INTRODUCTION

The growth and advances in irrigation have contributed greatly to the prosperity of semi-arid and arid areas around the globe, but they have also led to increased hazards, such as environmental degradation, trapping of livestock and wild animals in lined canals, drowning incidents involving the public, capsizing of boats, vehicle rollovers on canal banks, and other operational accidents. Without adequate safety measures, irrigation systems can lead to loss of life, property damage, production delays, and increased operational costs. Safety practices lessen their occurrence, and these practices must be upgraded to keep pace with the innovations and changes taking place in industry, construction, transportation, etc. There must be the best unified efforts on the part of operating and management personnel, and all others concerned must be utilized to help eliminate accidents around irrigation facilities and to protect the general public, the operation and maintenance personnel employed in the project, and the domestic livestock and wild animals in the area. There is much to be learnt concerning the value of safety and accident prevention practices in industry, construction, highways, and other activities in general.

Management personnel in irrigation enterprises should critically examine the value of safety programs in place and the benefits that accrue from them. A healthy safety program should constantly seek ways to improve. It must be emphasized that in older projects, where safety devices and other measures were not included in the original design and construction, they should be included. Fortunately, advancements over the years in irrigation technology and design practices have enabled us to include a variety of safety features in newer projects, such as automated gates, remote monitoring, and real-time hazard alerts. These innovations not only enhance the safety of systems but also reduce the workload and responsibilities of the operational personnel, contributing to more efficient and secure irrigation management.

Achieving and sustaining a safe irrigation environment requires coordinated efforts among operators, management personnel, regulatory bodies, and local communities. For purposes of discussion, the subject of safety is categorized into four main areas: (1) safety of the public, (2) safety of operation and maintenance personnel, (3) safety of livestock and wild animals, and (4) safety of the system. Additionally, the importance of emergency preparedness and response and systematic record keeping for identifying hazards, preventing incidents, and improving safety over time are discussed.

20.2 PUBLIC SAFETY

The facilities, constructed primarily for irrigation use, are increasingly utilized for recreational purposes, such as boating, fishing, and swimming. Storage reservoirs, diversion weirs, and canals have become informal leisure sites in many regions,

DOI: 10.1201/9781003747444-23

but their use often comes with serious risks. Drownings from overturned boats, hazardous undercurrents, and vehicle accidents on canal banks occur far too often. Increased use has led to an increase in the number of fatalities and injuries. People have drowned when boats are capsized or overturned due to storms or negligence on the part of the operator following unsafe practices. These accidents have grown faster than occupational accidents. The use of canals and waterways for recreation has resulted in more fatalities than occupational accidents. Accidental deaths and injuries from their use have resulted primarily from adults and children swimming, fishing, or playing in the waterways and from automobiles trespassing on canal banks or running into the numerous canals and laterals that parallel or cross many roads in project areas.

Accidents and drownings that occur are usually the subject of much public indignation and criticism. It is true that after an accidental death or drowning, the public is aroused to the dangers of swimming in the canal right-of-way. Unfortunately, they soon lapse into apathy to be aroused again only by a subsequent accident; nevertheless, the public demands more safety precautions not only for themselves but also for the protection of domestic and wild animal life. There is reason to believe that considerable expenses can be justified for these safety measures.

20.2.1 Safety Measures

With respect to the use of reservoirs and other bodies of water for recreational purposes, it is more or less evident that those in charge of recreation, whether it is the district or some other organization or agency, must formulate and enforce safety rules and regulations. Hazardous practices must be discouraged. For example, many casualties in past boating accidents could have been prevented with the mandatory use of life jackets and other safety devices on board.

Some measures that enhance the safety of boating and boats in reservoirs include the following: (1) Hazard marking: Installing markers for submerged rocks and other underwater hazards, similar to the practices of the Tennessee Valley Authority (TVA) (with maps and markers) (TVA, 2023) (Figure 20.1a); (2) Restricted zones: Placing guard booms near spillways and power plant intakes can significantly mitigate risks posed by underwater obstacles and strong currents (Figure 20.1b); (3) Safe access alerts: Ensuring the provision of safe launching and landing facilities; and (4) Weather alerts: Installing systems to warn of approaching storms or sudden releases from upstream structures. Many of the safety rules developed by the U.S. Coast Guard may be applicable (USCG, 2023).

In addition to recreational safety, specific areas within irrigation systems are much more hazardous than others. To optimize public protection in these zones, three things should be done: (1) educating the public (operating personnel may be included) about existing dangers, (2) preventing unauthorized access to these hazardous areas, and (3) providing well-planned means of self-escape, which ensures that individuals have the necessary routes or mechanisms for safe evacuation in the event of an emergency. Collectively, these integrated safety measures not only safeguard individuals engaging in recreational activities but also protect those working in or around potentially dangerous areas of irrigation systems.

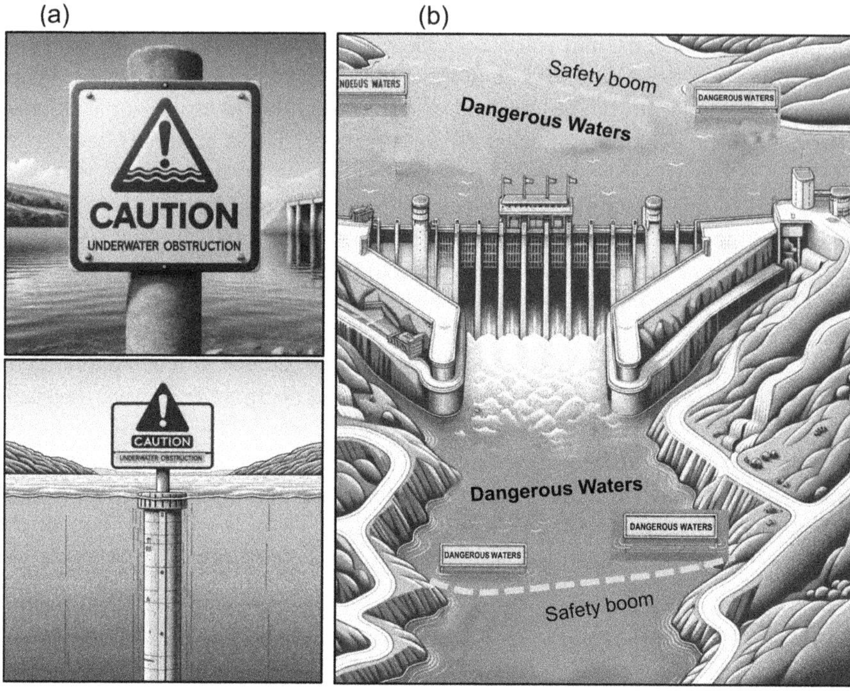

FIGURE 20.1 Illustration of (a) markers on submerged rock and similar hazards and (b) safety booms near spillways.

20.2.2 Education of the Public

One of the most effective means of protecting the public from the hazards of irrigation projects is to educate the public about these hazards. Conducting safety campaigns is one of the best ways to educate the public, which requires the collaborative engagement of district personnel, schools, and organizations near reservoirs and canals about the dangers inherent in all such works.

To effectively disseminate this critical information, educational campaigns must be directed toward all age groups. Parents are the primary audience, as their awareness and cooperation are essential in safeguarding youth from canal-related hazards. Consequently, dedicated efforts to educate school-aged children are necessary. These campaigns must clearly explain the dangers inherent in irrigation systems, especially hidden hazards, such as strong undercurrents near drop structures, siphon inlets, and gated spillways. Even calm-looking water can conceal powerful flows capable of overpowering experienced swimmers. To enhance understanding, safety programs should also demonstrate how to use emergency escape devices, such as ladders, grab ropes, and floating booms, ensuring that the public knows their location and purpose, while fostering respect for these installations to reduce vandalism.

Successful education efforts use multiple communication channels, including school presentations, community workshops, radio and television announcements,

FIGURE 20.2 Example of a school water safety poster with canal hazard warnings.

and printed materials such as posters and brochures. For example, the Imperial Irrigation District (IID) in California integrates its Dippy Duck mascot into school visits and seasonal media campaigns, reinforcing the "Stay Away from Canals" message for children and families (https://www.iid.com/customer-service/safety/canal-safety). Visual aids, such as illustrated hazard maps or safety posters, help convey messages effectively to diverse audiences (Figure 20.2). Collaboration with schools, recreational groups, and local governments can strengthen these efforts and facilitate the enforcement of trespassing regulations around high-risk zones.

Through a persistent educational campaign, a community can significantly increase its safety consciousness regarding irrigation projects.

20.2.3 Preventing Access

Restricting entry to hazardous areas is one of the most direct and effective means to protect the public from irrigation-related accidents. A combination of fencing, guardrails, locked gates, and warning signs can discourage trespassers, alert visitors to dangers, and prevent people or animals from entering restricted and hazardous areas. Among these measures, fencing is the most widely used physical barrier. Chain-link fences are particularly common for securing siphon inlets, canal banks near towns or schools, and large hydraulic structures where public access is frequent. In many irrigation districts, these fences completely encircle the canal right-of-way and protect critical points, such as inlets to siphons and other structures that are easily accessible from nearby communities (Figure 20.3a).

In more remote or agricultural zones, standard barbed-wire fencing may be used, although it is typically removed near population centers where greater protection is needed. Handrails, while useful for marking edges, are generally ineffective as standalone safety barriers at dangerous sites such as siphon inlets. For improved security, some projects enhance railings with wire mesh or additional rope lines between the rails. To facilitate maintenance, especially weed or moss removal, railings in such areas are designed to be removable.

Guardrails and guard posts play an important role in overall safety when constructing opposite ditch bank approaches, along narrow canal banks in hazardous reaches, on curves, and when canal crossings, culverts, and bridge or structure approaches are marked. To provide better visibility by day, the railing and posts should be painted and provided with reflectors or luminous painted stripes for visibility at night.

Although warning signs in hazardous areas may meet legal requirements, it is not known how many people are alerted to the dangers that exist and how many people actually obey them. Some individuals may fail to notice the signs, underestimate the risks, or choose to ignore them entirely. Nonetheless, warning signs play an important role in comprehensive safety strategies. While they are not a complete solution to the problem of unauthorized access, they serve as a visible hazard alert, provide legal notice, and complement physical barriers and education efforts.

Signs should be strategically installed at all known access points, road and foot crossings, and popular recreational hotspots, and they should be presented in multiple languages where applicable. For example, the IID's All-American Canal employs bilingual "No Trespassing – Dangerous Water" signs with clear drowning hazard pictograms to reinforce the safety message (Figure 20.3b). Although a small

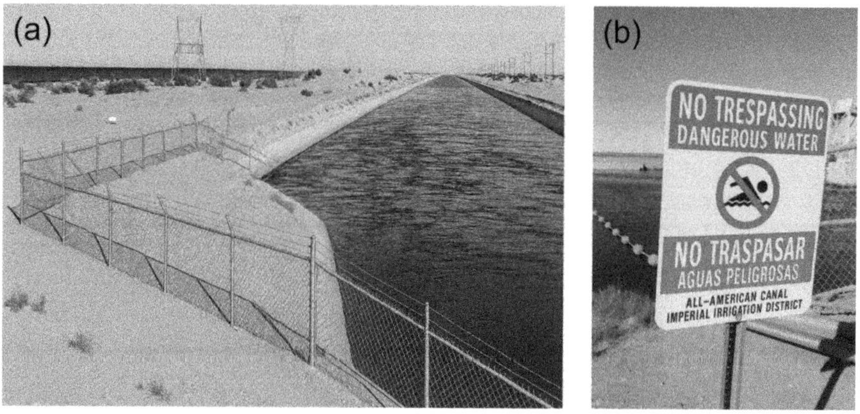

FIGURE 20.3 (a) Chain-link fence installed along a canal and (b) bilingual warning sign with a drowning hazard installed by the Imperial Irrigation District (IID) in California.

percentage of people may ignore such warnings, their consistent placement increases hazard awareness among the majority of visitors and strengthens the overall impact of public education campaigns.

20.2.4 Means of Self-Escape

Another important consideration in any safety program for canals and waterways is to provide a means of self-escape for both humans and animals in the case of an accidental fall into canals. Without such measures, swift currents, steep banks, and smooth canal linings can make self-rescue nearly impossible. Over the years, a variety of devices have been installed, including ladders, safety ropes, nets, booms, and step structures, each with its own benefits and maintenance considerations.

While these devices are intended to save lives, their design must balance safety, durability, and operational practicality. Poorly designed installations can hinder canal operations by obstructing flow, interfering with cleaning and inspection, or catching debris that reduces water quality. For example, escape devices that protrude too far into the canal may become snag points for floating vegetation or trash, requiring frequent maintenance.

Ladders are probably the most common installations used to provide self-escape for a person. These may be mounted on the walls of a structure or consist of steel rungs embedded in the structure's concrete or the lining of a canal. In concrete-lined canals, escape ladders are typically installed at 150m (500-ft) intervals alternating from one side to the other and are often positioned near siphon inlet transitions. The effectiveness of ladders alone in providing escape from a waterway is questionable. A person caught in a flowing canal midstream needs something to grab. For this reason, a net, cable, boom, rope, etc., in combination with a ladder is much more effective.

A typical installation consists of a cable placed across the canal with a steel-wire safety screen suspended from the cable. Some projects have adopted the use of nets made from treated hemp rope, and they offer certain advantages. Another type of interception device of this nature is a timber boom or float device in sections of 4–5 ft (1.25–1.5 m) in length and sufficient width to support a person. Each section becomes a segment of a more or less continuous walkway, which is held in position by a cable anchored on each side of the canal so that the device lies in an arc across the canal near the head of a hazardous structure. A newer type of escape device structure consists of concrete beams or steps installed horizontally across the inlet transition with a vertical opening to permit the passage of water. This type provides a means of escape for both humans and animals.

20.3 SAFETY OF OPERATION AND MAINTENANCE PERSONNEL

The safety of employees engaged in the operation and maintenance of irrigation systems is the most important concern for any organization. It is recognized that accidents not only pose a risk to the well-being of personnel but also result in operational disruptions, increased maintenance costs, and potential damage to materials, tools, and equipment. As such, a well-structured, proactive safety program is necessary. Preventing accidents is far more cost-effective than responding to them after they occur.

Safety programs must be methodically planned and seamlessly integrated into daily work routines so that safety awareness becomes part of the work culture. Although establishing such a program initially requires time and financial investment, the long-term human and economic benefits significantly outweigh these initial expenditures.

In managing safety within irrigation systems, it is essential to address two categories of accidents: (1) accidents arising from unsafe work procedures and (2) accidents involving unsafe structures, equipment, and apparatuses. Accidents arising from unsafe work procedures are incidents that occur due to lapses in following established safety protocols or a lack of proper safety training. Regular safety drills, clear communication of procedures, and ongoing education about the risks associated with specific tasks can reduce such incidents. Accidents involving unsafe structures, equipment, and apparatuses are due to equipment failure, structural issues, or inadequate maintenance of tools and machinery. Ensuring regular inspections, timely repairs, and adherence to maintenance schedules can mitigate these risks.

A proactive strategy focused on identifying and mitigating potential hazards is fundamental to maintaining a safe work environment. This may involve conducting risk assessments, updating equipment, reinforcing safe work practices, and ensuring that all safety measures align with the latest standards and technologies.

20.3.1 Educational Programs

The most effective means of minimizing accidents due to human error within irrigation projects is a well-structured educational system. This program must be well designed and include all personnel in the project, addressing the specific safety concerns and protocols relevant to their roles. Safety education should not be a one-time event but rather an ongoing process. Regular training sessions, updates on new safety procedures, and refreshers on existing protocols are important. A system needs to be in place to ensure that safety instructions and regulations are adhered to. This involves regular monitoring and follow-ups.

Safety efforts can succeed only if those with the final authority truly want to reduce accidents. The workforce must see that the management's commitment to safety is sincere, not merely a perfunctory gesture. Foremen and supervisors play important roles in reinforcing safety. They should be encouraged to view accident prevention as an integral part of their responsibilities. Their active participation in promoting safety can have a substantial effect on the attitudes and behaviors of their teams.

The ultimate aim of these educational programs is to cultivate a safety-conscious culture within the organization. This culture is characterized by a collective commitment to safety, where every member of the team understands the importance of safety measures and is proactive in maintaining a safe work environment.

20.3.2 Safety Installations

In general, many of the means employed to protect the safety of visitors also provide protection to operation and maintenance personnel. However, there are safety installations, such as guard rails, which may interfere with the normal activities of operating personnel. These conflicts can usually be resolved through careful design and planning. For example, guardrails can be designed to be removable or adjustable, allowing uninterrupted access for maintenance work.

Proper irrigation system design is essential for protecting personnel by minimizing exposure to hazards and ensuring safe operating conditions. Given the inherent risks in wet conditions, electrical circuits within irrigation systems must be meticulously designed and installed to prevent shock hazards. This includes the use of waterproof materials and proper grounding techniques. Irrigation systems should also be designed with accurate pressure ratings. The installation of pressure relief valves and other safe devices where high pressure is a concern is crucial to prevent system failure and ensure safe operation. Moreover, all components of the irrigation system must be installed according to specific safety and operational specifications. This adherence to standards ensures both the functionality of the system and the safety of the operators. Furthermore, power units and drive units must be properly sized, mounted, aligned, and shielded. This not only ensures the safe, long-term operation of these units but also reduces the risk of accidents caused by equipment malfunction or improper use.

20.3.3 Safety Procedures

Any construction activities associated with the project should be conducted in accordance with existing safety procedures. These procedures are well documented in the engineering safety manual and the literature and will not be repeated here (USBR, 1969). Notably, many maintenance activities closely resemble construction activities in terms of risks and required precautions. General rules include the use of hard hats, avoidance of loose clothing around moving equipment, adherence to safety signage, and maintaining situational awareness of coworkers' activities (USBR, 1958).

Working around water can be hazardous. Banks and structural surfaces are often wet, slippery, and uneven, increasing the likelihood of slips and falls. Many maintenance operations, such as gate servicing, debris removal, and inspection, require personnel to work at the water's edge or directly above the flow. The wet and often slippery conditions of banks and structures near water bodies require extra caution. The unpredictable nature of these surfaces increases the risk of slips and falls. The best practices to mitigate these risks include the following:

1. **Mandatory personal protective equipment**: Life jackets should be worn at all times near open water, and safety belts or harnesses should be securely attached to guardrails or other sturdy anchors to prevent falls into the water.
2. **Buddy system**: Where feasible, employing a buddy system (having personnel work in pairs) ensures that immediate help is available in the case of an accident. This practice is particularly beneficial in preventing and responding to water-related incidents.
3. **Safe access and escape points**: Incorporating fixed ladders, stepped beam escape structures, or floating timber booms at strategic intervals allows for rapid self-rescue.

4. **Improved traction surfaces**: Applying non-slip coatings or textured steel gratings to decks, platforms, and stair treads reduces slip hazards.
5. **Accessible anchor points**: Installing dedicated anchor rings or cables at routine work locations makes fall protection easier and faster to implement.
6. **Enhanced guardrails**: This method uses robust, corrosion-resistant rails that can be temporarily removed for equipment access without compromising safety.

20.3.4 Safe Driving of Vehicles

Vehicle accidents have long been a source of injury, property damage, and operational delays in irrigation projects. In many cases, the roads that project personnel must travel on are far from meeting highway standards and are often unpaved, narrow, and intersecting with canals, laterals, or steep embankments. Such conditions require constant vigilance, particularly when seasonal hazards such as mud, ice, or reduced visibility during storms occur.

To address these challenges, many irrigation districts implement proactive measures such as driver training and testing to ensure that only qualified operators handle project vehicles. Speed limits are established to reflect terrain and road conditions, whereas specialized driver education programs focus on off-road and canal-side driving techniques. These include controlled braking on loose surfaces, navigating narrow embankments, and recognizing high-risk areas during adverse weather. Vehicle selection also plays a role, with districts often choosing four-wheel-drive trucks equipped with reinforced suspensions, high-traction tires, and appropriate lighting for low-visibility conditions. Routine vehicle maintenance, including regular inspections of brakes, steering, and tires, helps reduce mechanical failures that could contribute to accidents.

To mitigate the risks associated with driving in such challenging conditions, several proactive measures are often adopted by projects: (1) Driver training and testing: Implementing a requirement for vehicle operators to have driving examinations ensures that only skilled and knowledgeable drivers are behind the wheel. This approach helps familiarize drivers with the specific challenges they may encounter in the project area. (2) Establishing speed limits: Given the rough and often unpredictable terrain, setting and enforcing speed limits is crucial. These limits should reflect the conditions of roads and trails, helping reduce the likelihood of accidents. (3) Driver education programs: Operating driver education schools or programs within the project can significantly increase the driving skills of personnel. These programs should focus on the unique aspects of driving under off-road conditions, including handling rough trails and avoiding skids. (4) Vehicle selection and maintenance: Choosing vehicles suited to rough terrain, such as four-wheel-drive trucks with reinforced suspension, and performing regular maintenance checks on brakes, tires, and steering systems are essential for safe operation.

The emphasis on cautious driving cannot be overstated. Experience and a mindful approach to driving in these areas are key to preventing accidents. Drivers should be constantly reminded of potential hazards, such as the risk of vehicles overturning on uneven paths or skidding into canals and laterals.

20.3.5 Machine Safety

Accidents involving the operation of power equipment and machinery have long been a concern in irrigation projects, with many incidents linked directly to operator actions. The operator of such equipment is the key to safe job performance. Equipment manufacturers have greatly improved the safety of their products, and many power machines today have "built-in" safeguards, such as emergency shut-off switches, rollover protection structures, and operator-presence control. These features are only effective when operators understand and use them correctly. Following the manufacturer's instructions, reading, comprehending, and applying them in daily operations is critical for maximizing safety benefits. Unsafe practices, such as overloading equipment, exceeding performance limits, or operating at excessive speeds, can override even the most advanced safety systems, leading to dangerous situations. The safest machine still must be operated with care and with knowledge of its performance capabilities. The man who has the ultimate responsibility is the operator, even though comprehensive safety programs are arranged by management.

One particularly severe hazard in irrigation projects involves cranes, excavators, or similar machines inadvertently swinging their booms into high-voltage transmission lines. Such contact can result in electrocution or severe burns not only to the operator but also to nearby personnel. Many modern machines are equipped with boom-proximity warning devices or insulating links on crane hooks to reduce this risk. These safety features, however, are not substitutes for maintaining safe clearance distances and conducting pre-operation hazard checks. Operators should also receive task-specific training in recognizing electrical hazards, interpreting overhead line markers, and implementing lock-out/tag-out procedures when working in energized areas.

20.3.6 Handling of Hazardous Materials

The use, handling, and storage of hazardous materials, including poisons, pesticides, herbicides, and explosives, present significant risks in irrigation projects. These substances can endanger human health, damage the environment, and compromise system operations if mishandled. Mitigating these risks requires strict adherence to established safety protocols and industry best practices.

The primary line of defense against accidents involving hazardous materials is the careful reading and understanding of instructions provided by manufacturers. These instructions are typically found on containers and cover critical aspects of safe handling and storage. In addition to individual responsibility, providing comprehensive training on the safe handling of these materials is essential for management. This training should cover proper usage, emergency procedures in cases of spills or exposure, and safe disposal practices. Given the potential for serious harm, regular refresher courses are recommended to ensure that all personnel are up to date on the latest safety procedures and regulatory changes regarding hazardous materials.

Hazardous materials should be stored in designated areas that are secure, well ventilated, and away from potential sources of ignition or contamination. All containers should be clearly labeled, and an inventory of hazardous materials should be maintained. This documentation aids in the quick identification of materials and their hazards, particularly in emergency situations. In addition, an emergency response plan should be in place, outlining the steps to be taken in the event of accidental spillage, leakage, or exposure. This plan must be easily accessible to all personnel responsible for the handling of hazardous materials.

20.3.7 Operational Safety

Operational safety in irrigation projects, particularly when working inside confined spaces, such as large penstocks, siphons, or closed conduits, presents unique and potentially life-threatening hazards. There are two hazards that need special attention: (1) The danger of someone inadvertently opening a valve or gate and allowing water to enter and (2) the danger of accumulated gases or a lack of oxygen in the tube. In the first instance, many projects use locking mechanisms with a single key for all valves and gates. The key is held by the person responsible for the work within the conduit and is returned only once it is confirmed that all personnel have safely exited. This approach ensures control over water flow and prevents accidental inundation. Second, ensuring proper ventilation is crucial for the safety of workers. Additionally, carrying gas detection devices can provide early warning of dangerous air quality conditions, allowing timely evacuation or remedial actions.

The personal element is by far the most important factor involved in preventing accidents. The safety of operation and maintenance personnel can best be ensured if each individual follows safe practices. Each individual's commitment to safety significantly reduces the likelihood of accidents. Regular training in safety procedures is essential. This training should include a wide range of scenarios, equipping staff with the knowledge and skills to handle different types of hazards effectively. Unsafe acts by individuals are often the primary cause of accidents, although a lack of safety devices, inadequate training, and unsafe structures contribute to these incidents.

20.3.8 Improvement of Working Conditions

Improving working conditions in irrigation projects is an ongoing process that directly enhances personnel safety and operational efficiency. Several practices can be implemented to create a safer work environment and reduce accidents. For example, the installation of metal foot plates or anti-slip treads on stairs and operating decks can greatly improve traction and reduce the risk of slips and falls especially under wet, icy, or snowy conditions. Using steel grating instead of traditional concrete or wood flooring on operating decks offers better grip strength and stability, minimizing the likelihood of slipping. These modifications are particularly beneficial in pump stations, spillway decks, and canal gate platforms, where moisture and algae growth can create hazardous surfaces.

Additional protective features can further safeguard personnel. The installation of handrails and safety nets around structures provides additional support and security, especially in elevated or potentially hazardous areas. In addition, installing cables to attach ropes to safety belts ensures that personnel can secure themselves safely, particularly when working in precarious positions. Automating the operation of large gates and valves helps reduce physical strain and the risk of injuries associated with manual operation.

Regular inspections of structures are important for detecting loose bolts, corroded surfaces, or weakened supports before they become hazards. Additionally, being familiar with the built-in safety features of power equipment is crucial. Operators must receive proper training to fully understand and effectively utilize these features. When such features are absent, cost-effective retrofitting, such as adding guard rails to exposed gears or installing splash shields on pumps, can offer substantial protection. Even simple, improvised safety measures, when thoughtfully designed, can significantly improve safety.

20.3.9 Self-Escape

Self-escape devices, which were originally designed for public protection, can be equally critical for safeguarding operation and maintenance personnel in hazardous areas of irrigation projects. These devices serve as a last line of defense, allowing individuals to exit dangerous situations quickly and safely. In high-risk environments, such as steep canal banks, confined spaces, or areas near high-flow spillways, the ability to self-rescue can mean the difference between a near miss and a fatality.

In many cases, the same self-escape mechanisms, such as ladders, ropes, and safety nets, that are installed for the public can be effectively used by operational staff. Their strategic placement in and around hazardous areas ensures that personnel have immediate and reliable means of escape in emergency situations. Recognizing the unique risks faced by operators, it may be necessary to install additional or specialized self-escape devices at hazardous locations within the project area. These installations should be customized to the specific hazards of the site and the nature of the work being conducted. Given their critical role in worker safety, these self-escape devices must be regularly inspected and maintained to ensure that workers are always in optimal working condition. This practice is crucial for guaranteeing their reliability and effectiveness in emergency scenarios.

20.4 SAFETY OF LIVESTOCK AND WILD ANIMALS

Irrigation canals and laterals can be death traps for livestock and large wild animals, especially those built with hard surface linings. The steps, often covered with slime, moss, or algae, make these waterways extremely slippery. Animals, and even humans, struggle to escape under these conditions. Furthermore, these waterways are designed for higher flow velocities than their unlined or earth-lined counterparts, increasing the risk of animals being swept away.

20.4.1 Prevention of Entrance

The most effective way to protect animals is to prevent their entry into these dangerous areas. For animals such as deer and antelopes, which can easily jump over obstacles, the height of the fences around waterways is a crucial factor. Fencing should be installed along high-risk reaches of canals, with special attention given to access

FIGURE 20.4 A wildlife-friendly canal crossing in Arizona.

points near rangelands or known wildlife migration routes. Where natural animal trails intersect with lined canals, wildlife-friendly bridges can be constructed to allow safe passage without disrupting movement patterns. These bridges, which are designed exclusively for animal use, can be lighter than vehicle-bearing structures, reducing costs while maintaining safety. For example, the Central Arizona Project has installed wildlife crossings to facilitate the movement of wildlife, including mule deer, javelina, sheep, and mountain lions (Figure 20.4).

20.4.2 Self-Escape

Self-escape devices can be highly effective in preventing the drowning of animals in irrigation canals and associated structures, although their design typically differs from that intended for human use. One design consists of a grid of reinforcing bars welded at approximately 6-in. (15 cm) centers and hung on the side of the canal lining to aid animals in reaching the concrete slope. Another more elaborate device, suitable for both humans and animals, was developed for installation at the entrance to inverted siphons along a canal in Utah, United States. This device provides wide steps (across the canal) from a level below the normal water surface to the top of the siphon inlet head wall. This device was designed with the aid of model studies, and the resistance to flow is not high.

Structures with sloping concrete floors, such as still pools and siphon entrances or exits, which are left undrained at the end of the irrigation season, are potential hazards to animals. Animals, often seeking water, can inadvertently slide into these structures because of their slippery surface. Once they are in the water, they may struggle to escape, leading to potential drowning or death from exhaustion. If it is not practical to drain such structures, they should be fenced or barricaded to prevent animals from accessing them.

20.5 SAFETY OF IRRIGATION SYSTEMS

A critical aspect of irrigation system safety is protecting it from vandalism and sabotage, which can lead to significant operational disruptions and damage. Measures should be taken to secure gates and other vital components of the system. Unauthorized or accidental opening of these gates can lead to flooding and compromise the integrity of canals and structures. Implementing monitoring systems or surveillance can deter malicious activities and enable a quick response to any incidents of vandalism or sabotage.

Designing a system that can withstand natural disasters and adverse climatic conditions is equally important. The design must account for potential natural threats, such as flooding, extreme weather conditions, and earthquakes. This involves robust engineering practices and the use of materials that can withstand such events. Features such as security spillways in canals are examples of design adaptations that can preemptively address potential problems. These spillways help manage excess water flow, thereby preventing overflows and structural damage.

Regular maintenance and prompt repairs are important to preventing deterioration within irrigation system. Regularly inspecting and maintaining the system ensures that it operates efficiently and safely. Promptly addressing any damage or malfunction in the system is vital for maintaining its integrity and functionality. Delaying repairs often results in greater structural deterioration and increased repair expenses.

20.6 EMERGENCY PREPAREDNESS AND RESPONSE

The development of a comprehensive emergency plan is essential to anticipate, manage, and respond to the unique operational and safety risks associated with irrigation projects. These plans should include two key components: (1) identifying specific risks and (2) preparing response strategies. Initially, it was important to evaluate the specific hazards associated with the project, such as flooding, equipment failure, or breaches in waterways. Following this assessment, detailed strategies for each identified risk should be developed. These strategies should include evacuation routes, communication protocols, and response team roles.

Equipping for an effective emergency response is crucial. This involves ensuring that first aid stations are not only accessible but also adequately stocked with essential medical supplies. Response teams should be equipped with reliable communication tools to coordinate efforts during an emergency. Additionally, the supply of essential rescue equipment, such as life jackets, ropes, and stretchers, should be maintained.

Effective emergency response requires coordination beyond the project's personnel. Connections with local emergency services and authorities should be established for support and assistance during emergencies. Local communities should be encouraged to raise awareness about risks and emergency procedures. Community involvement can enhance the effectiveness of response and aid in preventive measures.

20.7 RECORD KEEPING

It is important that an organization keeps records of all types of accidents. Only in this way can information be accumulated that points to hazardous locations in the project and hazardous practices being followed by individuals. Such recorded information, such as where the accident happened, what the employee or visitor was doing at the time, what caused the accident (if possible), and similar information, will be very helpful in time. For example, if the records show that a canal at a certain location is hazardous, it would indicate a need for guard posts or rails or perhaps changing the grade or alignment of the road at that location.

Maintaining detailed records of accidents is crucial for managing safety within irrigation projects. These records provide an essential basis for detecting patterns and trends that may not be immediately evident through routine observation. There are key elements that should be included in accident records: (1) Accident location. Documenting the exact location of each accident helps in identifying hazardous spots within the project area. (2) Activities at the time of the accident. Recording what the employee or visitor was doing when the accident occurred provides insights into potential risky behaviors or unsafe conditions. (3) Causes of the accident. While it may not always be possible to determine the cause of an accident, noting that the cause can be useful in preventing future incidents.

The accumulation of these data over time is essential for making informed decisions about safety enhancements. If records indicate a pattern of accidents at a specific location, this indicates a need for safety interventions, such as the installation of guard posts, rails, or modifications to the existing infrastructure, such as changing the grade or alignment of nearby roads. Regular analysis of accident records can lead to the revision of current safety protocols and the implementation of new safety measures.

20.8 CASE STUDIES

20.8.1 Imperial Irrigation District (California)

The IID operates as one of the largest irrigation delivery systems in the United States. Its safety programs integrate public education, operational safeguards, wildlife protection, and infrastructure maintenance to protect people, workers, ecosystems, and the irrigation system itself.

Public safety and education: Since 1959, IID has launched a water safety program to prevent drownings and educate the public about canal hazards, emphasizing its core message "Stay Away from Canals". IID formally declares May as a "water safety month" each year and delivers school outreach via its long-standing mascot, *Dippy Duck*, whose three cardinal rules are as follows: "Never swim in a canal, never play in a canal, never jump in a canal" (IID, 2025; KYMA, 2024). Safety signs are placed along canals to reinforce this message. The IID has also established signs near canals to remind the community that the best way to stay safe is to stay away, reinforcing that canals are not safe recreational sites.

As a harm-reduction strategy, IID funds summer pool-opening grants to local governments and community groups, providing safe recreational water alternatives. By 2023, the program had been awarded up to $41,500 collectively (Calexico Chronicle, 2023).

Drinking-water safety: Under the safe drinking water act, IID requires all customers using untreated canal water for homes or businesses to have an alternate potable supply (IID-SDWA, 2025). Following a 2021 enforcement settlement, IID committed to installing and maintaining at least 25 point-of-entry treatment systems for homes receiving canal water, addressing bacterial and contaminant risks (Water Resources Control Board, 2021). In addition, IID coordinates Title 22 source-water sampling with certified labs and shares results with local drinking-water systems to support consumer confidence reports.

Wildlife protection: To protect endangered desert pupfish, IID adjusts the timing of on-farm conservation measures that affect drainage flows, coordinating with state and federal agencies—an example of operational safety for ecosystems and regulatory compliance (AP News, 2024).

Operation and maintenance safety: IID's water department manages the open-channel delivery network and its Colorado River entitlement. To ensure canal bank and access safety, IID issues canal bank driving restrictions and inspection protocols to minimize the risks of vehicle rollovers and work-site accidents. Additionally, IID's safety guidance warns against storing haystacks or operating equipment near power lines, citing fire, outage, and liability risks, which are relevant to the safety of operation and maintenance personnel (IID, 2025).

Safety of the system: The IID safeguards the integrity of its irrigation system through a structured program of regular inspections, preventive maintenance, and prompt repairs. Routine activities, such as vegetation control, canal cleaning, concrete lining repair, and gate and pipe upkeep, are carried out under a scheduled "Canal Cutout" program to keep infrastructure in optimal condition. Sedimentation control and canal improvement projects are prioritized on the basis of structural assessments, ensuring long-term operational stability. Additionally, IID coordinates with local agencies to maintain critical crossings, culverts, and siphons, reducing the risk of failure and protecting both water delivery and public safety. IID maintains detailed engineering standards and drawings (e.g., for swing gates, utility crossings, drainage systems) to ensure that infrastructure construction adheres to safety and reliability requirements. These materials are accessible and well documented via official records management systems.

Emergency preparedness and response: The IID maintains an Office of Emergency Management to oversee disaster preparedness and coordinate district-wide emergency response efforts (Office of Emergency Management, 2025).

This office develops and regularly evaluates emergency strategies, conducts drills, collaborates with external agencies, and manages real-world incident responses, ensuring a structured, well-documented, and proactive approach to safety. To facilitate rapid communication, IID publishes both toll-free and local emergency contact numbers, specifically for energy-related incidents.

Record keeping: The IID operates a dedicated records administration department responsible for document control, public records access, and the systematic tracking of safety-related incidents. This function supports institutional learning and continuous improvement in safety practices. Additionally, the district offers a mobile platform, i.e., IID Customer Connect, that enables the public to report canal and drainage issues and check the status of submitted concerns (IID Customer Connect, 2025).

IID maintains a department responsible for records administration, document control, and public records access, which is critical for safety incident tracking and institutional learning. The public can report and check canal and drainage issues through an App called IID Customer Connect (IID Customer Connect, 2025).

20.8.2 Hume Dam (Murray–Darling Basin Authority, Australia)

The Hume Dam, located just downstream of the confluence of the Murray and Mitta Rivers in the Riverina region of New South Wales, is a critical multipurpose infrastructure. It supports flood mitigation, hydroelectric power generation, irrigation, water supply, and water conservation. Its safety management integrates proactive public risk reduction, rigorous operational standards, and infrastructure resilience. Under the leadership of the Murray–Darling Basin Authority (MDBA), in collaboration with partner agencies, dam operations align with dam safety legislation, strategic flood management practices, and structured maintenance programs. These coordinated measures protect the public, safeguard operations staff, and ensure the reliable performance of the system.

Safety of the public: Hume dam managers provide clear safety advice at recreational zones, such as wearing life jackets, supervising children, avoiding underwater hazards, and maintaining a safe distance when boating, to reduce public risk around the dam and its waterways (WaterNSW, 2025). During high-inflow events, the MDBA issues flood management warnings and increases dam releases to manage downstream flood risks. For example, in November 2022, when the dam reached 96% capacity, releases increased to 9.5×10^{10} GL/day to create airspace and mitigate potential flooding (ABC News, 2022).

Safety of operation and maintenance personnel: Personnel safety is governed by compliance with state dam safety legislation and best-practice operational standards (MDBA, 2024a). The significant safety upgrades completed in 2013 included installing a parapet wall along the crest, enhancing spillway structural safeguards, and improving drainage systems. These measures protect workers while also improving the long-term structural performance of the dam (Australian Water Partnership, 2024).

Safety of the system (infrastructure and reliability): Regular structural inspections of the dam wall, spillways, embankments, and outlet works are conducted to detect any signs of movement, seepage, or deterioration. Following earlier concerns about structural movement and leakage, extensive remedial works between the late 1990s and the early 2010s, including the construction of a secondary earth wall and improved drainage, were undertaken to increase Hume Dam system safety (WaterNSW, 2025). Ongoing maintenance and service of mechanical and electrical components, such as gates, penstocks, and trash racks, are needed to ensure that they operate as designed during both normal and emergency conditions (IPART, 2024).

Emergency Preparedness and Response: The MDBA applies operational strategies, such as pre-releases, to create "airspace" for incoming flows, reducing flood risk. This was demonstrated in the 2022 flood event, where proactive water level management helped protect downstream communities (ABC News, 2022).

Record Keeping: Operations are supported by the MDBA's River Murray Operations Joint Program, which encompasses asset management planning, scheduled maintenance, and performance reporting. Comprehensive public safety documentation, such as hazard signage and reporting procedures, enhances transparency, supports emergency readiness, and reinforces community trust (MDBA, 2024b; WaterNSW, 2025).

REFERENCES

ABC News (2022). River Communities on Flood Watch as Hume Dam Releases Increase. Australian Broadcasting Corporation. https://www.abc.net.au/news/2022-11-11/river-communities-on-flood-watch-as-hume-dam-releases-increase/101645538

AP News (2024). Tiny, Endangered Fish Hinders California's Colorado River Conservation Plan. https://apnews.com/article/california-colorado-river-imperial-valley-irrigation-pupfish-d4296229cdd265afe74cbfdde5cad527

Australian Water Partnership (2024). Australian Handbook for Dams. https://waterpartnership.org.au/wp-content/uploads/2024/05/24216-AWP-Handbook-for-Dams-Web.pdf

Calexico Chronicle (2023). IID Awards Water Safety Summer Activities Grants. https://calexicochronicle.com/2023/06/21/iid-awards-water-safety-summer-activities-grants/

IID (2025). *Water Safety*. Imperial Irrigation District. https://www.dippyduck.com/water-safety/home/

IID Customer Connect (2025). Imperial Irrigation District. https://www.iid.com/?utm

IID-SDWA (2025). Safe Drinking Water Act of Imperial Irrigation District. https://www.iid.com/water/service-pipe-customers/safe-drinking-water-act

IPART (2024). 2024 Pricing Proposal: WaterNSW Attachment 16: MDBA Pass-Through Charges. Independent Pricing and Regulatory Tribunal. https://www.ipart.nsw.gov.au/sites/default/files/cm9_documents/2024-Pricing-Proposal-WaterNSW-Attachment-16-MDBA-pass-through-charges.PDF

KYMA (2024). News: May Declared as Water Safety Month by IID Board. https://kyma.com/news/imperial-county/2024/05/08/may-declared-as-water-safety-month-by-iid-board/

MDBA (Murray-Darling Basin Authority) (2024a). Murray–Darling Basin Authority Annual Report 2023–24. Government of Australia. Transparency Portal. https://www.transparency.gov.au/publications/agriculture/murray-darling-basin-authority/murray-darling-basin-authority-annual-report-2023-24

MDBA (Murray-Darling Basin Authority) (2024b). Hume Dam. Murray–Darling Basin Authority. https://www.mdba.gov.au/water-management/infrastructure/hume-dam

Office of Emergency Management (2025). Imperial Irrigation District. https://www.iid.com/about-iid/department-directory/office-of-emergency-management

TVA (Tennessee Valley Authority) (2023). Boating & Water Safety. https://www.tva.com/environment/lake-levels/water-safety

USBR (1958). Canal Safety: Safety in the Operation of Canals and Other Waterway Structures. U.S. Department of the Interior, Bureau of Reclamation, Denver, CO.

USBR (1969). *Construction Safety Standards.* 1st ed. U.S. Department of the Interior, Bureau of Reclamation, Denver, CO, June 1, 1968 (Revisions to January 1, 1969)

USCG (U.S. Coast Guard) (2023). Recreational Boaters. https://uscgboating.org/recreational-boaters/index.php?m=rb

WaterNSW (2025). Hume Dam. https://www.waternsw.com.au/nsw-dams/regional-nsw-dams/hume-dam

21 Recreation and Fish and Wildlife Enhancement

21.1 INTRODUCTION

Recent population growth, coupled with rise in disposable income, mobility, and leisure time combined, especially among people living in urban areas, have magnified the importance of recreational activities. This shift has particularly increased the demand for water-based recreation facilities, a trend that is expected to persist in the coming years. The increasing popularity of activities offered by reservoirs not only meets this growing demand but also has a considerable economic impact. The U.S. Bureau of Labor Statistics reported that outdoor recreation contributed $459.8 billion to the GDP in 2019, which increased by 22% compared with 2016 (OIA, 2019).

Irrigation projects in semi-arid areas, where normally there is a deficiency of lakes and suitable water courses to provide facilities for water-based recreation, immediately become attractive to a large number of people. Storage reservoirs and pools above diversion dams can provide opportunities for boating, swimming, picnicking, camping, fishing, etc. Large canals may be used for fishing, boating, swimming, and other types of recreation. This rapid growth in use and demand for recreation has created important new problems in the planning, design, operation, and maintenance of irrigation projects.

In the past, such activities were not planned for and were generally discouraged (NRLSC, 1999). However, the present trend is to make the facilities available for such purposes. In fact, one of the uses of multiple-purpose reservoirs is now considered to be fishing and recreation. Costs allocated for this use are non-reimbursable in projects financed by the federal government. Furthermore, it is now well accepted that the large amounts of money spent by individuals for recreation can be important to the economy of the area.

The integration of recreational activities into irrigation project facilities, while beneficial, introduces various management challenges. These challenges tend to be more pronounced in older projects where recreational use was not originally envisaged. There are numerous aspects that require careful management. Facilities, such as access routes to recreational areas, camping and sanitary facilities, campgrounds, and boat docks, require operation and maintenance. Safety of visitors must be assured insofar as it is practical. This includes guarding against vandalism, preventing fires, controlling pollution, and minimizing interference with the project's normal operations. These additional responsibilities significantly broaden the scope of duties for management personnel.

Recreational use of project facilities can and does present problems for management personnel of irrigation enterprises. These problems are not as acute in projects which are planned to include this purpose as in older projects where such usage was

DOI: 10.1201/9781003747444-24

not anticipated in the planning. In either case, access to recreational areas, camping and sanitary facilities, camp grounds, boat docks, and many such improvements must be operated and maintained. Vandalism, fires, pollution, and interference with normal operations increase the responsibility and duties of management personnel. This chapter presents some general considerations and discusses well-defined problems related to the recreational use in irrigation projects. Fundamental to recreation is the issue of access, a short discussion of which is in order.

21.2 ACCESS TO PROJECT FACILITIES

When determining public access to irrigation project facilities for recreational purposes, there are three basic choices that may be made concerning visitations to project facilities: (1) prohibition of visitations and recreational use; (2) allowance for access to parts of the system, and (3) encouragement of recreational use of all, or nearly all, project facilities.

21.2.1 Denial of Access

It is difficult to deny the use of reservoirs and waterways in irrigation systems for recreational purposes. Water simply attracts people living in semi-arid areas and they are drawn to it. Fences, locked gates, signs, and possibly armed guards will be circumvented and some people will gain access. On multiple-purpose projects where one of the uses of the water has been established to be recreation, such action is not possible. Public opinion will normally be against such practice. Public relations between project personnel and the people of the community will be strained. In a number of cases, legal action has been taken to force the opening of facilities to the public.

21.2.2 Limited Access

Allowing access to selected areas of the irrigation project, such as reservoirs and regions above diversion dams and parts of the conveyance system, can create valuable recreational spaces without endangering visitor safety or disrupting the system's operations. This approach also typically involves minimal financial investment. However, limited access should be carefully considered, especially in areas where the right-of-way access is limited, or where there are safety hazards, operational interferences, or other critical concerns.

Prioritizing visitor safety is crucial. This involves identifying potential risks in accessible areas and taking steps to mitigate these hazards. Safety measures must be implemented to ensure a secure environment for recreational activities. In addition, recreational activities should not interrupt the routine operational and maintenance procedures of the irrigation system. Transparent communication with the community about restricted access areas is also important. Informing the community about the reasons for restricted access is also important. Through educational initiatives, the public can understand and respect the boundaries and limitations set for safety and operational efficiency. A detailed understanding of safety considerations and public education strategies related to limited access to irrigation projects can be found in Chapter 20.

21.2.3 Full Access

Encouraging comprehensive recreational use of all or most project facilities can pose significant financial challenges for the management. This approach requires careful consideration of various costs, including securing right-of-way, constructing necessary improvements for accessibility and safety, and covering ongoing operating and maintenance expenses. These costs, if solely borne by the district without substantial financial support from external organizations or the users themselves, can lead to financial hardship.

Financial analysis should be conducted to evaluate whether the benefits of enhanced public relations and community engagement justify the associated costs. In multipurpose projects where recreation is an integral purpose, the inclusion of recreational facilities and operational plans is standard. This integration significantly reduces the financial burden, as these considerations are incorporated from the project's inception. Generally, public opinion and, to some extent, legislative frameworks favor comprehensive recreational access. This preference must be weighed against the practical and financial implications for the irrigation enterprise.

Full access typically results in increased duties, responsibilities, and expenditures for the irrigation project. This includes ensuring safety, maintaining facilities, and managing the increased use of resources. While full access to irrigation project facilities for recreation can offer substantial community benefits, it demands careful planning and financial consideration. The decision to full access should be based on a balanced assessment of benefits, costs, and the project's capacity to manage the additional requirements, especially safety.

21.3 OPERATION OF RECREATIONAL FACILITIES

General concerns related to the management and operation of recreational facilities include (1) determining who will provide and operate these facilities, (2) financing construction costs, (3) establishing user fees, and (4) handling liabilities. There are three general procedures for handling recreation in an irrigation project: (1) The district may assume the entire obligation; (2) by agreement, some other organization (such as those of County, State, or Federal governments, or recreation districts) may take over completely; and (3) a combination of the previous two where major facilities like recreation at reservoirs are managed by other organizations, while the district assumes responsibility for recreational areas. This model is often favored, especially in newer projects, as it balances responsibilities and resources. In either of the two latter procedures, agreements must be reached that clearly define the responsibilities of each organization concerned.

21.4 SPECIFIC PROBLEMS

21.4.1 Right-of-Way

One of the major problems in a district where facilities were not planned for recreational use is the matter of right-of-way. This is particularly acute at reservoirs and above diversion dams. If the land immediately above the high-water line is not controlled by the irrigation enterprise and is in private ownership, access to shoreline recreation areas is very difficult, and the matter of trespass becomes a major concern. Public relations become strained, and possible legal action might follow. Acquisition of necessary right-of-way for access very probably will be costly and slow. Management personnel will need to exercise extraordinary judgment to reach even a compromise solution.

21.4.2 Safety of Recreationists

Ensuring the safety of recreationists is a primary concern for management. Activities like fishing and swimming in canals, particularly near structures, pose severe hazards, including drowning risks. Implementing warning signs, constructing protective works, and initiating public awareness programs are essential strategies to reduce these accidents and enhance safety.

21.4.3 Sanitary Regulations

Establishing and enforcing sanitary regulations is crucial for maintaining a healthy recreational environment. This challenge may necessitate the construction, operation, and maintenance of sanitary facilities, adding to overall costs. Clear guidelines and strict adherence are essential to maintain health standards and mitigate environmental impacts.

21.4.4 Other Problems

There are many other problems connected with recreation but with careful planning squarely, many may be solved to the satisfaction of all concerned. Others will remain a source of annoyance and items of added costs. Recreation and fish and wildlife enhancement present many problems for the operators of irrigation enterprises. Solutions to these problems are not always easy and are usually the result of cooperative efforts on the part of many people representing different disciplines and interests. Compromise solutions can be found in most instances by concerted effort and understanding of the problem.

21.5 FISH AND WILDLIFE ENHANCEMENT

Another important consideration in the operation and management of irrigation enterprises is the enhancement of fish and wildlife. Encouragement of the propagation and preservation of game fish and various species of wildlife is now an added responsibility of management. This enhancement may be accomplished many times without making an undue effort, but in some instances, considerable trouble is involved.

21.5.1 Migration of Fish

In the case of fish, stocking reservoirs with desirable species is not too difficult or costly. However, keeping fish out of pumps and turbines and preventing their release from reservoirs usually calls for added construction and consequently added items of operation and maintenance. Dams prohibit migration of fish in public streams; therefore, fish ladders, fish passages, spawning basins, or other facilities may be necessary for passage. Usually, the design and construction of such facilities are not the responsibility of operation, and maintenance is accomplished by an agency concerned with fish management. There are many instances where operators of projects must be involved in planning, designing, constructing, and operating and maintaining the necessary improvements. Solutions to the problems are generally reached by cooperative effort on the part of engineers, fish biologists, and others.

21.5.2 Fish Screen

Perhaps the most common problem is the operation and maintenance of "fish screens" at diversion dams, as shown in Figure 21.1. In some areas in the United States,

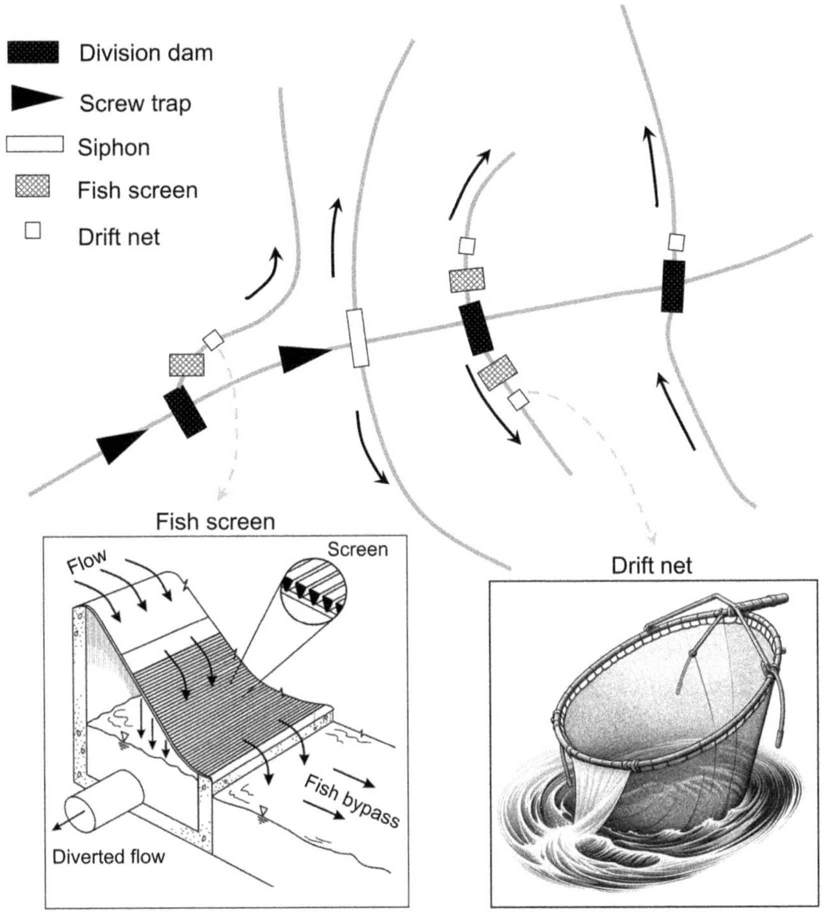

FIGURE 21.1 Illustration of the locations of irrigation canal, diversion dams, fish screens, drift nets, a siphon, and screw traps in a river. (Note: The fish screen figure was adapted from USDA (2013).)

these are required by law to prevent fish in public streams from entering the irrigation system (WDFW, 2000). Usually, these facilities are of a nature such that they greatly magnify the weed and floating debris problems. Further discussion on this subject can be found in Houk (1951). In addition, a technical document from the U.S. Department of Agriculture, Natural Resources Conservation Service provides instructions and useful advice on how to install screens that prevent fish from entering irrigation systems, such as pipelines, canals, and ditches (USDA-NRCS, 2013).

Essential factors to consider when designing a fish screen include the following: (1) approach velocity: The water flow speed toward the screen, considered along with the swimming speed of the target fish; (2) screen sweeping velocity: The water flow speed along the screen, taking into account the fish's ability to swim before needing to rest; (3) approach velocity to sweeping velocity ratio; (4) open area percentage (generally from 27% to 50%); (5) screen opening size, which is determined based on the size of the target fish; and (6) screen location, which is positioned either in the stream or river, or in the outtake with a bypass channel back into the river (USDA-NRCS, 2013).

21.5.3 Fish Enhancement

The establishment of irrigation systems can disrupt the natural equilibrium of streams, negatively affecting the habitats that fish depend on for spawning and feeding. It is essential for planners and designers to anticipate these potential impacts and to integrate effective mitigation strategies into the planning and design process. In practice, it often falls to operation and maintenance teams to uphold and improve these habitats. Effective restoration and maintenance of natural or near-natural stream conditions require collaboration between local authorities, conservationists, and recreational users. By working together, these groups can implement innovative solutions to recreate conditions conducive to the fish's lifecycle and well-being, aiming to satisfy the needs and concerns of all parties involved.

Before the practice of irrigation is introduced into an area, natural streams are somewhat stabilized. Natural pools and feeding and spawning areas for fish are established by nature. When these streams are used to convey irrigation water, many of the natural conditions may be destroyed. It should be the responsibility of planners and designers to foresee this problem. However, operation and maintenance personnel often become responsible for providing and maintaining improvements. The procedures adopted to maintain or restore natural or near natural conditions should be a joint effort among representatives of the district, fish conservationists, and, to a degree, the recreationists. By working together, these groups can implement innovative solutions to recreate conditions suitable for the propagation and preservation of fish, aiming to satisfy the needs and concerns of all parties involved.

21.5.4 Wildlife Enhancement

Similar statements apply to the enhancement of wildlife, such as birds and animals. Areas along the right-of-way that have been seeded to grass may provide suitable habitats. Usually, there are swampy or other wet areas in the project that can be improved and set aside for waterfowl. Non-agricultural lands within the project boundaries may be planted with suitable vegetation to provide habitats for land fowl and small animals.

In the United States, federal funds are now available, under certain conditions, to provide necessary facilities for fish and wildlife enhancement (U.S. Fish & Wildlife Service, 2023a). There is a public expectation to see these efforts in action.

Judicious use of pesticides and herbicides in the project will go a long way toward the preservation of fish and wildlife. Our current understanding of the complex interdependencies of ecosystems is insufficient to predict the long-term consequences of these chemicals. Therefore, it is important to carefully use these chemicals until more comprehensive data becomes available. Historical ecological disasters, such as the extensive fish kills in Germany's Rhine River due to industrial waste, have raised significant concerns among scientists about chemical pollution (Wilken, 2005). The U.S. Federal Government, through the Secretary of Health, Education, and Welfare, initiated steps to phase out DDT nationwide (USEPA, 1975). In addition, the Colorado State Game and Fish Commission implemented monitoring for mercury contamination in game birds, a response to the discovery of mercury in birds that consumed treated wheat seeds. Similarly, in Alberta, Canada, the hunting seasons for pheasant and Hungarian partridge were suspended in 1959, reflecting a commitment to wildlife protection in the face of chemical exposure risks. These measures showcase the need for regulation and careful monitoring of environmental toxins to protect our ecological systems.

Irrigation canals, especially those with impermeable linings, can become death traps for wild animals (e.g., deer) and domestic livestock. These animals enter canals seeking drinking water, but may fall into the canals and become unable to escape or get trapped while trying to cross the obstruction in their paths. To mitigate these risks, the installation of escape mechanisms within the canals, along with wildlife-friendly overpasses, fencing, and other structural enhancements, has proven to be effective in saving lives of many animals. Detailed strategies and devices are discussed in Chapter 20.

21.6 ESTABLISHMENT OF WETLANDS FOR FISH AND WILDLIFE ENHANCEMENT

A wetland is an area of land where the soil is permanently or seasonally saturated with water. These areas may be partially or entirely covered by shallow standing water. Wetlands include swamps and marshes, and their waters may be fresh, saline, or brackish. They support highly diverse ecosystems, with abundant plant communities and animal life, including insects, amphibians, reptiles, birds, and furbearers.

Irrigation districts may have some extra land that can be used to enhance habitats for different species by using the return water or wastewater. There are several partnerships between different institutions and irrigation districts to enhance habitats that may have an economic impact on the region. For example, Ducks Unlimited in Canada has collaborated with irrigation districts to establish wetlands specifically for waterfowl, optimizing water levels for both ecological and agricultural benefits (Ducks Unlimited Canada, 2022). Similar initiatives in the Lower Rio Grande Valley in Texas involve partnerships among irrigation districts, municipalities, and governmental entities to create wetlands that serve as sanctuaries for wildlife, illustrating a commitment to ecological stewardship and conservation (U.S. Fish & Wildlife Service, 2023b).

These efforts highlight an increasing awareness of the importance of wetlands in ecological systems, particularly in relation to agricultural practices and wildlife conservation. The U.S. Fish & Wildlife Service, through the North American Wetlands Conservation Act (NAWCA), provides matching grants to public–private partnerships in the United States, Canada, and Mexico (U.S. Fish & Wildlife Service, 2023c). These grants support projects that align with the North American Waterfowl Management Plan, aiming to protect, restore, and manage wetland ecosystems and habitats crucial for migratory birds and other wildlife in North America. Additionally, the U.S. Department of Agriculture (USDA) has invested significantly in wetland restoration. They allocated $28 million to six new Wetland Reserve Enhancement Partnership (WREP) projects and four ongoing ones (USDA, 2021). This initiative enables conservation partners and producers to collaborate on returning critical wetland functions to agricultural landscapes. These efforts range from restoring forested wetlands in the Mississippi Alluvial Valley to improving wildlife habitat and water quality in various regions across the United States. Moreover, the Wetland Reserve Enhancement Partnership (WREP) is a voluntary program under the USDA that works with eligible partners to leverage resources for high-priority wetland protection, restoration, enhancement, and wildlife habitat improvement (USDA-NRCS, 2023). This program allows for effective integration of wetland restoration in working agricultural landscapes, offering benefits not only to the environment but also to farmers and ranchers who participate in the program.

21.7 CASE STUDIES

21.7.1 Imperial Irrigation District (California)

Under the quantification settlement agreement (QSA), Imperial Irrigation District (IID) committed to fund and implement environmental mitigation projects to offset impacts from Colorado River water transfers, by integrating recreation and fish and wildlife enhancement into its water operations, especially around the Salton Sea. The initiatives balance habitat preservation with safe public recreation, such as fishing, boating, and birdwatching.

Habitat enhancement: In partnership with the California Natural Resources Agency and California Department of Water Resources, IID is supporting the species conservation habitat (SCH) Project (2020–2025) to create over 4,000 acres of habitat at the Salton Sea's southern edge to provide roosting, foraging, and breeding areas for fish-eating birds like pelicans and cormorants (CNRA, 2025). Meanwhile, IID manages flows to the Salton Sea, which help sustain wetland and shoreline habitats critical for migratory birds on the Pacific Flyway (Bureau of Reclamation, 2022). To mitigate habitat losses from water transfers, IID operates and maintains 900 acres of managed marsh near the Salton Sea, providing nesting and feeding areas for waterfowl and shorebirds (IID, 2025). The district undertakes vegetation planting and water-level management to improve conditions for fish and wildlife while accommodating public recreation such as birdwatching and photography. In order to preserve desert pupfish habitat in its drainwater and wetland systems, IID also collaborates with state and federal agencies to integrate fish passage structures and timing maintenance to avoid spawning disruptions (FWS, 1993).

Public recreation access and facilities: Public access points near IID-managed wetlands and along sections of the Salton Sea support activities, such as fishing, birdwatching, and nature photography, though IID coordinates with local agencies to balance access with habitat protection. IID also works with Imperial County to ensure safe boating practices on canals and reservoir areas, with designated launch points and posted safety signage.

21.7.2 Delta Lake Irrigation District (Texas)

The Delta Lake Irrigation District (DLID) actively enhances recreation and wildlife through habitat-focused management (drawdowns, stocking), infrastructure upgrades (bank stabilization, bypass canal), park development (Delta Lake Park), and wildlife-friendly wetlands (marshes for viewing and day-use). These efforts make the reservoir a multifaceted resource, e.g., for agriculture, habitat, and public enjoyment.

Species habitat enhancement: The reservoir supports sport fish, e.g., channel catfish, white bass, largemouth bass, white crappie, and exotic species like grass carp and suckermouth catfish (IFD, 2012). DLID implemented winter drawdowns and spring refilling to boost spawning and recruitment and to control invasive flora and fauna by exposing them to cold temperatures or leaving them on dry banks. Additionally, DLID conducted fingerling stocking of Bluegill, Channel Catfish, and Largemouth Bass. Press releases highlighted Blue Catfish angling opportunities after its successful appearance in the lake. During 2004–2006, DLID drained the reservoir to carry out a bank stabilization project and construct an east-side bypass canal, which improved resilience during droughts and irrigation flexibility. In post-refurbishment, the reservoir was carefully refilled to support aquatic habitat (IFD, 2012).

Recreational access and facilities: Established in 1997, the 65-acre Delta Lake County Park offers amenities, such as barbecue grills, picnic tables, an open pavilion, softball field, playground, jogging/biking trails, and birdwatching zones, raising public engagement with the reservoir environment. DLID operates the park along the reservoir shoreline featuring marsh and pothole habitats with cattails and bulrushes (TCWP, 2025). Visitors can enjoy wildlife viewing, hiking, and picnicking. Day use is free, and group camping is available by arrangement. Delta Lake is a known spot for fishing species like alligator gar, channel catfish, and blue catfish, with regular user reports verifying its recreational use for angling (Fishbrain, 2025).

REFERENCES

Bureau of Reclamation (2022). Salton Sea Program. https://www.usbr.gov/lc/region/programs/saltonsea.html

California Natural Resources Agency (CNRA) (2025). Species Conservation Habitat (SCH) Project. https://projects.saltonsea.ca.gov/Project/Detail/17267

Ducks Unlimited Canada (2022). Ducks Unlimited Canada and Its Partners Begin Second Phase of Detailed Wetland Mapping in the Outaouais. Conserving Canada's Wetlands. https://www.ducks.ca/news/national/mapping-in-the-outaouais/

Fishbrain (2025). Fishbrain Community Data on Delta Lake's Popular Fish Species. https://fishbrain.com/fishing-waters/UnAXTGFq/delta-lake

FWS (Fish & Wildlife Service) (1993). Recovery Plan for Desert Pupfish. https://ecos.fws.gov/docs/recovery_plan/Draft%20Desert%20Pupfish%20RP%20Amendment.pdf

Houk, I.E. (1951). *Irrigation Engineering. II. Agricultural and Hydrological Phases*. Vol. 10. John Wiley & Sons, Inc. II, New York.

IFD (Inland Fisheries Division) (2012). 2012 Fisheries Management Survey Report: Delta Lake. Texas. https://tpwd.texas.gov/publications/pwdpubs/lake_survey/pwd_rp_t3200_1285/#download

IID (Imperial Irrigation District) (2025). Managed Marsh – QSA Related Documents. https://www.iid.com/water/library/qsa-water-transfer/mitigation-implementation/managed-marsh-qsa-related-documents

NRLSC (National Recreation Lakes Study Commission) (1999). Reservoirs of Opportunity. Report of the National Recreation Lakes Study Commission. The Commission.

OIA (Outdoor Industry Association) (2019). Outdoor Participation Report. Outdoor Foundation, pp. 1–41. https://vegetableproject.org/wp-content/uploads/2020/02/2019-outdoor-activity-participation-report.pdf

TCWP (Texas Community Watershed Partners) (2025). Delta Lake Park. https://tcwp.tamu.edu/wetland-restoration/lower-coast-wetland-sites/

USDA (2021). USDA Invests $28 Million in New Projects to Help Restore Lost Wetland Functions, Benefits on Agricultural Landscape. U.S. Department of Agriculture, Washington DC. https://www.usda.gov/media/press-releases/2021/03/15/usda-invests-28-million-new-projects-help-restore-lost-wetland

USDA-NRCS (2013). Technical note: Guide to Fish Screens. Pocket Guide to Screening Small Water Diversions. A Guide for Planning and Selection of Fish Screens for Small Diversions. U.S. Department of Agriculture, Natural Resources Conservation Service, Washington, DC.

USDA-NRCS (2023). Wetland Reserve Enhancement Partnership. U.S. Department of Agriculture, Natural Resources Conservation Service, Washington, DC. https://www.nrcs.usda.gov/programs-initiatives/wrep-wetland-reserve-enhancement-partnership

USEPA (1975). DDT Regulatory History: A Brief Survey. U.S. Environmental Protection Agency. https://www.epa.gov/archive/epa/aboutepa/ddt-regulatory-history-brief-survey-1975.html

U.S. Fish & Wildlife Service (2023a). Wildlife Restoration. https://www.fws.gov/program/wildlife-restoration

U.S. Fish & Wildlife Service (2023b). Lower Rio Grande Valley National Wildlife Refuge. https://www.fws.gov/refuge/lower-rio-grande-valley

U.S. Fish & Wildlife Service (2023c). North American Wetlands Conservation. https://www.fws.gov/program/north-american-wetlands-conservation

WDFW (Washington Department of Fish and Wildlife) (2000). Fish Protection Screen Guidelines for Washington State, April 2000. https://wdfw.wa.gov/sites/default/files/publications/00050/wdfw00050.pdf

Wilken, R.D. (2005) The recovered rhine and its history. In: T.P. Knepper, Eds., *The Rhine*. The Handbook of Environmental Chemistry, vol. 5L. Springer, Berlin, Heidelberg. https://doi.org/10.1007/698_5_036

22 Public Relations

22.1 INTRODUCTION

Webster defines public relations as "the art or science of developing reciprocal understanding and goodwill between a person, firm, or institution and the public". The public ordinarily refers to the people as a whole but can relate to people with a common interest. In the context of irrigation management, this involves not only the general public but also specific groups with different interests, such as water users, employees, community members, partner agencies, and government institutions. There is no better way for management to develop "reciprocal understanding and goodwill" than to communicate with all those concerned. It is, therefore, essential to keep them informed and up to date on what has happened, what is currently happening, and what is projected for the future. These future projections can serve as a "sounding board" as to whether the general public or others will accept the proposed actions well or whether opposition may be expected. When encountering opposition, this may be because the proposal is not understood. The key to resolution in such situations is open and transparent communication. Communication methods, including public hearings, press releases, and interactive media platforms, can significantly aid in addressing and resolving misunderstandings.

The management of an irrigation enterprise may take numerous courses of action, which may be within legal bounds. However, in taking such actions, it is possible that relationships between water users and district management or between the local population and district management might be strained. It is also possible that relationships with the county or state governments could be affected. The management personnel of irrigation enterprises must be thoroughly aware of the legal framework within these bounds. They must at all times maintain public relations at the highest level of practicality. Therefore, keen adjustments must be made to avoid legal issues and, at the same time, arrive at decisions that will not antagonize the public or personnel within the district. They must not take action that the law empowers them to carry out legally if it appears that public relations will be strained or broken. Some examples include decisions on matters, such as drainage, bridges across conveyances, recreational activities, mosquito control, fences, trespass rights, and right-of-way. The operation and maintenance of the drainage system are usually included in the operating budget. Additionally, a minor amount may be allowed each year for the construction of needed drainage systems. If sinking funds are accumulated for the purchase of major equipment items, rehabilitation, and betterment, or other reasons, the money set aside each year for these funds must be included in the budget.

It is well to remember that Charles Simmons, an American literary figure, has said, "Public opinion, or public sentiment, is able to sustain, or pull down any law of the common wealth". This highlights the power of public perception and its impact on legal and administrative frameworks. Therefore, maintaining good public

DOI: 10.1201/9781003747444-25

relations should be one of the principal objectives of the management personnel of an irrigation district. To achieve this, a strategic and continuous program is essential. A "hit-and-miss" effort will do little more than mend strained situations that develop. Proactivity, consistent engagement, and communication are key to building and preserving a favorable public image, which, in turn, supports the successful and sustainable operation of irrigation enterprises.

Whatever is done in the field of public relations must be done well. It is crucial to understand that public relations activities are not routine tasks; instead, they reflect the enterprise's overall commitment to quality and efficiency. High-quality publications, well-planned tours, open houses, and other events are essential. If these activities are executed haphazardly, it could lead the public to form a negative perception of the enterprise's overall operations and business practices.

The qualities that are exhibited in public relations influence how others perceive the quality of enterprise. Public relations must be maintained between irrigation enterprises, the general public, other agencies and organizations, water users, and personnel employed in the enterprise. One of the best means to foster this relationship is to communicate with all concerned individuals. Keeping all parties well-informed about past achievements, current initiatives, and future plans is essential for fostering mutual understanding and trust. Without this ongoing communication, there is a risk of misunderstanding and suspicion, which can hamper the organization's goals and operations.

A well-executed public relations strategy does much more than manage the image of the enterprise. It plays a fundamental role in the successful management of the organization, facilitating smoother operations and contributing to overall success. It involves a strategic, ongoing effort to engage, inform, and collaborate with all relevant stakeholders, ensuring that the actions of the organization are understood and supported. Such a comprehensive public relations strategy is integral to establishing and sustaining a positive institutional reputation and ensuring long-term success.

22.2 STRATEGIES FOR PROMOTION OF PUBLIC RELATIONS

Effective public relations are important to the success of irrigation enterprises. A well-structured program helps build trust, enhances relationships with stakeholders, and promotes the efficient and reliable operation of the system. There are some ways to establish and maintain good public relations. Many of them can be carried out with little effort and expense. Others will be more difficult, but should not be omitted from the program. The establishment and maintenance of robust public relations can be achieved through various strategies.

Several low-cost strategies include (1) leveraging social media, (2) networking, (3) community involvement, and (4) media relations. Social media platforms, such as Facebook, X (Twitter), or local WhatsApp groups, are cost-effective channels through which to share updates, announce water delivery schedules, promote conservation practices, or highlight achievements. A consistent online presence also counters misinformation by providing accurate, real-time updates. Networking, such as participation in professional conferences, agricultural fairs, and community forums, increases visibility and credibility. Such interactions allow managers to exchange ideas and promote the enterprise's contributions to water management and local development. Community involvement is also significant. Participation in local events, sponsorship of youth programs, or volunteering in conservation projects demonstrate that the irrigation enterprise investing in the welfare of the community it serves. This approach enhances goodwill and fosters loyalty among water users and residents. Establishing and maintaining positive relationships with the media is important to ensure accurate messaging and strong public engagement. Developing relationships with local journalists and providing timely press releases ensures that the enterprise's perspective is accurately represented. Positive media coverage can strengthen public confidence and can highlight an organization's success.

Enhanced investment strategies for an irrigation enterprise's public relations program may include (1) dedicated professional development, (2) comprehensive crisis communication planning, and (3) in-depth research and analytics. Training staff in communication and customer service skills enables them to handle inquiries effectively, manage conflicts, and act as ambassadors of the enterprise. Address how to manage public relations during crises, such as droughts or controversies over water rights. Strategies for open communication during these times should be provided to maintain trust and credibility. Using surveys, focus groups, and digital analytics tools helps managers understand public perceptions and measure the impact of public relations efforts. These insights allow for data-driven refinement of strategies.

In an era of rapidly spreading misinformation, irrigation enterprises must prioritize accuracy, honesty, and inclusiveness. Open hearings, community feedback channels, and participatory forums help ensure that diverse viewpoints are reflected in decision-making. Ethical communication fosters not only compliance with legal requirements but also long-term trust and cooperation.

In addition to traditional approaches, many irrigation districts now employ digital innovations, including (1) stakeholder dashboards that provide real-time updates on water allocations; (2) mobile apps that notify users about water delivery schedules or outages; and (3) virtual town halls to reach dispersed stakeholders who cannot attend in person. These tools expand the reach of public relations efforts and promote transparency across a broad audience.

Example of Central Arizona Project (CAP): The CAP faced increasing scrutiny over water allocation priorities during recent drought negotiations. To address public concerns, CAP launched a comprehensive communication program that included (1) an online stakeholder portal called AquaPortal, with real-time data on water delivery and storage (CAP AquaPortal, 2025); (2) a public education campaign branded as "CAP University", offering tours, workshops, and webinars (CAP University, 2025); and (3) regular updates through social media and press releases in both English and Spanish. These efforts reduced misinformation, improved transparency, and fostered greater trust among CAPs, farmers, urban users, and tribal communities.

22.3 ENGAGEMENT WITH WATER USERS

Engagement with water users is crucial for managing an irrigation enterprise (Hugh, 1995). Successful operations depend on cultivating trust, understanding, and collaboration with individuals and organizations that rely on water services.

When users feel heard and informed, conflicts are minimized, and cooperative problem-solving becomes possible. Several strategies can strengthen these relationships: (1) Responsive communication. All calls, emails, and in-person inquiries should be welcomed and addressed promptly. Active listening demonstrates that the organization not only values user perspectives but is also committed to understanding and responding to their needs. (2) Equal opportunity for interaction. Every water user, regardless of farm size, economic status, or crop type, should have fair access to management. Equitable treatment builds confidence that decisions are made transparently and without bias. (3) Demonstrations and open houses (Axthelm, 1968). The enterprise should regularly provide demonstrations of new developments, such as advanced equipment, innovative water measurement devices, or planned infrastructure upgrades, allowing users to see progress firsthand. Open houses following the completion of new structures or the acquisition of equipment foster transparency and pride in collective achievements. (4) Cultivating positive relationships. Every employee, from field technicians to office staff, represents the enterprise. Training in communication and customer service ensures that staff interactions consistently reflect professionalism and respect. An organizational culture that values courtesy and responsiveness reinforces goodwill at all levels. (5) Feedback and continuous improvement. Encouraging users to provide suggestions or raise concerns through surveys, community meetings, or digital platforms helps managers identify issues early. Integrating user feedback into decision-making enhances both service quality and legitimacy.

22.4 EMPLOYEE RELATIONS

The success and efficiency of an irrigation enterprise depend not only on sound infrastructure and financial stability but also on the motivation, satisfaction, and commitment of its employees. Public relations, therefore, must extend both inward and outward. Positive employee relations foster a cohesive workplace culture, reduce turnover, and ensure that employees serve as credible ambassadors for the organization when interacting with water users and the public. The key strategies for maintaining healthy employee relations include the following: (1) Personal engagement. The enterprise should show interest in employees' personal problems and situations, fostering a supportive work environment where employees feel valued not just as workers but also as individuals. Understanding and addressing personal issues can enhance employee satisfaction and strengthen organizational loyalty. (2) Cultivating team pride: Instill a sense of pride in being part of the team. This can be achieved through team-building activities, recognizing and celebrating individual and collective achievements, and routinely highlighting the team's role in the company's success. Strong team spirit enhances collaboration and a sense of belonging. (3) Providing quality work conditions. The working environment should be safe, comfortable, and conducive to productivity. This includes providing adequate, modern equipment and facilities that enable employees to perform their duties efficiently. A satisfactory work environment demonstrates a company's commitment to its employees' well-being. (4) Transparent communication of policies. All employees should be kept well-informed about policies related to wages, promotions, and

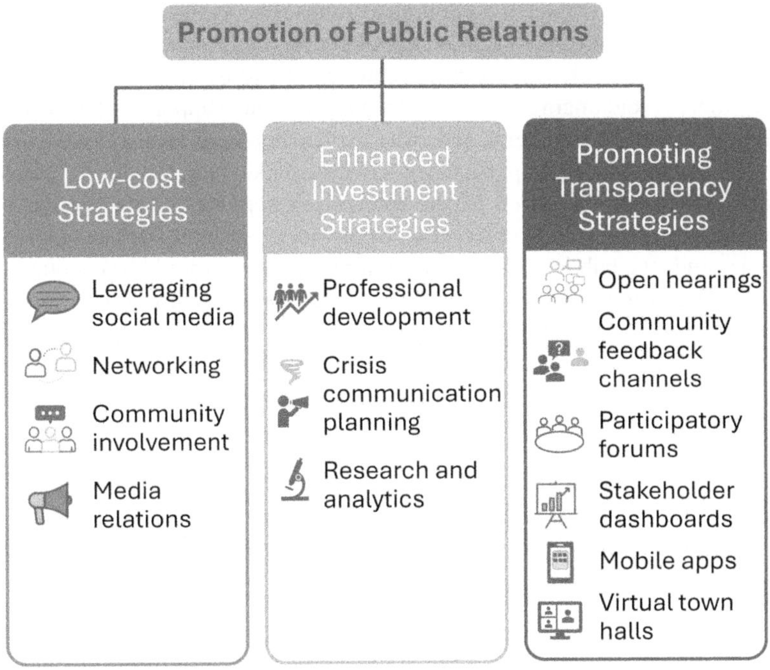

FIGURE 22.1 Strategies for promoting public relations.

other employment matters. Transparency in communication builds trust and helps in managing expectations. Regular updates and open discussion forums provide an effective means of maintaining clear communication and encouraging stakeholder engagement. (5) Recognition and awards. An award program should be initiated to acknowledge and reward jobs well done. Recognizing individual and team achievements increases morale and motivates employees to maintain high performance standards. This can range from formal award ceremonies to informal acknowledgments in team meetings.

22.5 PUBLIC

Engaging with the public is an important aspect of maintaining a positive image and transparent operations for an irrigation enterprise. Effective public engagement informs and builds trust and understanding about the activities and objectives of the organization. The following strategies can be used to improve public relations: (1) Diverse media communication. The organization should utilize various news media channels, including newspapers, television, radio, and personal contacts to communicate with the public. This broad approach ensures wide reach and caters to different preferences in how the public consumes information. (2) Hosting public events. Open houses, demonstrations, and field days should be organized to showcase the enterprise's operations and development. Such events offer a hands-on and transparent view of the organization's work, fostering public interest and engagement.

(3) Outreach to educational and community groups. The management should actively participate in public forums, such as school events, club meetings, and other community gatherings. These opportunities can be used to explain ongoing and future projects, thereby enhancing public understanding and support. (4) Informational materials. Informative brochures and other media should be created and distributed. A monthly newsletter can be particularly effective in keeping the public, employees, and other stakeholders updated on the enterprise's activities and achievements. (5) Welcoming visitors. Visits to enterprises from various groups, such as business professionals, students, public officials, and community clubs, should be encouraged and facilitated. Providing guided tours is an excellent way to demonstrate transparency and engage directly with different segments of the public. (6) Openness to feedback. Maintaining an open and receptive stance to public comments and feedback. Listening to and considering different perspectives can provide valuable insights and assist in adapting services and communications to more effectively align with public needs and expectations.

22.6 COLLABORATION WITH AGENCIES AND ORGANIZATIONS

Effective collaboration with external agencies and organizations can establish a robust network of support and shared knowledge. This cooperation can lead to better decision-making and enhanced operational efficiency. The following approaches can be used to build and maintain these collaborative relationships: (1) Engagement in consultation and assistance. The organization should actively seek comments or assistance from relevant agencies and organizations when their interests intersect with a proposal or project. This not only brings diverse expertise to the table but also ensures that all stakeholders are considered in the decision-making process. (2) Offering mutual support. These entities should proactively cooperate in addressing the challenges faced by these entities. This support can strengthen relationships and open up opportunities for future collaboration. (3) Showing genuine interest. The management should regularly visit their offices or operational sites to demonstrate an interest in their activities and initiatives. This helps better understand their operations and fosters a sense of camaraderie and mutual respect. (4) Inviting representatives in relevant discussions: When discussing matters that concern these agencies and organizations, ensure that their representatives are invited to meetings. This inclusion not only provides valuable insights but also reinforces the importance of their contributions to the enterprise. (5) Sharing information and resources: The management should include these entities on the mailing list for the distribution of publications, reports, and other relevant materials. Keeping them informed about the enterprise's activities and findings promotes transparency and ongoing knowledge exchange.

22.7 CASE STUDIES ON PUBLIC RELATIONS

22.7.1 Imperial Irrigation District (IID), California

The Imperial Irrigation District (IID), which serves over 500,000 acres in southern California, has long faced challenges involving water rights, environmental concerns, and cross-border issues with Mexico. To address growing public scrutiny, the district launched the IID Water Conservation Program, a large-scale outreach initiative that included bilingual newsletters, community workshops, and farmer demonstration plots. These proactive efforts reduced tension during negotiations with the state and improved cooperation with farmers.

The IID emphasizes transparency by making public records available under the California Public Records Act (CPRA). This allows stakeholders, including water users, agencies, and the public, to formally request and access key documents. The IID has institutionalized its public relations strategies through board-level policies. A notable example is *IID Resolution No. 6-2025*, which formally directed staff to produce a detailed report on the district's community outreach and communications plan (IID Board Resolutions, 2025). In addition to allowing the public to request records from the district by directly submitting a request via official email or mail box, the IID maintains a *News & Resources* channel on its official website to routinely release annual reports, budget plans, and audited finance statements, as well as video records and livestreams of IID meetings (IID News & Resources, 2025). The *News & Resources* channel is also actively used to update ongoing projects, policies, operations, and alerts, keeping stakeholders informed in a timely manner. During periods of high temperatures or strain on the electrical grid, IID issues "Conserve Alerts" to encourage voluntary energy conservation and prevent potential power outages through various channels, including the official IID website and social media platforms such as Facebook, the IID Connect Mobile App, and Customer Service Center.

Following controversial decisions, such as the proposed rate hikes, IID has demonstrated responsiveness to stakeholder concerns by enhancing outreach efforts and reviewing the effectiveness of its communication strategies. Following public backlash in early 2025, IID's board instructed staff to expand community engagement efforts and ensure that information was delivered more clearly and effectively. Similarly, when facing a significant budget deficit in late 2024, IID emphasized the importance of transparent communication and improved outreach as part of its fiscal response strategy.

The IID also engages the community proactively through public outreach and funding support, particularly by enhancing safety, sustainability, and community well-being. For example, in July 2025, the IID announced the provision of $300,000 in public benefit funds to support 12 energy-efficient lighting projects at local schools and municipalities across IID's service area, which is part of IID's new GLOW (Green Lighting for Outdoor Wellness) grant program. At the same time, the IID plans to award over $500,000 in public green space grants to support much-needed shade and environmental improvement projects across the Imperial and Coachella Valleys. To support safe water activities in the 2025 summer, the IID Board of Directors approved a $40,000 fund to help offset expenses associated with keeping pools open, providing free or discounted swimming lessons, and offering safe water activities for youth and adults, such as water slides in the Imperial Valley.

The IID engages in several practices to foster positive relationships with its employees. For example, IID's commitment to safety is evident in their award-winning programs and safety culture, which fosters a secure environment for their team members. IID supports employee growth through initiatives such as a Professional

Development Program for graduating electrical engineering students, offering opportunities to learn and contribute to a growing organization (IID Professional Development Programs, 2025).

The IID involves significant collaboration with other agencies and organizations, particularly in areas such as water resources management, energy initiatives, and community programs. For example, the IID is a partner in the development of a multi-species management plan for the Lower Basin (Water Education Foundation, 2014), working alongside the Bureau of Reclamation and other federal, state, and local partners. This plan aims to protect threatened and endangered species while ensuring the continued delivery of water and power benefits. Additionally, the IID is a balancing authority within California and collaborates with other entities to manage its diverse resource portfolio, which includes native generation, partnerships with the Southern California Public Power Authority (SCPPA), and power purchases to meet its Renewables Portfolio Standard (SCPPA, 2025).

22.7.2 Lower Rio Grande Valley, Texas

The irrigation districts in the lower Rio Grande Valley face recurring droughts, seawater intrusion, and disputes over Mexico–U.S. water sharing agreements. To maintain trust, districts adopted transparent water accounting systems and mobile phone alerts for farmers about water delivery schedules. They also hosted joint forums with universities and local governments, which helped reassure farmers about the district's efforts while also enhancing credibility with policymakers. For example, institutional staff in the irrigation districts in the valley regularly engage with regulatory bodies, such as the Texas Commission on Environmental Quality, reflecting sustained collaboration and communication with oversight agencies (Lower Rio Grande Valley Development Council, 2024). Additionally, a Texas Water Resources Institute-led initiative, supported by a 10-year grant, expanded from a metering calibration facility into the Rio Grande Center for Agricultural Water, serving as a public resource to educate and support irrigation stakeholders on efficient water practices (TWDB, 2017).

The 2016 Regional Water Plan states that "educational programs for farmers, irrigation district boards of directors, and irrigation district employees are recommended and should be supported by the TWDB, TCEQ, and the universities in Texas" (TWDB, 2016). To support this recommendation, the Texas Water Development Board (TWDB) launched the Lower Rio Grande Valley Irrigation Education and Outreach Program, which provides growers and irrigation district personnel with pertinent and timely educational resources on irrigation water conservation (Gregory et al., 2020). The irrigation districts in the valley are engaged in a variety of public communication activities. For example, public hearings are held for significant milestones, such as the release of draft regional water plans. Workshops address the role of irrigation district infrastructure and operating practices in on-farm water conservation. The Rio Grande Watermaster Program communicates daily with the International Boundary and Water Commission (IBWC) and water rights holders regarding river and reservoir conditions and water releases (Rio Grande Watermaster Program, 2025).

22.7.3 Central Arizona Project (CAP), Arizona

The CAP manages a 336-mile aqueduct delivering Colorado River water to Arizona. Public relations challenges arose during debates about water allocation priorities between agriculture, cities, and tribes. CAP responded by creating the CAP University program, where citizens, students, and stakeholders attend seminars, facility tours, and workshops to learn about the system (CAP University, 2025). This initiative significantly improved the understanding of the CAP's role and reduced misinformation in public debates.

The CAP employs several public relations practices designed to foster community awareness and promote transparent stakeholder engagement. It operates public stakeholder briefings on topics such as water operations, rate setting, and budgeting, which help maintain open lines of communication with diverse constituencies, including tribal communities (CAP Community, 2025). CAP also ensures accessibility through public board meetings, which are available via live streams, enabling community members to observe governance proceedings remotely. In addition to reinforcing its outreach capacity, the CAP's leadership, including the Director of Public Affairs, oversees internal and external communications, media relations, and stakeholder outreach programs (CAP Management Council, 2025).

22.7.4 South San Joaquin Irrigation District (SSJID), California

The canal systems of the South San Joaquin Irrigation District (SSJID) have been modernized, and pressurized pipelines have been implemented. To foster acceptance, it conducted pilot demonstrations on farms, invited growers to test new water measurement devices, and provided cost–benefit analyses. These measures encouraged farmers to adopt innovations and demonstrated that the district valued transparency and user feedback.

The SSJID has implemented public relations efforts to reinforce community messaging and transparency. The district launched a public awareness campaign with the slogan "Water=Food=Life" that used billboards and messaging to highlight the critical connection between water rights and regional food security during periods of drought, aiming at both urban and agricultural audiences. The SSJID also leverages board meeting technology, offering public access via Zoom for regular board sessions, thereby enabling broader community participation and transparency in governance (SSJID Board Minutes, 2024). Moreover, the public and government relations manager is embedded in project communications, including updates and groundbreaking events for major infrastructure projects (Tri-Dam Power Board Packet, 2025).

REFERENCES

Axthelm, D. (1968). *Guidelines for New Directors of Irrigation Districts*, E.C. 68-774. Extension Service, University of Nebraska, College of Agriculture and Home Economics, Lincoln, NE, 2268.

CAP Aquaportal (2025). Central Arizona Project. https://aquaportal.cap-az.com/Data

CAP Community (2025). Central Arizona Project community. https://www.cap-az.com/community/

CAP Management Council (2025). Bridget Schwartz-Manock, Public Affairs Director, Central Arizona Project. https://www.cap-az.com/about/management-council/bridget-schwartz-manock/

CAP University (2025). Central Arizona Project University. https://www.cap-az.com/community/cap-university/

Gregory, L., Nelson, S., Simpson, C., Young, M., Prewett, R., and Ford, K. (2020). Lower Rio Grande Valley Irrigation Education and Outreach Program. Texas Water Development Board Project Contract No. 1613581996. https://www.twdb.texas.gov/publications/reports/contracted_reports/doc/1613581996.pdf

Hugh, T. (1995). Recent Trends in Irrigation Management Changing Directions for the Public Sector. Natural Resource Perspectives, no. 5. https://hdl.handle.net/10535/4232

IID Board Resolutions (2025). Imperial Irrigation District Resolution No. 6-2025: Directing a Report on Community Outreach and Communications. https://www.iid.com/about-iid/department-directory/cal-public-records-act-requests

IID News & Resources (2025). Imperial Irrigation District News & Resources. https://www.iid.com/about-iid/news-resources/news-releases

IID Professional Development Programs (2025). Imperial Irrigation District Professional Development Programs. https://www.iid.com/Careers/Professional-Development-Programs

Lower Rio Grande Valley Development Council (2024). Board Meeting Packet. https://www.lrgvdc.org/downloads/packets/LRGVDCBoardMeetingPacket03-27-24.pdf

Rio Grande Watermaster Program (2025). Rio Grande Watermaster Program. Texas Commission on Environmental Quality. https://www.tceq.texas.gov/permitting/water_rights/wmaster/rgwr#duties-of-the-watermaster

Southern California Public Power Authority (SCPPA) (2025). Irrigation District – SCPPA. https://scppa.org/about-us/member-imperial/.

SSJID Board Minutes (2024). Regular Meeting Minutes, South San Joaquin Irrigation District. https://www.ssjid.gov/wp-content/uploads/SSJID_Minutes_121024.pdf

Texas Water Development Board (TWDB) (2016). 2016 Regional Water Plan Project Prioritization. https://www.twdb.texas.gov/waterplanning/rwp/plans/2016/

Texas Water Development Board (TWDB) (2017). Agricultural Water Conservation Demonstration Initiative Report. https://www.twdb.texas.gov/publications/reports/contracted_reports/doc/503580013_FinalReport.pdf

Tri-Dam Power Board Packet (2025). TRIDAM Project Updates. https://www.tridamproject.com/files/b8e4f4f26/E-Packet%2B07.17.2025.pdf

Water Education Foundation (2014). Lower Colorado River Multispecies Conservation Program. https://www.watereducation.org/aquapedia/lower-colorado-river-multi-species-conservation-program

23 Coping with Change

23.1 INTRODUCTION

One of the most valuable assets for the management personnel of irrigation enterprises is likely the ability to look ahead and visualize the effects of change on operations. To become lost in the details of current activities, with no time to observe external trends or plan for the future, will stagnate the organization and negatively impact its performance. This status will have a degenerating effect on all activities of the enterprise.

The fast-moving advances in technology, the increased demands for food and fiber to sustain a rapidly expanding population, and the growing competition among users of limited water resources for purposes other than irrigation necessitate careful and continuous assessment of current conditions and forecast future possibilities. Organizations cannot remain static. They must either move ahead or else will slip behind other similar endeavors.

The conditions in an irrigated area are never static. Changes in land use resulting from urbanization, industrial development, the construction of freeways, new crops, changing cropping patterns, etc., may be expected. Cropping patterns shift with markets, climate, consumer preferences, and technology. Water may be allocated to higher-priority uses. Some agricultural land may be converted to recreational or environmental uses. Acreage can be lost to channel enlargement, new drains, soil salinization, or the retirement of land of poor quality from crop production. Because systems are usually designed to serve maximum acreage and specific service standards, such changes can stress hydraulics and finances. Small changes may require minor adjustments, whereas major changes may necessitate rehabilitation or reoperation strategies. The results are likely to have a significant impact on fiscal considerations and could lead to severe financial difficulties. Management must be alert to signs of looming change and prepare through structured, anticipatory planning.

Specific guides in this subject area are challenging to establish because of the wide range of activities involved in an irrigation enterprise. However, some guidelines can be proposed. Bearing in mind that the primary objective of the organization is to deliver water equitably, efficiently, and at a reasonable cost to the water users, it is more or less evident that any pending development or trend that will affect this objective will also affect the standing of the organization and perhaps the ultimate success of the project.

23.2 CHANGES IN LAND USE

The dynamics of land use within an irrigation project area have profound implications for its operations and sustainability. These changes, often gradual but sometimes abrupt, necessitate a vigilant and adaptive management approach. The key

DOI: 10.1201/9781003747444-26

factors influencing this change include (1) urbanization, (2) shifts in cropping, (3) transportation and industrial infrastructure development, (4) environmental and recreational land conversion, and (5) regulatory changes.

The encroachment of urban development into traditionally agricultural areas is a significant factor. Urban expansion often requires a comparable amount of water per acre as agricultural use; however, the quality standards, delivery methods, and usage patterns vary significantly. Urban water use typically requires higher quality and often involves more complex distribution systems. As cities expand, irrigation enterprises must adapt their water delivery and treatment processes to meet these new demands, while balancing the needs of remaining agricultural users.

Trends in agricultural production can also reshape irrigation needs. A shift from water-intensive crops to less demanding ones, or vice versa, can significantly alter water usage patterns. These shifts could be driven by market demands, the impacts of climate change, or advances in agricultural technology (Singh and Su, 2022a). Flexible operations policies, such as service windows, minimum order sizes, and advance notice rules, help districts adapt without sacrificing conveyance efficiency.

The loss of agricultural land due to infrastructure development, such as roads, drains, freeways, and industrial plants, can result in a reduction in the total service area of an irrigation project. This not only impacts the direct availability of water for agricultural purposes but can also impose financial strains on the remaining agricultural lands. The reduced user base might lead to increased costs for the remaining users, challenging the economic viability of the irrigation enterprise.

In some cases, agricultural lands, such as wildlife refuges or recreational uses, are converted for environmental conservation. While these changes can be beneficial for ecological balance and community recreation, they alter the water usage landscape. The irrigation system must adapt to these new uses, which may have different water quantities and quality requirements than traditional agriculture does.

Zoning laws and environmental regulations also influence land use. Zoning codes, floodplain ordinances, riparian and wetland setbacks, source-water and habitat protection overlays, groundwater sustainability mandates, and total maximum daily load (TMDL)-driven water quality requirements condition where when, and how irrigation can occur. Overlays can narrow the irrigable footprint, alter the required delivery standards, require additional monitoring, or prompt rotational fallowing/managed retirement of sensitive tracts. In some jurisdictions, groundwater sustainability plans and source-water protection also drive land fallowing or crop switching in vulnerable basins. These effects allow districts to plan canal reratings, turnout consolidations, or pipeline conversions before conflicts materialize.

To stay ahead of land use change, irrigation enterprises should maintain an authoritative GIS showing parcels, easements, and all regulatory overlays; refresh it at least annually with planning/permitting agencies; and set up an early-notification protocol for subdivision maps and major permits. Track and publish service-level commitments (pressure, flow, and delivery windows) by zone, and update the service charter as conditions evolve. Adopt clear connection, relocation, and abandonment policies that recover stranded and betterment costs through capacity fees and standardized agreements. Use developer agreements to fund relocations, appurtenances, and metering upgrades, and to record easements and access rights. Scenario-test proposed zoning or regulatory changes for hydraulic performance, e.g., peak day/peak week, freeboard, spill risk, water-quality compliance, and revenue stability; then brief the board with mitigation pathways and triggers, e.g., when perennial acreage exceeds a threshold. Finally, expected changes should be communicated early to water users and municipal partners so that planting decisions, subdivision design, and capital projects can be coordinated rather than contested.

23.3 DRAINAGE AND SALINITY

As urban areas expand and competition for high-quality freshwater intensifies, irrigation districts increasingly rely on the reuse and careful management of saline and lower-quality waters. These sources include drainage water, irrigation return flows, treated municipal effluent, stormwater, brackish groundwater, and livestock wastewaters. The central question is no longer whether such waters will be used but how to use them safely and profitably while protecting soils, aquifers, and receiving streams. Successful examples of the use of saline water for agricultural purposes are documented in Hoffman and Shalhevet (2007). These examples include utilizing water with total dissolved solids (TDS) of approximately 1,500–5,000 mg/L in Colorado's Arkansas Valley for crops such as alfalfa, sorghum, and wheat; employing water with 2,500 mg/L TDS in Texas's Pecos Valley for alfalfa, melons, and cotton; and leveraging monsoon rains to irrigate wheat with water containing up to 10,000 mg/L in regions of India. Similarly, in Uzbekistan, water with 5,000–6,000 mg/L has been used to cultivate cotton with notable yields, and in semi-arid regions of China, saline water ranging from 2,000 to 5,000 mg/L has been applied since 1970 for crops, such as wheat, corn, and cotton. In each case, success depended on sound drainage, realistic leaching targets, careful crop/variety selection, and timing/blending strategies that match water quality to crop sensitivity.

Looking ahead, climate variability will intensify both sides of the problem. More intense, short-duration storms exacerbate waterlogging and increase erosion and sediment export, whereas longer, hotter dry spells intensify the evaporative concentration of salts in the root zone. Sustainable agriculture, therefore, requires controllable drainage rather than fixed outlets. Districts should retrofit key laterals with low-head weirs or adjustable risers to raise or lower the water table by the season and forecast. This approach holds water higher during cool, noncritical periods to conserve soil moisture and draws down ahead of planting or large storms to preserve trafficability and root aeration. Because retention alone does not ensure acceptable water quality, controllable drainage should be paired with edge-of-field practices such as denitrifying bioreactors, two-stage ditches, and constructed wetlands to intercept and transform nitrogen, reduce the amount of suspended solids, and polish return flows before they enter canals and streams (USDA-ARS, 2021).

Salinity management itself must evolve. The leaching requirement should be determined on the basis of the salinity of the applied water (electronic conductivity, EC), measured soil salinity (EC of the saturated paste extract), crop-specific salt and ion tolerances, and stage of growth, rather than from blanket, conservative tables. Evidence-based leaching targets can provide equal or better control with less water and lower off-site salt and nutrient loads. In practice, districts should formalize blending and cyclic-use rules (e.g., applying more saline water early in the season and switching to fresher water during sensitive stages), schedule cool-season flushing windows when off-site risks are lower, and establish a simple, auditable monitoring program. For example, EM-directed soil salinity mapping (electromagnetic-induction surveys to guide and interpret soil salinity sampling), continuous or routine EC and flow measurements at selected tiles/outfalls, and targeted nutrient plus, where reuse is practiced, can be used. Housing these data in a cloud system with quality control and clear dashboards enables timely operational decisions (USDA-ARS, 2021).

Because storage remains a limiting factor, many districts will also combine reuse with managed aquifer recharge (MAR) and aquifer storage and recovery (ASR), often leveraging soil-aquifer treatment during subsurface passages. Where geology and policy allow, capturing storm peaks, treated effluent, or tailwater for infiltration basins or river-bank galleries can buffer supply and dilute salinity. Safeguards are important: pretreatment to limit clogging and pathogens, monitoring wells to track attenuation and breakthrough, operating rules for rest and recovery cycles, and source/quality criteria that protect crops at points of recovery. Accounting should clearly distinguish between consumptive use, return flows, and recharge credits so that benefits can be recognized in water rights and water quality programs.

Sound economics and environmental responsibility are integral to coping with change. Every drainage or salinity intervention, including controlled drainage, lining, piping, wetlands/bioreactors, blending facilities, and MAR/ASR, should be evaluated on both unit water and salt outcomes (e.g., $ per acre-foot conserved or reused; $ per ton salt prevented from export) and farm-gate metrics (yields, crop water productivity, energy per acre-foot). Hydrologic studies to identify restricting layers, perched zones, and lateral connectivity help target investments where waterlogging and salinity risks are highest and where edge-of-field treatment intercepts the largest loads. To maintain accountability and enable adaptive management, an annual monitoring and evaluation report should be published, summarizing the following: (1) the shear of fields within crop soil salinity tolerance; (2) median seasonal water table depth; (3) trends in electrical conductivity and nutrient concentrations at drainage outlets; and (4) volume captured, reused, or recharged. Each indicator should have predefined performance thresholds with corresponding, pre-authorized management actions.

23.4 APPLICATION OF WATER

As the cost of water and energy increases, irrigation districts are increasingly motivated to implement conservation strategies to deliver the right amount of water at the right time while reducing losses. In practice, success depends on three linked capacities: reliable measurement and telemetry at diversions and turnouts; decision support that translates weather–soil–plant information into executable set-points for timing and depth; and delivery systems (sprinkler, drip, modernized surface) capable of implementing spatially and temporally variable prescriptions. Modernization, therefore, includes lining or piping to reduce seepage, redesigning checks and spillways for stable control, and implementing targeted automation with accurate flow measurement. These network upgrades need to be coordinated with on-farm shifts toward pressurized systems and protected production, which typically require smaller, more frequent irrigations. Because operations are shared across the canal and the farm, districts achieve the best results when they involve users in setting service windows, minimum run times, and data-sharing practices, ensuring that on-farm schedules can actually be met.

The operational shift is from rules of thumb to data-based prescriptions. Center pivots, linear moves, and microirrigation can already vary applications by zone; the challenge is turning diverse data streams into simple, reliable instructions that an operator can trust. A practical decision support system (DSS) combines field zones, soil-moisture readings, crop canopy information (e.g., temperature or indices from imagery), local weather, and short-term forecasts and then outputs controller-ready plans, including pivot speed or sector rates for sprinklers, pulse width or pressure set-points for drip, so that equipment follows the plan rather than improvises (Ale et al., 2023; Taghvaeian et al., 2020).

Despite steady innovation, the adoption of science-based methods remains low. In the United States, more than 70% of irrigation scheduling still relies on traditional methods, such as "visual crop conditions" or "feel of soil", whereas less than 20% of decisions use soil sensors, ET-based methods, plant sensors, and crop models (USDA-NASS, 2019) (Figure 23.1). Barriers include high costs for hardware and software, the need for specialized skills, heavy data requirements, uncertain accuracy under deficit irrigation, dependence on nearby weather networks, and limited integration of economic evaluations that farmers can use at the field scale (Barnes et al., 2020; Ale et al., 2023). Similar challenges arise in countries, such as China, India, Pakistan, Iran, Mexico, Indonesia, and Bangladesh, where smallholder systems predominate, characterized by limited capital and scarce technical support.

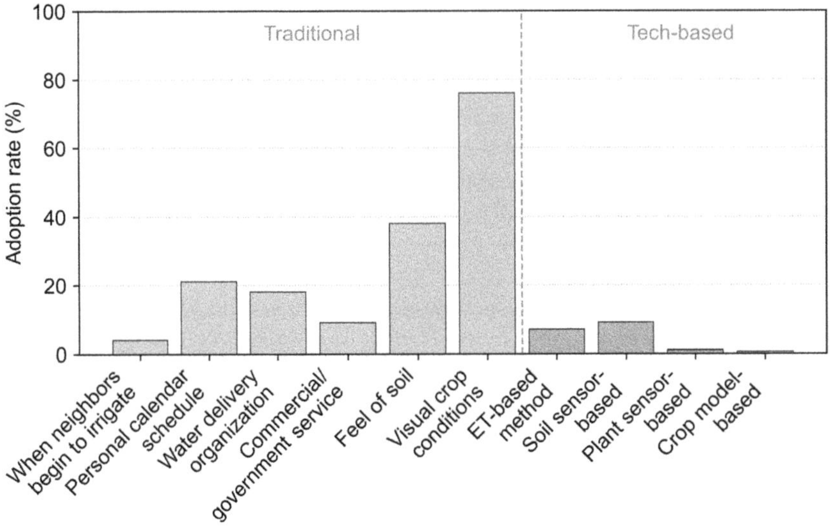

FIGURE 23.1 Irrigation scheduling methods in the United States. Note: The total percentage exceeds 100% since a field may use more than one method. (Data are from USDA-NASS, 2019.)

To improve adoption, practical pathways include lower-cost sensors and loggers; simpler user interfaces that require minimal training; local calibrations for soils, crops, and climates; and public/NGO programs that provide training and partial cost shares (Su & Singh, 2024).

To keep pace, irrigation districts must modernize, leveraging computerization and remote-control technologies to stay aligned with these evolving conditions and demands. This transition reflects a broader trend toward more sustainable and technologically integrated water management practices in agriculture, emphasizing efficiency, conservation, and collaborative decision-making.

23.5 ADVANCES IN TECHNOLOGY

Many systems were initially designed for certain crops and fixed delivery patterns. As cropping patterns, irrigation methods, and management goals have changed, some systems no longer align with today's needs. Technology provides a practical path to modernize without rebuilding everything: districts can add communication, sensing, analytics, and automated control to make existing works more observable and more responsive. A proper upgrade process involves connecting the system (telemetry), observing it (via remote meters and sensors), and then controlling it (through automated gates/valves), allowing water to be delivered when and where it is needed, even as demands shift throughout the season. This integration helps districts keep pace with pressurized on-farm methods and protected production, which require smaller, more frequent applications, and supports collaborative operations between canal managers and water users.

Digital infrastructure and automation are the backbone of this shift. Field staff use computers and mobile devices to interact with remote-read flow meters and level/pressure sensors at critical locations, while automated control gates/valves respond to changing orders. Together, these tools provide real-time visibility of flows, enable faster, smoother adjustments, reduce operational spills, and align canal services with on-farm schedules. Over time, districts can link devices to simple, auditable decision rules, such as minimum run times, order windows, pool set-points, and alarm thresholds, so routine decisions are consistent and transparent. Designing for reliability means adding basic redundancies (backup power/communications) and clear manual-override procedures so that automation fails safely rather than interrupting service.

Advances in sensing and remote sensing have increased visibility from canals to the field. Contemporary programs combine soil–water sensors, canopy stress indicators (thermal or spectral), and microclimate stations with low-cost proximal imaging and satellite-based evapotranspiration (ET) to construct a coherent, multiscale picture of field water status. Optical imagery helps distinguish irrigated areas from rainfed areas, thermal imagery with energy-balance models provides field-scale ET, and microwave sensors detect topsoil moisture to identify irrigation timing and check whether schedules match soil conditions. When data accumulate over seasons, quality-controlled time series of ET, soil moisture, yields, weather, deliveries, and pressures support model calibration and transparent reporting across fields and laterals. At the system edge, the practical steps are straightforward: meter and telemeter turnouts to ensure that deliveries and pressures are known; encourage zone-based sensing in fields; and adopt simple data standards to enable third-party decision tools and controllers to interoperate.

New data-driven decision tools translate diverse data streams into simple actions that operators can trust. Machine learning (ML) methods can fuse weather, soil sensors, and imagery to estimate soil moisture and ET, predict short-term crop stress, and suggest irrigation scheduling. Because ML learns from both historical and real-time data, it can improve the timing and depth of events and reduce waste, especially when paired with well-sited field sensors.

Finally, technology works best when paired with targeted rehabilitation, which removes the bottlenecks that the sensors expose. This often includes selective lining or piping, improved checks and spillways for stable control, pressure regulation, and filtration for on-farm compatibility, as well as upgrades to communications and power. The goal is a cohesive, resilient system in which instruments, models, and controls reinforce each other: telemetry makes conditions visible; analytics convert data into clear controller targets (set-points) for flow, depth, pressure, and timing; automation holds those set-points within tolerance; and operators retain simple, safe manual overrides when needed. In this way, modernization becomes a source of reliability and flexibility, helping districts adapt to changing crops, methods, and climates while sustaining equitable and efficient services (USDA-ARS, 2021).

23.6 CHANGES IN COSTS

In the dynamic landscape of irrigation management, it is crucial to account for fluctuating costs in labor, materials, and water. These changes have a significant impact on the operation and maintenance of irrigation systems. For example, while the need for ditch riders may decrease in some scenarios, enhancing their efficiency with improved communication tools and decision-making resources becomes essential.

The adoption and modification of maintenance practices are also heavily influenced by varying costs, as well as the emergence of new knowledge and technological advancements. A notable example is the shift from mechanical to biological methods for weed control, which can significantly reduce costs while maintaining effectiveness (Harker and O'Donovan, 2013). Furthermore, the cost of water itself plays a pivotal role in shaping investment decisions and in determining the prioritization of projects. Evaluating the cost per unit of water conserved emerges as a key metric for selecting and justifying specific rehabilitation or upgrading projects. This approach ensures that investments are not only economically sound but also align with the overarching goal of efficient and sustainable water resources management. In summary, adapting to changes in costs requires a multifaceted approach that considers financial, technological, and operational aspects to ensure the continued effectiveness and sustainability of irrigation systems.

23.7 CHANGES IN LAWS AND REGULATIONS

Another aspect that irrigation districts need to consider is the changes in laws and regulations related to (a) water rights, (b) water quality, (c) the use of insecticides and herbicides, (d) the conservation of natural resources, (e) recreation, and (f) fish and wildlife enhancement. Each of these factors has a significant influence on the operational dynamics of irrigation systems. For example, stricter water quality and irrigation regulations may necessitate changes to reduce runoff. This could involve expanding wetland areas to capture and clean runoff or requiring farmers to implement vegetative buffer strips between agricultural fields and drainage systems.

Additionally, some districts might face restrictions on water usage for irrigation to ensure the maintenance of adequate reservoir levels or to preserve ecological flows in rivers. This would call for more efficient water use strategies, possibly including the implementation of advanced irrigation technologies or revised water distribution protocols.

Navigating these regulatory changes requires a flexible and informed approach from irrigation districts. By maintaining awareness of legal requirements and proactively adapting management strategies, they can achieve compliance, while simultaneously serving the diverse needs of their stakeholders.

23.8 USE OF WATER

Dynamic shifts in water usage across the urban, agricultural, and industrial sectors profoundly impact investment strategies, rehabilitation projects, and the overall management of irrigation water (Singh and Su, 2022b,c). These evolving priorities present both challenges and opportunities for smarter, more resilient water resource management.

Allocation under change requires explicit operating rules for both normal and shortage years. Effective operating rules do two things. First, they identify clear triggers, such as reservoir storage, forecast runoff, groundwater levels, and require environmental flow or salinity limits. Second, they pair each trigger with a preapproved action, such as adjusting allocation volumes, tightening delivery windows, enabling short-term exchanges, offering voluntary follow-up, or switching to alternative sources. Because agriculture increasingly shares canals and turnouts with urban and environmental uses, operations must also manage peaking (pressure/flow guarantees for mixed customers) and quality coordination (e.g., meeting instream EC targets while serving irrigators).

The key to effectively navigating these changes lies in proper strategic planning. By anticipating potential shifts and probable trends in water usage, irrigation managers can develop adaptable and resilient strategies. This process involves not only understanding current water demands but also projecting future needs and identifying how these needs might impact existing irrigation infrastructure.

Proper consideration and prior planning are essential for ensuring the continued success of irrigation projects. It allows for timely adjustments in terms of resource allocation, infrastructure development, and management practices. By adopting such an approach, irrigation systems can better protect against unforeseen challenges and efficiently meet the diverse and changing needs of urban, agricultural, and industrial water users.

Demand and supply tools must advance together. On the demand side, volumetric billing, tiered rates, seasonal quotas, and short-term exchanges can allocate water to higher-value uses in dry years while funding on-farm upgrades, provided that flow is measured and service is predictable. On the supply side, districts increasingly blend treated effluent, stormwater, drainage, and brackish groundwater, guided by source-specific quality criteria, blending and cyclic-use rules by crop stage, and revised leaching requirements, based on the salinity of applied water and measured soil salinity. Where timing is the constraint, on-farm storage, interties, and MAR/ASR provide buffering, which is paired with pretreatment, clogging control, and monitoring wells to protect groundwater.

A practical way to plan under growth and climate variability is to forecast regional water stress and test scenarios before committing to long-lived assets. For example, Su and Karthikeyan (2023) integrated a socioeconomic system dynamics model with a hydrologic supply model to estimate the water stress index (WSI), i.e., the ratio of total withdrawals to available supply, at the sub-basin scale across Texas for the period 2015–2050 (Figure 23.2). Using consistent class breaks (e.g., low, moderate, medium-high, and high when the WSI exceeds a threshold), the analysis projected sharp increases in domestic withdrawals in metropolitan areas and rising irrigation withdrawals across many agricultural basins. In contrast, manufacturing withdrawals declined with increasing efficiency. Together, these shifts expanded the footprint of high water stress, particularly in intensive agricultural regions (High Plains,

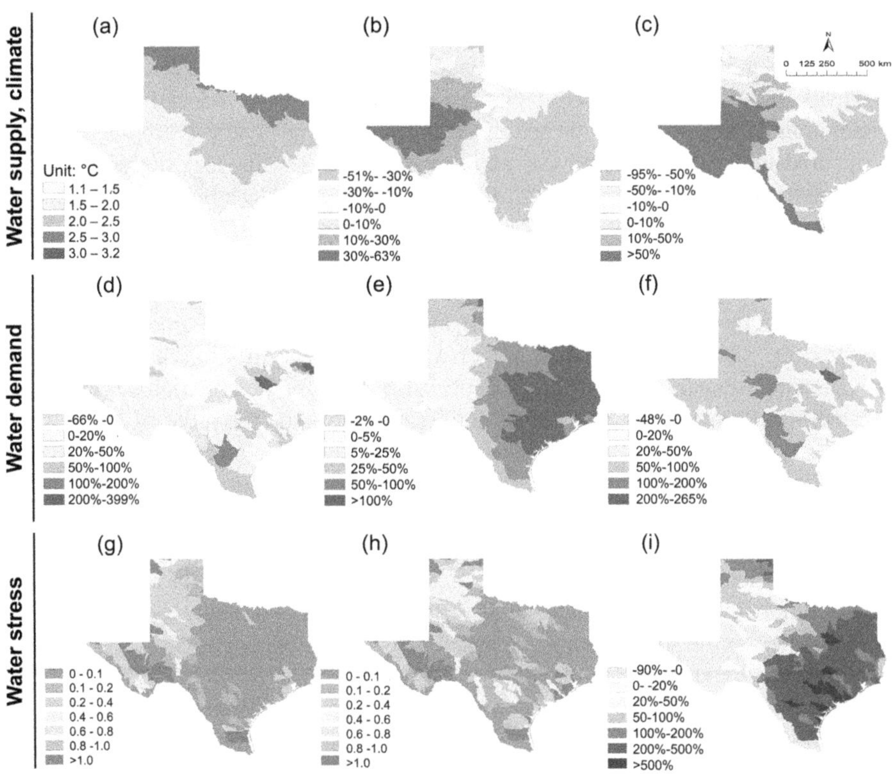

FIGURE 23.2 Predicted 2050 vs. 2015 changes in water supply, demand, and stress in Texas. (a) Changes in mean air temperature; percent changes in (b) precipitation and (c) water supply; percent changes in water use from (d) domestic, (e) irrigation, and (f) all sectors; current (g) and future (h) water stress index (WSI); and (i) percent changes in the WSI from 2015 to 2050. (Adapted from Su and Karthikeyan, 2023.)

Rolling Plains) and fast-growing metros (Dallas–Fort Worth, Houston, Austin, San Antonio). Sensitivity tests revealed that faster population growth significantly exacerbates urban stress, whereas higher irrigation rates intensify stress in agricultural basins. This highlights the need to pair any expansion of irrigated areas with efficiency measures and diversified sources. In practice, districts can use a simple stress metric (e.g., WSI) to flag hotspots, test what-if scenarios before investing, and then tie each scenario to pre-approved actions. This means conservation pricing and supply diversification in fast-growing metropolitan areas, as well as crop/area shifts, efficiency upgrades, and quality-aware blending/leaching in farm regions.

23.9 CLIMATE CHANGE IN THE WATER CYCLE AND CROP YIELD PRODUCTIVITY

Climate change is intensifying the global hydrological cycle, manifesting as an increase in the frequency and severity of floods, droughts, and rainfall variability. These changes are already negatively impacting global agriculture, threatening water availability, crop yields, and ultimately, food security. Between 1981 and 2010, changes in precipitation patterns led to reductions in global yields for maize (−4.1%), soybeans (−4.5%), rice (−1.8%), and wheat (−1.8%) (IPCC, 2023). Furthermore, droughts and extreme weather events are estimated to have reduced global cereal production by 9%–10%. Floods are also a significant contributor to crop loss. While some cold-temperate zones have experienced localized gains, e.g., higher maize or soybean yields with increased summer rainfall, the negative effects on production and quality predominate at the global scale (high confidence) (IPCC, 2023).

Projections suggest that these pressures will intensify as global warming continues. Hydrologic hazards, such as droughts, floods, and compound events, are increasing in frequency and magnitude in all assessed regions over mid- to long-term timeframes, with a greater risk at higher warming levels. At ~2°C of global warming, snowmelt-dependent basins may experience a decline in irrigation-season water availability of up to 20%, and global glacier mass loss, on the order of 18% ± 13%, will further reduce dry-season flows that support agriculture, hydropower, and settlements (medium to high confidence). Direct flood damage is projected to be ~1.4–2 times greater at 2°C and ~2.5–3.9 times greater at 3°C than at 1.5°C without adaptation (medium confidence). At ~4°C, approximately 10% of global land could experience increases in both extremely high and low river flows in the same locations, complicating planning across all water-use sectors (medium confidence). The groundwater on the small islands is particularly vulnerable to climate-driven salinization (high confidence) (IPCC, 2023).

These hydrologic shifts translate directly into crop performance. Without adaptation, rain-fed yields in many regions, especially for maize, are projected to decline by approximately 20%–33% by the end of the century. Many regions currently supporting multiple crops per year may become unsuitable for rain-fed agriculture or change to a single-crop system. The vulnerability of agriculture is compounded by its status as the largest consumptive water user globally (50%–90% of withdrawals), meaning that much of the water applied is lost as ET and is not available for rapid downstream reuse. Irrigation, often the first line of defense against water stress, will itself face tighter limits in already water-stressed basins and in regions where groundwater is depleted, e.g., parts of India, North China, and the northwestern United States. The net effect is an increasing gap between crop water demand and reliable supply, coupled with greater year-to-year variability in yield, quality, and farm income.

Farmers are already adapting. Common responses include expanding or improving irrigation systems; conserving soil water through the use of mulch, residue cover, reduced tillage, and contouring; adjusting planting dates and cropping patterns; adopting drought-, flood-, or heat-tolerant cultivars; and refining agronomic practices (e.g., precision nutrient management to minimize stress interactions). Indigenous and local knowledge increasingly informs choices about what to grow, when to plant, and how much to irrigate. Economic tools, such as index-based insurance, can buffer income against droughts and floods. Additionally, training and capacity building, when co-designed with farmers, help translate new information and technology into practical applications. Many of these measures provide multiple benefits (water savings, yield stability, income gains), but the degree to which they reduce climate risk varies by place, crop, and management and remains a key knowledge gap.

Adaptation has limits and trade-offs that tighten with increasing warming. Most agricultural adaptations perform better at ~1.5°C than at 2°C or 3°C; beyond ~2°C, substantial residual risks persist even with improved practices (high confidence). Irrigation stabilizes yields but can accelerate groundwater depletion and salinization if not managed within sustainable recharge and leaching limits; at large scales, irrigation can also modify local temperature and precipitation patterns (high confidence). Intensification strategies may meet near-term food and income goals while creating equity, biodiversity, and ecosystem-service trade-offs. Genetic improvement, from conventional breeding to modern biotechnologies, can increase climate resilience, but uptake is shaped by biophysical ceilings and by socioeconomic and institutional constraints, especially for smallholders.

23.10 CROPS AND CROPPING PATTERNS

Cropping patterns are already shifting in response to climate change, and the pace of change is expected to accelerate in the future. Warmer springs advance phenology, such as earlier budbreak, flowering, and harvest, whereas hotter summers increase ET, thereby increasing crop water demand and heat stress. Longer dry periods increase the salinity pressure and the need for high-frequency irrigation. Extreme weather events, such as heatwaves around flowering/fruit set, multi-week droughts, and intense storms, are increasingly responsible for sudden yield shortfalls. Earlier phenology can also increase frost exposure for perennials, and declining winter chill affects cultivar suitability for some fruit and nut crops. Together, these changes indicate that crops need water earlier in the season and more water during the hottest months. There is less room for missed or late deliveries, and small, frequent irrigations become much more important.

Coping with these shifts begins with stage-specific water management. The crop coefficient method, i.e., crop water use = reference ET × crop coefficient, remains important for estimating crop water use; however, the stages most sensitive to warming are changing. Where physiology allows, regulated deficit irrigation remains useful in noncritical stages, such as for grapes, olives, and some tree fruits, whereas row crops often require tighter protection at establishment and flowering. Districts can support climate-aware scheduling by aligning deliveries with stage sensitivity and, where alternative waters are available, applying blending and cyclic-use rules. This involves using higher-EC water during tolerant phases, switching to fresher supplies during sensitive phases, and scheduling cool-season leaching when off-site risks are lower. Because longer dry spells increase salinity pressure, crop and variety choices should reflect salt and ion tolerance, such as chloride/boron sensitivity, as well as the drainage capacity required to maintain soil salinity within safe limits.

Climate change also changes the cropping mix. In hotter, drier districts, some acreage will shift toward heat- and drought-tolerant cereals (e.g., sorghum, millets), oilseeds, or short-season varieties that avoid late-summer heat. Perennial blocks may shift to cooler aspects or deeper soils with more plant-available water or adopt shade nets, reflective mulches, and targeted evaporative cooling. When winter chill declines, growers may regraft to low-chill cultivars or transition to species with compatible chilling needs. Protected cultivation is likely to expand for high-value crops, enabling smaller, more frequent irrigations with tight pressure and quality tolerances.

Irrigation districts can take practical steps to make these transitions easier. Reliable measurement and telemetry at turnouts are provided, allowing growers to trust ET- and sensor-based scheduling and adjust set points when heat/ET anomalies occur. Offer service tiers that pair a firm base allocation with an interruptible block and publish order windows and minimum run times that automation can benefit, especially for microirrigation and protected cultivation, which depend on small, frequent applications. Reuse is part of the portfolio; crop-stage-specific blending and leaching guidance tied to local soil salinity conditions and drainage capacity are published; and pilot trials, such as variable-rate irrigation, regulated deficit irrigation, heat-avoidance planting dates, shade/cover strategies, and reuse blends, are run on representative soils and crops. Finally, progress in an annual monitoring and evaluation summary, tracking crop mix, ET, and delivery trends, reuse volumes, salinity outcomes, and service reliability is reported so that boards and growers can see how well the district is adapting.

Because change is continuous, resilience comes from habits, not one-off projects. Districts that pair agronomic insight (stage sensitivity, salinity tolerance, and water productivity) with operational flexibility (service tiers, reliable telemetry, variable-rate irrigation, and safe reuse) will meet diverse crop needs, protect soils and water quality, and stay financially sound as patterns evolve. A change-ready district should (1) perform measurements first, such as turnout metering/telemetry and discharge water-quality monitoring; (2) use explicit operating rules with clear triggers and pre-authorized actions; (3) invest in controllable drainage and quality-aware reuse; (4) use science-based scheduling and integrate with sensors and short-term forecasts; (5) modernize with automation while preserving safe manual overrides; (6) manage rates and finances to cover fixed costs while rewarding conservation; (7) test scenarios (e.g., "what-ifs") before committing capital; (8) support crop transitions with service tiers and stage-specific guidance; (9) report annually through a monitoring and evaluation summary to stay accountable and adaptive; and (10) cultivate a learning culture—pilot, evaluate, and scale what works. Districts that institutionalize these practices deliver equitable, reliable service while navigating the uncertainties of climate, markets, and technology.

REFERENCES

Ale, S., Su, Q., Singh, J., Himanshu, S., Fan, Y., Stoker, B., Gonzalez, E., Sapkota, B.R., Adams, C., Biggers, K., Kimura, E., and Wall, J. (2023). Development and evaluation of a decision support mobile application for cotton irrigation management. *Smart Agricultural Technology*, 5, 100270.

Barnes, E.M., Campbell, B.T., Vellidis, G., Porter, W.M., Payero, J.O., Leib, B.G., Sui, R., Fisher, D.K., Anapalli, S., Colaizzi, P.D., Bordovsky, J.P., Porter, D.O., Ale, S., Mahan, J., Taghvaeian, S., and Thorp, K.R. (2020). Forty years of increasing cotton's water productivity and why the trend will continue. *Applied Engineering in Agriculture*, 36(4), 457–478. https://doi.org/10.13031/aea.13911

Harker, K.N. and O'Donovan, J.T. (2013). Recent weed control, weed management, and integrated weed management. *Weed Technology*, 27(1), 1–11. https://doi.org/10.1614/WT-D-12-00109.1

Hoffman, G.J. and Shalhevet, J. (2007). Controlling salinity. In: G.J. Hoffman, R.G. Evans, M.E. Jensen, D.L. Martin, and R.L. Elliot, Eds., *Design and Operation of Farm Irrigation Systems*. ASAE, St Joseph, MI, pp. 160–201. 49085-9659.

Intergovernmental Panel on Climate Change (IPCC) (2023). *Climate Change 2022: Impacts, Adaptation and Vulnerability*. Working Group II Contribution to the Sixth Assessment Report of the Intergovernmental Panel on Climate Change. Cambridge University Press, Cambridge.

Singh, V.P. and Su, Q. (2022a). Irrigated agriculture under climate change. In: T.R. Lee, Ed., *Case Studies: Insights on Agriculture Innovation 2021*. IAAS Series. Vital Wellspring Education Pte. Ltd, Singapore, pp. 41–49.

Singh, V.P. and Su, Q. (2022b). Water: How secure are we under climate change? In: B. Yadav, M.P. Mohanty, A. Pandey, V.P. Singh, and R.D. Singh, Eds., *Sustainability of Water Resources: Impacts and Management*. Water Science and Technology Library, # 116. Springer International Publishing, Cham.

Singh, V.P. and Su, Q. (2022c). Water-environment-energy-food nexus: Challenges and opportunities under climate change. *Indian Journal of Soil Conservation*, 50(3), 177–189.

Su, Q. and Karthikeyan, R. (2023). Regional water stress forecasting: Effects of climate change, socioeconomic development, and irrigated agriculture: A Texas case study. *Sustainability*, 15, 9290. https://doi.org/10.3390/su15129290

Su, Q. and Singh, V.P. (2024). Advancing irrigation management: Integrating technology and sustainability to address global food security. *Environmental Monitoring and Assessment*, 196, 1018. https://doi.org/10.1007/s10661-024-13145-5

Taghvaeian, S., Andales, A.A., Allen, L.N., Kisekka, I., O'Shaughnessy, S.A., Porter, D.O., Sui, R., Irmak, S., Fulton, A., and Aguilar, J. (2020). Irrigation scheduling for agriculture in the United States: The progress made and the path forward. *Transactions of the ASABE*, 63(5), 1603–1618.

USDA-ARS (2021). USDA-ARS: National Program 211 Water Availability & Watershed Management 2021–2025. United States Department of Agriculture, Agricultural Research Service. Washington, DC. https://www.ars.usda.gov/ARSUSERFILES/NP211/NP211%20ACTION%20PLAN%202021-2025FINAL.PDF

USDA-NASS (2019). *2018 Irrigation and Water Management Survey*. United States Department of Agriculture, Washington, DC. https://www.nass.usda.gov/Publications/AgCensus/2017/Online_Resources/Farm_and_Ranch_Irrigation_Survey/index.php

Index

D

E

F

M

N

O

P

R

S

T

U

V

W

Y

Z

For Product Safety Concerns and Information please contact our EU
representative GPSR@taylorandfrancis.com
Taylor & Francis Verlag GmbH, Kaufingerstraße 24, 80331 München, Germany

www.ingramcontent.com/pod-product-compliance
Ingram Content Group UK Ltd.
Pitfield, Milton Keynes, MK11 3LW, UK
UKHW022213190926
475702UK00003B/331

* 9 7 8 1 0 4 1 2 5 6 0 8 3 *